Teacher's Pack, **Higher 1**
Delivering the AQA Specification

NEW GCSE MATHS
AQA Linear

Matches the 2010 GCSE Specification

Brian Speed • Keith Gordon • Kevin Evans • Trevor Senior • Chris Pearce

Contents

Introduction	4–7

1 Number: Number skills — 8
- 1.1 Solving real-life problems — 10
- 1.2 Multiplication and division with decimals — 12
- 1.3 Approximation of calculations — 14
- Functional Maths: Planning a dinner party — 16
- Answers — 18

2 Number: More number — 20
- 2.1 Multiples, factors, prime numbers, powers and roots — 22
- 2.2 Prime factors, LCM and HCF — 24
- 2.3 Negative numbers — 26
- Functional Maths: Creating your own allotment — 28
- Answers — 30

3 Number: Fractions, percentages and ratios — 32
- 3.1 One quantity as a fraction of another — 34
- 3.2 Adding and subtracting fractions — 36
- 3.3 Increasing and decreasing quantities by a percentage — 38
- 3.4 Expressing one quantity as a percentage of another — 40
- 3.5 Compound interest and repeated percentage change — 42
- 3.6 Reverse percentage (working out the original quantity) — 44
- 3.7 Ratio — 46
- Functional Maths: Understanding VAT — 48
- Answers — 50

4 Number: Proportions — 54
- 4.1 Speed, time and distance — 56
- 4.2 Direct proportion problems — 58
- 4.3 Best buys — 60
- 4.4 Density — 62
- Functional Maths: Organising your birthday dinner — 64
- Answers — 66

5 Geometry: Shapes — 68
- 5.1 Circumference and area of a circle — 70
- 5.2 Area of a trapezium — 72
- 5.3 Sectors — 74
- 5.4 Volume of a prism — 76
- 5.5 Cylinders — 78
- Functional Maths: Fitting a carpet — 80
- Answers — 82

6 Algebra: Expressions and equations 1 — 84
- 6.1 Basic algebra — 86
- 6.2 Factorisation — 92
- 6.3 Solving linear equations — 94
- 6.4 Setting up equations — 100
- 6.5 Trial and improvement — 102
- Functional Maths: Temperature scales — 104
- Answers — 106

7 Algebra: Expressions and equations 2 — 110
- 7.1 Simultaneous equations — 112
- 7.2 Rearranging formulae — 114
- Functional Maths: Walking using Naismith's rule — 116
- Answers — 118

8 Geometry: Pythagoras and trigonometry — 120
- 8.1 Pythagoras' theorem — 122
- 8.2 Finding a shorter side — 124
- 8.3 Applying Pythagoras' theorem in real situations — 126
- 8.4 Pythagoras' theorem in three dimensions — 128
- Problem Solving: Map work using Pythagoras — 130
- Answers — 132

9 Geometry: Angles — 134
- 9.1 Special triangles and quadrilaterals — 136
- 9.2 Angles in polygons — 138
- Functional Maths: Tessellations — 140
- Answers — 142

© HarperCollins*Publishers* 2012

10 Geometry: Constructions **144**
    10.1 Constructing triangles   146
    10.2 Bisectors   148
    10.3 Defining a locus   150
    10.4 Loci problems   152
    Problem Solving: Planning a football pitch   154
    Answers   157

11 Geometry: Transformation geometry **160**
    11.1 Congruent triangles   162
    11.2 Translations   164
    11.3 Reflections   166
    11.4 Rotations   168
    11.5 Enlargements   170
    11.6 Combined transformations   172
    Problem Solving: Developing photographs   174
    Answers   176

12 Statistics: Data handling **182**
    12.1 Averages   184
    12.2 Frequency tables   186
    12.3 Grouped data   188
    12.4 Frequency diagrams   190
    12.5 Histograms with bars of unequal width   192
    12.6 Surveys   194
    12.7 Questionnaires   196
    12.8 The data-handling cycle   198
    12.9 Other uses of statistics   200
    12.10 Sampling   202
    Functional Maths: Fishing competition on the Avon   204
    Answers   206

13 Algebra: Real-life graphs **212**
    13.1 Straight-line distance–time graphs   214
    13.2 Other types of graphs   216
    Functional Maths: Planning a motorbike trip to France   218
    Answers   220

14 Statistics: Statistical representation **222**
    14.1 Line graphs   224
    14.2 Stem-and-leaf diagrams   226
    14.3 Scatter diagrams   228
    Functional Maths: Reporting the weather   230
    Answers   232

15 Probability: Calculating probabilities **234**
    15.1 Experimental probability   236
    15.2 Mutually exclusive and exhaustive events   238
    15.3 Expectation   240
    15.4 Two-way tables   242
    15.5 Addition rule for events   244
    15.6 Combined events   246
    Functional Maths: Fairground games   248
    Answers   250

16 Algebra: Algebraic methods **254**
    16.1 Number sequences   256
    16.2 Finding the nth term of a linear Sequence   258
    16.3 Special sequences   260
    16.4 General rules from given patterns   262
    Problem Solving: Patterns in the pyramids   264
    Answers   266

17 Algebra: Linear graphs and equations **268**
    17.1 Linear graphs   270
    17.2 Drawing graphs by the gradient-intercept method   272
    17.3 Finding the equation of a line from its graph   274
    17.4 3D coordinates   276
    Problem Solving: Traverse the network   278
    Answers   280

18 Algebra: More graphs and equations **284**
    18.1 Quadratic graphs   286
    18.2 The significant points of a quadratic graph   288
    Problem Solving: Quadratics in bridges   290
    Answers   292

19 Algebra: Inequalities and regions **294**
    19.1 Solving inequalities   296
    19.2 Graphical inequalities   298
    Functional Maths: Linear programming   300
    Answers   302

Schemes of Work   **304**

Introduction

Welcome to Collins *New GCSE Maths* Higher 1, which has been written for the AQA Mathematics (Linear) B Specification for GCSE Mathematics (4365). This Teacher's Pack accompanies Student Book Higher 1, which covers all the content required for the first year of teaching the course.

The new GCSE contains some types of questions that have not appeared in GCSE mathematic exams before. This guide and its lesson plans will help you to prepare students to tackle the new aspects with confidence.

Functional maths
In the Higher tier about 20–30% of the questions will have a functional element. This is all about 'real-life' mathematics. In the Student Book, functional maths questions are highlighted with this icon **FM** to show students when they are practising their functional skills.
One way to decide if a question is functional is to ask if anyone would want to know the information in a real-life context; for example, Work out 30% of £15 is **not** functional.

Functional maths example 1 (Grade D)
A school outing has been planned for 142 students. Two coaches can carry 104 students. The remaining students will travel by car – maximum five per car. How many cars will be needed?
(3 marks)

Functional maths example 2 (Grade C)
Lin belongs to a local gym.
This table shows the kilojoules used on four exercise machines at the gym.

Exercise machine	Kilojoules per minute Intensity		
	Low	Med	High
Rowing machine	33	46	59
Treadmill	25	38	50
Bicycle	21	33	46
Cross-trainer	42	59	75

Lin goes to the gym for an hour. She likes to do 30 minutes of low-intensity exercise and no more than 10 minutes of high-intensity exercise.
Draw up a training programme for Karina that uses each machine, and will allow her to burn up at least 400 calories. Hint: 4.184 kilojoules = 1 calorie (5 marks)

New Assessment Objectives
In the previous GCSE, assessment objectives were topics such as Number and Algebra but now assessment objectives are related to the processes of doing mathematics. There are three types:
AO1   recall and use knowledge of the prescribed content
AO2   select and apply mathematical methods in a range of contexts
AO3   interpret and analyse problems and generate strategies to solve them.

AO1 will be assessed in about 50% of the examination with straightforward questions that test if students can do mathematics. These questions will be very familiar from past papers.

    AO1 example 1 (Grade C)
    Expand and simplify $4(4e + 3) - 2(5e - 4)$ (3 marks)

© HarperCollins*Publishers* 2012

AO1 example 2 (Grade B)
Leon spends £220 a month on food and cleaning products. This is 24% of his monthly take-home salary.
Food costs him 80% of this amount. The remainder is for cleaning products.
How much does he spend on cleaning products?
What is his monthly take-home salary? (3 marks)

AO2 will be assessed in approximately 30% of the examination to test if students understand the topics and can apply mathematics in slightly more involved situations. These types of question are not totally new because Using and Applying Mathematics has been assessed before. What is new are questions that assess students' understanding. These are flagged in the Student Book exercises with the icon (AU).

AO2 example 1 (Grade C)
A bag contains 3.5 kg of soil, to the nearest 100 g.
What is the least amount of soil in the bag?
Give your answer in kilograms and grams. (3 marks)

AO2 example 2 (Grade A)
Find the area of the triangle enclosed by these three equations.
$y - x = 2$ $x + y = 6$ $3x + y = 6$ (3 marks)

AO3 will be assessed in about 20% of the examination to test if students can solve problems. This is a new type of question in GCSE Maths and there is lots of practice in the Student Books with questions marked (PS).

AO3 example 1 (Grade B)
In the game called Pontoon, you are dealt two cards. If one card is an Ace and the other is a King, Queen or Jack you have been dealt a 'Royal Pontoon'.
What is the probability of being dealt a Royal Pontoon? Give your answer to 3 decimal places. (4 marks)

AO3 example 2 (Grade C)
A group of boys and girls wait for school buses. The first bus arrives and 25 girls get on. The ratio of boys to girls at the bus stop is now 3 : 2. The second bus arrives and 15 boys get on. Now the same number of boys and girls are at the bus stop.
How many students were originally at the bus stop? (3 marks)

Quality of Written Communication
Another new requirement in GCSE Maths exams is to have questions that will assess the Quality of Written Communication (**QWC**). The examination board will indicate in some way which questions are allocated marks for QWC, for example, with an asterisk. Students will be assessed on their ability to write a logical answer and correctly use mathematical vocabulary, symbols and notation.

QWC example (Grade C)
Each of three restaurant chains claims to have the lowest average price increase over the year. The table below summarises the average price increases.

Price increase (p)	1-5	6-10	11-15	16-20	21-25	26-30	31-35
Wutherly's	4	10	14	23	19	8	2
Harvestright	5	11	12	19	25	9	6
O'Nanny's	3	8	15	31	21	7	3

Using their average price increases, make a comparison of the restaurant chains and write a report on which chain, in your opinion, has the lowest price increases over the year. Justify your answers.

© HarperCollins*Publishers* 2012

How to use this book

Chapter overview
Each chapter starts with an outline of the content covered in the entire chapter so you can plan ahead easily.
- **Overview** shows the topics in each section at a quick glance.
- **Context** provides a summary of the work students will encounter and offers context for the key mathematical ideas.
- **Curriculum references** show how the material meets the requirements of the new curriculum with references to the 2010 Specification, KS4 Programme of Study, Functional Skills Standards, Personal, Learning and Thinking Skills (PLTS) and
- Assessing Pupils' Progress (APP).
- **Route mapping chart** for each exercise indicates the level of work students will meet. The chart shows, for example, that questions 1–9 in Exercise 1A are targeted at Grade D.
- **Overview test** is a quick diagnostic test you can use to establish if students have the appropriate level of understanding to tackle the topic.
- **Why this chapter matters** gives suggestions on how to use the corresponding chapter opener in the Student Books to help students make links across other subjects and cultures, see how maths has changed through history and how it can be applied to everyday life.

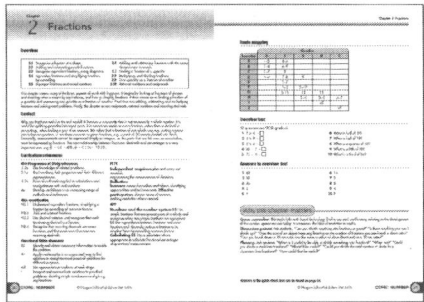

Lesson plans
Each section in the Student Book is supported by a double-page lesson plan, and each one follows the same format making it easy to use and prepare lessons.

On the left-hand page:
- **Learning objectives** indicate clearly what the lesson is about and the level of the content; they are useful to measure the success of a lesson.
- **Learning outcomes** provide a graded indication of what basic knowledge students **must** gain, what they **should** know and what they **could** extend to by the end of the lesson.
- **Key words** highlight important mathematical vocabulary.
- **References to the curriculum** and **Collins resources** show how the course is integrated.
- **Prior knowledge** highlights the underpinning maths that students need for the lesson.
- **Common mistakes** and remediation are pinpointed so they can be quickly recognised and rectified.
- **Useful tips** help students remember key concepts easily.
- **Functional maths** and **problem-solving** help support new areas of the curriculum with extra tips and advice on how to tackle the FM and PS questions in the Student Book.
- **Differentiation** suggestions are highlighted throughout in bold so you can direct students to the work that best suits them.

© HarperCollins*Publishers* 2012

6

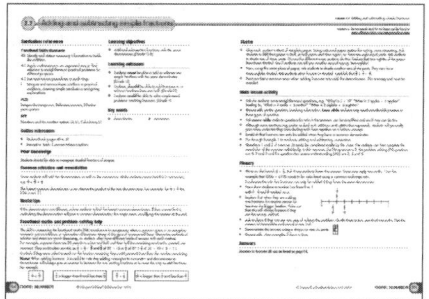

On the right-hand page:
- **Starter** suggestions involve the whole class and give you ideas on how to capture the attention and interest of students.
- **Main lesson activities** help you lead students into exercise questions.
- **Plenary** offers guidance on how to round off the three-part lesson.

Functional maths and problem-solving lesson plans
- These accompany the Student Book double-page activities at the end of most chapters and help you deliver functional skills lessons with confidence.
- They follow a similar format to the lesson plans for the exercises with curriculum references, learning objectives and learning outcomes.
- The lessons are broken down into 3 steps:
  – **Build** includes discussion points, structured questions and possible student answers
  – **Apply** concentrates on how students apply their understanding of maths with helpful prompts for you to use in the classroom.
  – **Master** shows you how to use the functional activity to assess students' understanding, by giving them minimal guidance.
- Answers and solutions for the functional maths activities are contained within each lesson..

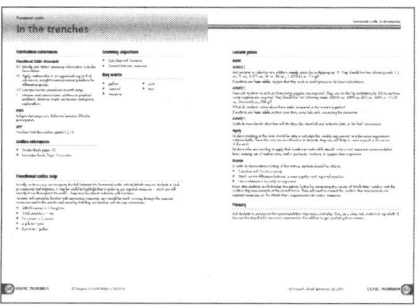

Answers to questions
Answers to the Student Book questions are at the back of the Student Book and at the end of each chapter in this Teacher's Pack. Answers to the Quick Check questions are available in the Teacher's Pack. Answers to the Examination questions can be found in both books.

Scheme of Work
A flexible 2-year and 3-year Scheme of Work is at the back of the book and in Word on the Teacher's Pack CD-ROM. It suggests two routes through the content, with an overview for medium term planning and a detailed breakdown showing timings for lessons.

Teacher's Pack CD-ROM
- The CD-ROM contains all of the Teacher's Pack material in Word to enable you to customise the lessons as you wish.
- **Grade progression maps** for all the Strands (Number, Geometry and Measures, Statistics and Probability, and Algebra) show students what they need to do to move up a grade. These can be printed off and used by students for self-assessment or used for display purposes to track progress.

© HarperCollins*Publishers* 2012

Chapter 1 Number: Number skills

Overview

> 1.1 Solving real-life problems
> 1.2 Multiplication and division with decimals
> 1.3 Approximation of calculations

This chapter covers a variety of issues, starting with solving real-life problems without using a calculator. It progresses to multiplying and dividing with decimals, rounding to a given number of significant figures and using estimations to approximate various calculations.

Context
The ability to solve real-life problems is an essential skill in preparation for the world of work. It is important that students can relate topics to everyday situations and this is a key theme of the book. Using estimation for checking answers or simply to gauge the size of an answer is another fundamental skill, which students should develop.

KS4 Programme of Study references
1.1a Applying suitable mathematics accurately within the classroom and beyond.
1.3b Understanding that mathematics is used as a tool in a wide range of contexts.
2.1a Identify the mathematical aspects of the situation or problem.
2.1d Select mathematical information, methods, tools and models to use.
2.2p Estimate, approximate and check working.
3.1a Real numbers, their properties and their different representations.
3.1b Rules of arithmetic applied to calculations and manipulations with real numbers.
4a Develop confidence in an increasing range of methods and techniques.

Linear specification
N1.2 Add, subtract, multiply and divide any number.
N1.4h Approximate to specified or appropriate degrees of accuracy, including a given number of decimal places and significant figures.
N1.8 Index notation for squares, cubes and powers of 10.

Functional Skills standards
R1 Understand routine and non-routine problems in familiar and unfamiliar contexts and situations.
R2 Identify the situation or problems and identify the mathematical methods needed to solve them.
R3 Select a range of mathematics to find solutions.
A1 Apply a range of mathematics to find solutions.
A2 Use appropriate checking procedures and evaluate their effectiveness at each stage.
I1 Interpret and communicate solutions to multistage practical problems in familiar and unfamiliar contexts and situations.

PLTS
Independent enquirers identify questions to answer and problems to resolve. **Creative thinkers** ask questions to extend their thinking. **Reflective learners** assess themselves and others, identifying opportunities and achievements. **Team workers** collaborate with others to work towards common goals. **Self-managers** organise time and resources, prioritising actions. **Effective participators** discuss issues of concern, seeking resolution where needed.

APP
Numbers and the number system L4.2 Recognise and describe number relationships including multiple, factor and square. **L5.1** Use understanding of place value to multiply and divide whole numbers and decimals by 10, 100 and 1000 and explain the effect. **L5.2** Round decimals to the

Chapter 1 Number: Number skills

nearest decimal place and order negative numbers in context. **L5.3** Recognise and use number patterns and relationships.
Calculating L5.3 Understand and use an appropriate non-calculator method for solving problems that involve multiplying and dividing any three digit number by any two-digit number. **L7.3** Add, subtract, multiply and divide fractions.
Using and applying mathematics L7.3 Justify generalisations, arguments or solutions.

Route mapping

Exercise	Grades		
	D	**C**	**B**
A	1–9	10–12	
B	1–3, 11	4–10, 12	
C	1	2–5, 9	6–8
D	1–2	3–5	
E	1–9	10–16	
F	1–5	6–12	

Overview test
Questions 1, 2a and b are grade G and question 2c is grade F. Answer the following questions **without** using a calculator.

1 Write down the answers to the following.
a $21.75 \times 10 =$ **b** $6.35 \times 1000 =$
c $81.1 \times 100 =$ **d** $0.056 \times 1000 =$
2 Use your skill with BIDMAS/BODMAS to work out the following.
a $5 + 5 \times 5 =$ **b** $5 + 4^2 - 3 =$
c $(5 + 5) \times 5 =$

Answers to overview test
1 a 217.5 **b** 6350 **c** 8110 **d** 56
2 a 30 **b** 18 **c** 50

Why this chapter matters
Cross-curricular: This work links to the sciences, Design and Technology, Geography and any subject that requires 'numbers' knowledge.
Introduction: Numbers are the framework of everyday life, so being able to multiply and divide decimal numbers and to know the approximate answer to a calculation before embarking on working it out is a great advantage in life.
Discussion points: How will knowledge of numbers be an asset in the building trade, working in a shop, flying an aircraft or working in a sales team? Could an engineer or designer get by without arithmetic?
Plenary: Prompt the class for some number-related everyday problems and discuss the number skills they require. Make sure students include questions that require rounding of answers.

Answers to the quick check test can be found at the end of the chapter.

© HarperCollins*Publishers* 2012

1.1 Solving real-life problems

Curriculum references

Functional Skills standards
R1 Understand routine and non-routine problems in familiar and unfamiliar contexts and situations.

PLTS
Independent enquirers, Creative thinkers, Reflective learners, Team workers, Self-managers, Effective participators

APP
Calculating L5.3

Collins references
- Student Book pages 8–10
- Interactive Book: Multiple Choice Questions

Learning objectives
- Solve problems set in a real-life context. [Grade D–C]

Learning outcomes
- Students **must** be able to decide whether a problem in words involves multiplication or division. [Grade D]
- Students **should** be able to accurately carry out a multiplication or division to solve a problem expressed in words. [Grade D]
- Students **could** be able to accurately carry out multi-step calculations to solve word problems involving multiplication or division. [Grade C]

Key words
- column method (or traditional method)
- grid method (or box method)
- long division
- long multiplication
- problem
- strategy

Prior knowledge
Students should know the multiplication tables and be comfortable when doing simple mental calculations. They should be able to multiply numbers such as 24 × 36, 162 × 78, £24.50 × 15, and divide numbers such as 1035 ÷ 55 without using a calculator. Students should be familiar with short division methods. They should know how to work out simple percentages of a quantity, such as 10%.

Common mistakes and remediation
Students may fail to answer a question due to not rounding answers logically, e.g. rather than 18.81 coaches, it should be 19. Students may misinterpret the remainders in the functional maths (FM) questions, but they can avoid this by giving fuller answers as outlined under 'Functional maths and problem-solving help'. Students with a weak grasp of multiplication tables may make basic errors. Address this by regularly testing students' multiplication table skills.

Useful tips
Less able students can use 'carried figures' if they need practice in laying out addition and subtraction.

Functional maths and problem-solving help
This topic has an overall element of functional maths (FM) and problem solving (PS). Exercise 1A includes real-life problems; students need to decide whether to multiply or divide, and so should read the PS questions carefully. The quality of written communication is assessed in GCSE mathematics examinations, so it is important for students to give answers in good English and clear mathematics. For example, in question 1, simply stating '6000' does not convey the answer clearly. A better response is: 'There were 6000 cans of soup in this delivery.' Ensure that students understand that they need to round up (as in question 3a) or round down (as in question 5), depending on the context of the question. It is good practice for students to give fuller answers than required. A fuller answer to question 3a would be '38 coaches with 24 spare seats' and to question 5, '18 with 45 pence left over'.

© HarperCollins*Publishers* 2012

Lesson 1.1 Solving real-life problems

Starter
- Ask students some mental multiplications, e.g. 20 × 5 (100), 7 × 400 (2800) and 30 × 500 (15 000)
- Combine two similar multiplications, e.g.
  30 × 8 (240), 2 × 8 (16) and 32 × 8 (240 + 16 = 256).

  Repeat for other numbers, increasing the difficulty according to ability. Now ask some mental divisions using the multiplication tables up to 10 × 10. Include examples that have a remainder, e.g. 40 ÷ 6 = 6, remainder 4.

Main lesson activity
- The focus of this lesson is to ensure that students have methods of multiplying and dividing with which they are comfortable. Depending on prior knowledge, they may be able to explain their methods. **Less able** groups should, at least in this lesson, concentrate on using just one method of multiplying and one method of dividing.

  Multiplication
- Demonstrate the grid (or box) method, e.g. with 213 × 54.
- Adding each answer to find the total gives 11 502. Now demonstrate the same calculation using the column method.
- Ask students which method they prefer. If there is a preference then stick to that method.
- Repeat with other calculations for **less able** groups. Go through E.g. 1 and show both methods.

  ```
      2 1 3
    ×   5 4
    ─────────
      8 5 2
    1 0 6 5 0
    ─────────
    1 1 5 0 2
  ```

×	200	10	3
50	10000	500	150
4	800	40	12

  Division
- Display 430 ÷ 2 and ask students how they would work it out. (215) They are likely to say they did a mental calculation. Increase the level of difficulty, e.g. ask students how they would work out 275 ÷ 25. Expect some to say that there are four 25s in 100, so 4 (100) + 4 (100) + 3 (75) + 11. Point out that this is a chunking method. Write out the calculation:

  ```
  4 × 25 = 100
  4 × 25 = 100
  3 × 25 =  75
  11 × 25 = 275, so 275 ÷ 25 = 11
  ```

- Now demonstrate a short division giving an answer with a remainder, e.g. 100 ÷ 7 (14, remainder 2) or 300 ÷ 21 (14, remainder 6). Explain that some questions will have remainders as the numbers do not divide exactly. Also explain that in some real-life or FM questions the remainder will need interpreting.
- Repeat the same question, using long division. Explain that long division is similar to short division but requires them to show more working, which may reduce errors.
- Ask: There are 30 students in the dinner queue. They sit at tables with four seats. How many tables are needed? (8). Ask students why eight tables are needed (there are two spare seats).
- Go through Example 2. Demonstrate the solution, using their method, e.g. for chunking 10 × 53 = 530, this leaves 672 − 530 = 142, 2 × 53 = 106, this leaves 142 − 106 = 36 as the remainder, so 640 ÷ 53 − 12, remainder 36. So 13 coaches are needed.
- Explain that if they have a variation on any of these methods or their own successful method, they can continue to use it. Encourage students to be precise with their calculations and to check answers to see if they seem sensible.
- Students can now do Exercise 1A.

Plenary
- Ask for the answer to 400 ÷ 24. Then ask students to write this as a question, e.g. In total, 400 people are on a train, with 24 people in every carriage but the last one. How many are in the last carriage? (16)

Answers
Answers to Exercise 1A can be found at the end of this chapter.

© HarperCollins*Publishers* 2012

1.2 Multiplication and division with decimals

Curriculum references

Functional Skills standards
A1 Apply a range of mathematics to find solutions.

PLTS
Independent enquirers, Creative thinkers, Reflective learners, Team workers, Self-managers, Effective participators

APP
Calculating L5.3

Collins references
- Student Book pages 11–12
- Interactive Book: Worked Exam Questions

Learning objectives
- Multiply a decimal number by another decimal number. [Grade D]
- Divide by decimals by changing the calculation to division by an integer. [Grade D]

Learning outcomes
- Students **must** be able to multiply two decimal numbers. [Grade D]
- Students **should** be able to divide one decimal number by another. [Grade D]
- Students **could** be able to carry out the appropriate calculation in a problem involving multiplication or division of decimals. [Grade D]

Key words
- decimal places
- integer
- decimal point

Prior knowledge
Students must be able to multiply three-digit integers by two-digit integers. They should also be able to multiply decimal numbers by 10, 100, 1000, etc.
Students should be able to use pencil and paper methods for division of integers.

Common mistakes and remediation

Multiplication: Students may forget to insert the decimal point or put it in the wrong position.

Division: Students may change both numbers into integers, and then have the decimal point in the wrong place. With repeated emphasis on the rules for multiplying and dividing with decimals, students should reduce making these errors.

Useful tips
As in Lesson 1.1, if **less able** students lack practice in laying out addition and subtraction, they could use 'carried figures' until they become confident. However, they should aim to develop the skill of carrying these mentally.

Functional maths and problem-solving help
This topic has a functional maths (FM) element overall, in its everyday application.
Advise students that a ream of paper (500 sheets) is about 5 cm thick; the paper in question 8 is probably thinner than normal. Challenge **more able** students to use their answer to question 8 to say how thick a sheet of paper from a ream 6 cm thick would be. (0.008 cm × 1.5 = 0.012 cm) The problem-solving (PS) questions are structured and straightforward. **Note:** For **less able** students this lesson could be split into two sessions, multiplication and division. To cater for this, the main lesson activity has been split into two parts.

© HarperCollins*Publishers* 2012

Lesson 1.2 Multiplication and division with decimals

Starter
Multiplication and division
- Ask students quick-fire, linked, mental multiplication questions, gradually increasing the level of difficulty, e.g. 6 × 7 (42) followed by 12 × 7 (84). Ask how they obtained the second answer. (Either from the 12 times table, or others may realise it is double the previous answer.)
- Now use other numbers, e.g. 9 × 8 (72), then 90 × 80 (7200). Again ask how they obtain the second answer.
- Then give several divisions questions with integer values and answers, e.g. 342 ÷ 3 (114), 8561 ÷ 7 (1223), 2784 ÷ 12 (232). Discuss the methods used: chunking, short division and long division.

Main lesson activity
Multiplying two decimal numbers together

$$\begin{array}{r} 1\,1 \\ \times\ \ 7 \\ \hline 7\,7 \end{array} \qquad \begin{array}{r} 1\,.\,1 \\ \times\ \ \ 7 \\ \hline 7\,.\,7 \end{array}$$

- Ask students to give the answer to 11 × 7 (77). Then ask for the answer to 1.1 × 7 (7.7). Show students the effect of the decimal point by setting out the calculations as shown.
- Now show students a more complex multiplication, e.g. 3.56 × 27.

$$\begin{array}{r} 3.5\,6 \\ \times\ \ 2.7 \\ \hline 2\,4\,9\,2 \\ 7\,1\,2\,0 \\ \hline 9.6\,1\,2 \end{array}$$

- Explain that the calculations now consist of multiplying two decimals together.
- Use the example above again but change 27 to 2.7.
- Explain that 3.56 × 2.7 is ten times smaller than 3.56 × 27. So 3.56 × 2.7 = 9.612.
- Explain that 3.56 has two decimal places and that 2.7 has one decimal place, so the answer has 2 + 1 = three decimal places.
- Work through Example 3. Then students can do multiplication questions 1–4 in Exercise 1B.

$$\begin{array}{r} 3.5\,6 \\ \times\ \ \ 2\,7 \\ \hline 2\,4\,9\,2 \\ 7\,1\,2\,0 \\ \hline 9\,6.1\,2 \end{array}$$

Dividing by a decimal
- Write 36 ÷ 2, 3.6 ÷ 2, 3.6 ÷ 0.2 and 36 ÷ 0.2. Then write the same calculations as fractions.

$$\frac{36}{2},\ \frac{3.6}{2},\ \frac{3.6}{0.2},\ \frac{36}{0.2}$$

- Ask for the answers (18, 1.8, 18, 180). Explain that dividing by 2 is easier than dividing by 0.2. Point out that dividing 36 by 2 is the same as dividing 3.6 by 0.2 because $\frac{3.6}{0.2} \times \frac{10}{10} = \frac{36}{2} = 18$. multiplying by 10 is multiplying by a whole one and so the answer is unchanged.
- Show an example using division by a two-decimal place number: 3.4 ÷ 0.05. Write it as a fraction: $\frac{3.4}{0.05}$ Explain that to change the format this time it is necessary to multiply by 100: $\frac{100}{100} \cdot \frac{3.4}{0.05} \times \frac{100}{100} = \frac{340}{5} = 68$
- Work through Example 4 with the class. Then students can complete the remainder of Exercise 1B.

Plenary
- Use the plenary as a lead-in to the next lesson on approximation. Use the exercise questions and estimate the answers, e.g. in 1a ask students what answer they have (67.2 × 35 = 2352) and say it is about 70 × 40 = 2800 (but less as both numbers are rounded), so answers between 2000 and 3000 seem reasonable.
- Repeat this for other parts, gradually handing over to students and asking them to do the estimate of their answers to check if they seem reasonable.

Answers
Answers to Exercise 1B can be found at the end of this chapter.

© HarperCollins*Publishers* 2012

1.3 Approximation of calculations

Curriculum references

Functional Skills standards
A2 Use appropriate checking procedures and evaluate their effectiveness at each stage.

PLTS
Independent enquirers, Creative thinkers, Reflective learners, Team workers, Self-managers, Effective participators

APP
Numbers and the number system L5.1, 2; L7.3

Collins references
- Student Book pages 13–20
- Interactive Book: Paper Animation

Learning objectives
- Round to a given number of significant figures. [Grade D–C]
- Approximate the result before multiplying two numbers together. [Grade D–C]
- Approximate the result before dividing two numbers. [Grade D–C]
- Round a calculation, at the end of a problem, to give what is considered to be a sensible answer. [Grade D–C]

Learning outcomes
- Students **must** be able to round a number to one, two or three significant figures. [Grade D]
- Students **should** be able to find an approximate answer to a multi-step calculation. [Grade C]
- Students **could** be able to round off the answer to a calculation in a practical context to an appropriate number of significant figures. [Grade C]

Key words
- approximate
- round
- estimation
- significant figures

Prior knowledge
Being able to round to a given number of decimal places (10, 100, 1000) will help students learn how to approximate.

Common mistakes and remediation
These are best shown with two examples, both rounded to two significant figures (2 sf):
275 986 = 28, rather than 280 000 (the zeros are forgotten) 0.000 371 04 = 0.00 or 37, rather than 0.000 37 (2 dp instead of 2 sf, or zeros forgotten). Students often have difficulty with the power of 10 in the answer. Sometimes students try to be too accurate in approximating calculations, rounding to more than 1 sf, or not rounding at all when using a calculator, assuming a totally correct answer is better than an approximation (when sometimes answers are better as 'roughly' or 'nearly').

Useful tips
Stress the value of estimating answers before carrying out calculations, or estimating to save time at every opportunity. Understanding the importance of this will enhance students' skills. Encourage students also to use rounding and estimation as a way of checking that their answers are sensible.

Functional maths and problem-solving help
The functional maths (FM) aspect of this topic is shown in question 10 of Exercise 1E, and questions 2 and 4 of Exercise 1F. The problem-solving (PS) questions are 7 and 8 of Exercise 1C, 4 of 1D, 14 of 1E and 10 of 1F. Less able students may need to take it in stages, possibly by covering multiplication first, then division, and completing the relevant questions as appropriate.

© HarperCollins*Publishers* 2012

Lesson 1.3 Approximation of calculations

Starter
Significant figures
- Display: 0.04, 0.4, 4, 40, 400. Ask students what they have in common. (all have a 4 and zeros)
- Write: 0.065, 0.65, 6.5, 65, 650. Ask how this set differs from the first set. (two non-zero digits)
- Ask students for a set of numbers with three non-zero digits, e.g. 0.0123, 0.123, 1.23, 12.3, 123, 1230.
  Multiplying and dividing by multiples of 10
- Start with quick-fire questions involving multiplying integers by 10, 100 and 1000.
- Display this table of numbers. Start with 31 and ask questions such as, "What is 31 × 10? What is 31 × 1000? What is 31 ÷ 10?" Continue, using a different starting number, e.g., "What is 3.1 × 10? What is 3100 ÷ 10?"

Main lesson activity
Rounding to significant figures
- Refer to the numbers in the starter. The first set had one non-zero digit (4). Those in the second set had two (65) and those in the third, three (123). The non-zero digits are significant figures or sf. Zeros before or after these digits maintain the value of the number but are not sf. Highlight the number 0.00301 in the first table in the Student Book. Explain that this has three significant figures, since the 0 in the middle is holding a place.
- Ask students to explain why, in the second table, 45 281 is 50 000 to one sf. (Only one non-zero digit, but place value must be correct.) Writing 45 281 to one sf requires the nearest number to 45 281 with only one non-zero digit is 50 000; to two sf it is 45 000, and so on. Ask quick-fire questions to reinforce the method.
- Now work backwards. If a number is 80 to one sf, what could the original number have been? (Any number from 75 to 84.999...) Students can now do Exercise 1C.
  Multiplying and dividing by multiples of 10
- Use the Student Book examples for practice. Ask students to multiply e.g. 3 × 4, 3 × 40, 30 × 40. Ask them to state any rules they notice. Point out that they need to be careful with zeros, e.g. 40 × 50 = 2000.
- Follow the same procedure for division, reviewing the starter activity. Demonstrate how to set out divisions in fraction form and cancel zeros. Students can now do Exercise 1D.
Approximations of calculations and sensible rounding
- Explain that approximations or estimations can be made more easy by rounding numbers to one sf. Give examples, such as $19.8 \times 3.01 \approx 20 \times 3 = 60$, $491 \div 9.6 \approx 500 \div 10 = 50$
- Work through Example 5, then students can start on questions 1–9 in Exercise 1E.
- Now work on more complex examples involving division by a decimal between 0 and 1. Write 86 ÷ 0.21 on the board. Explain that divisions are often better written as fractions: $\frac{86}{0.21}$ This is approximated as $\frac{90}{0.2}$
- Point out that multiplying a fraction by 1 leaves it unchanged and that $\frac{10}{10}$ is the same as 1, so $\frac{90}{0.2} = \frac{90 \times 10}{0.2 \times 10} = \frac{900}{2} = 450$
- Students can now complete Exercise 1E. Discuss sensible rounding with students before they complete Exercise 1F.

Plenary
- Write on the board: 571.8 to 1 sf = 6, 0.003276 to 2 sf = 0.00, 25.37 to 1 sf = 30.00

Answers
Answers to Exercises 1C, 1D, 1E and 1F can be found at the end of this chapter.

© HarperCollins*Publishers* 2012

Functional maths
Planning a dinner party

Curriculum references

Functional Skills standards
R2 Identify the situation or problems and identify the mathematical methods needed to solve them.
I1 Draw conclusions and provide mathematical justifications.

PLTS
Independent enquirers, Reflective learners, Self-managers

APP
Using and applying mathematics L7.3

Collins references
- Student Book pages 24–25
- Interactive Book: Real Life Video

Learning objectives
- Consolidate number work on ratio and proportion.
- Convert between metric units of measure.
- Break down a complex task into smaller manageable tasks.

Key words
- ratio

Functional maths help
Encourage all students to ask questions and then solve their own problems. They should write down the question to be answered and use it to check if they have achieved what they set out to do. Also encourage students to look at how the amounts in the recipes can be interlinked when deciding what ingredients to buy. In other words, they need to think about the practical aspect of multiplying or dividing recipe ingredients to vary the number of servings.
Students should record their findings in a systematic way; they may need support with this.
Make sure students know and can use Pythagoras' theorem, which they will require for Activity 4. If necessary, go over it with them: $a^2 + b^2 = c^2$.

Lesson plans
Activity 1
Before students complete the questions in Activity 2, let them answer 'Getting started' questions. (● six 150 g portions ● seven 75 g bars of chocolate ● one 1-litre bottle at £1.80 is cheaper because you would have to buy four 330 g cans, £2.36, to get one litre ● seven packets of biscuits)

Activity 2
Refer students to the ingredients cost list and the recipes in the Student Book. Ask these questions.

- How much of each ingredient will I need to make crème caramel for 10 people? (Double the ingredients. This serves 12, but it would be impractical to take two-thirds of each ingredient, e.g. $\frac{2}{3}$ of two eggs.)
- If I buy 800 g of caster sugar, how many chocolate brownies can I make? (Three batches of 15 = 45, because 3 225 g = 675 g, with 125 g remaining)
- How much will it cost to make mango sorbet to serve 8? If I buy the minimum ingredients needed, how much of each ingredient will be left over? (Cost: 4 mangos = £6, 2 limes = £0.40, sugar, with 1.85 over = £1.90. Total £8.30)
- How much of each ingredient will I need to make chocolate brownies to serve 6? (To be practical, halve the recipe: 25 g flour, 55 g butter, 1 egg, 112 g sugar, 80 g walnuts, $\frac{1}{2}$ tsp baking powder.)
- How much will it cost to make mango sorbet to serve 20? (To be practical, multiply each ingredient by 2.5: 10 mangos, 5 limes, 1.125 litres sugar syrup, using 375 g sugar = £15 + £1 + £1.90 = £17.90 with 1.6 kg of sugar left over.)

© HarperCollins*Publishers* 2012

Functional maths Planning a dinner party

In the text under 'Your task', students are required to decide on several numbers of dinner party guests. For each number of guests, they should plan three different combinations of desserts to serve, and also work out how much money these combinations would cost altogether and per person.

Activity 3
1. Students are required to find a recipe for their favourite dessert and then work out the cost of the dessert per person for each number of guests they decided on. (Students' costings will vary but make sure their working out is correct.)
2. Now they are required to create table decorations for the dinner parties, as in the outlines provided in the Student Book.

Apply
While students work on the first activity, guide them to think in practical terms of buying and using ingredients. Activity 3 can be entered into in a variety of ways. Students can use one or more of these points and move the task in the direction they prefer.
- For each recipe, find out the amounts needed for different servings. This ranges from very easy, multiples of servings, to more difficult, non-multiples of servings.
- Cost each recipe.
- Look at the number of servings that would best meet the amounts to be bought.
- For 100 guests, what would be the cheapest option?
- For 50 guests, list at least two options, with costings.

When students try to find their favourite dessert recipe, encourage using the internet. Guide students towards using a table to work out the costings. Check their completed calculations.

Extension
Before they start making table decorations, tell students that the diagram on the left is made up of right-angled triangles, and the diagram on the right is made up of rectangles. Briefly, go through the first table decoration with students. Then let them apply what they have learnt to complete the second one on their own, with minimal support.
More able students could work with those who are **less able** as this is a challenging activity. (e.g. one missing length is 6.8 cm)

Master
More able students will be able to access the task and develop it in the way they prefer.
If **less able** students find this difficult, suggest one of the entry points listed above.
In order to demonstrate mastery of the learning objectives, students should be able to:
- Accurately work out costs for different numbers of servings for at least two of the given recipes.
- Identify at least one efficient way of providing a range of desserts within the £50 budget.
- Calculate missing lengths using properties of similar shapes.
- Ask challenging questions and present the answers to these in a clear and accurate manner.

Plenary
Ask students to share their findings with the class. Their findings should include the following.
- Did they have enough information? What question did they ask originally?
- How did the task develop as it went on? What part did they find hardest? What part did they find easiest? What mathematics did they use during the lesson?
- Can they think of any other ways to extend the task?

© HarperCollins*Publishers* 2012

Answers Lessons 1.1 – 1.3

Quick check
1 a 468 b 366 c 54
  d 300 e 102 f 95
2 a 3841 b 41 c 625
3 a 17 b 25 c 5

1.1 Solving real-life problems
Exercise 1A
1 a 6000
  b 5 cans cost £1.95, so 6 cans cost £1.95. 32 = 5 × 6 ÷ 2. Cost is £10.53.
2 a 288 b 16
3 a 38
  b Coach price for adults = £8, coach price for juniors = £4, money for coaches raised by tickets = £12 400, cost of coaches = £12 160, profit = £240
4 £34.80
5 (18.81...) Kirsty can buy 18 models.
6 (7.58...) Eunice must work for 8 weeks.
7 £8.40 per year, 70p per copy
8 £450
9 15
10 3 weeks
11 £48.75
12 Gavin pays 2296.25 – 1840 = £456.25

1.2 Multiplication and division with decimals
Exercise 1B
1 a 0.028 b 0.09 c 0.192 d 3.0264
  e 7.134 f 50.96 g 3.0625 h 46.512
2 a 35, 35.04, 0.04 b 16, 18.24, 2.24
  c 60, 59.67, 0.33
  d 180, 172.86, 7.14
  e 12, 12.18, 0.18
  f 24, 26.016, 2.016
  g 40, 40.664, 0.664
  h 140, 140.58, 0.58
3 a 572 b i 5.72 ii 1.43 iii 22.88
4 a Incorrect as should end in the digit 2
  b Incorrect since 9 × 5 = 45, so answer must be less than 45
5 26.66 ÷ 3.1 (answer 8.6) since approximately 27 ÷ 3 = 9
6 a 18 b 140 c 1.4 d 12 e 21.3 f 6.9
  g 2790 h 12.1 i 18.9
7 a 280 b 12 c 0.18 d 450 e 0.62 f 380 g 0.26 h 240 i 12
8 750
9 300
10 a 27 b i 27 ii 0.027 iii 0.27
11 £54.20
12 Mark bought a DVD, some jeans and a pen.

1.3 Approximation of calculations
Exercise 1C
1 a 50 000 b 60 000 c 30 000
  d 90 000 e 90 000 f 0.5 g 0.3
  h 0.006 i 0.05 j 0.0009 k 10 l 90
  m 90 n 200 o 1000
2 a 56 000 b 27 000 c 80 000
  d 31 000 e 14 000 f 1.7 g 4.1 h 2.7
  i 8.0 j 42 k 0.80 l 0.46 m 0.066
  n 1.0 o 0.0098
3 a 60 000 b 5300 c 89.7 d 110 e 9
  f 1.1 g 0.3 h 0.7 i 0.4 j 0.8 k 0.2
  l 0.7
4 a 65, 74 b 95, 149 c 950, 1499
5 Any correct multiplication such as 200 × 6000, 1000 × 1200, 50 × 24 000, 25 × 48 000
6 Elsecar 750, 849, Hoyland 1150, 1249, Barnsley 164 500, 165 499
7 15, 16 or 17
8 1, because there could be 450 then 449
9 Donte has rounded to 2 significant figures or nearest 10 000

Exercise 1D
1 a 60 000 b 120 000 c 10 000 d 15
  e 140 f 100 g 200 h 0.028 i 0.09
  j 400 k 8000 l 0.16 m 45 n 0.08
  o 0.25 p 4 000 000 q 360 000
2 a 5 b 50 c 25 d 600 e 3000
  f 5000 g 2000 h 2000 i 400
  j 8000 k 4 000 000 l 3 200 000
3 a 54 400 b 16 000
4 30 × 90 000 = 2 700 000
  600 × 8000 = 4 800 000
  5000 × 4000 = 20 000 000
  200 000 × 700 = 140 000 000
5 1400 million

© HarperCollins*Publishers* 2012

Answers Lessons 1.1 – 1.3

Exercise 1E
1 a 35 000 b 15 000 c 960 d 5
  e 1200 f 500
2 a 39 700 b 17 000 c 933 d 4.44
  e 1130 f 550
3 a 4000 b 10 c 1 d 20 e 3 f 18
4 a 4190 b 8.79 c 1.01 d 20.7
  e 3.07 f 18.5
5 a £3000 b £2000 c £1500 d £700
6 a £15 000 b £18 000 c £17 500
7 £20 000
8 8p
9 $1000
10 a 40 miles per hour b 10 gallons
  c £70
11 a 80 000 b 2000 c 1000 d 30 000
  e 5000 f 200 000 g 75 h 140 i 100
  j 3000
12 a 86 900 b 1760 c 1030 d 29 100
  e 3930 f 237 000 g 84.8 h 163
  i 96.9 j 2440
13 Approximately 500
14 1000 or 1200
15 400 or 500
16 a i 27.571 428 57 ii 27.6
  b i 16.896 516 39 ii 16.9
  c i 704.419 889 5 ii 704

Exercise 1F
1 a 1.74 m b 6 minutes c 240 g
  d 83°C e 35 000 people
  f 15 miles g 14 m^2
2 82°F, 5 km, 110 min,
  43 000 people, 6.2 seconds, 67th,
  1788, 15 practice walks, 5 seconds.
  The answers will depend on the
  approximations made. Your
  answers should be to the same
  order as these.
3 40
4 300 miles
5 40 × £20 = £800
6 40 minutes
7 60 stamps
8 270 fans
9 80 000 kg (80 tonnes)
10 22.5° C − 18.2° C = 4.3 Celsius
  degrees
11 149 000 000 ÷ 300 000 = 496.66
  ≈ 500 seconds
12 Macau's population density is
  approximately 710 000 times the
  population density of Greenland.

Examination questions
1 6 weeks
2 17 boxes
3 13
4 a 30.946 944 26 b 30.95
5 a 3.586 440 678 b 3.59
6 4200
7 Briony
8 20 cartridges
9 £0.79 × 500 = £395; £14.95 × 24 =
  £358.80; so the two-year contract is
  cheaper.
10 20^2 ÷ (5 × 10) = 400 ÷ 50 = 8
11 a 0.873 581 8   b 0.874
12 Sal £5, Bill £7
13 18 km ≈ 3 hours 36 minutes.
  1400 m climbed ≈ 2 hour 20
  minutes.
  3 h 36 min + 2 h 20 min
  = 5 h 56 min which is about 6
  hours.

© HarperCollins*Publishers* 2012

Chapter 2 Number: More number

Overview

> 2.1 Multiples, factors, prime numbers, powers 2.3 Negative numbers
>
> 2.2 Prime factors, LCM and HCF and roots

This chapter covers multiples, factors, prime numbers, powers and roots, which are revised in preparation for writing numbers as products of prime factors, least common multiples (LCM) and highest common factors (HCF). The final lesson deals with negative numbers and the four rules (+, −, ×, ÷), with emphasis on multiplying and dividing negative numbers.

Context

The ability to be able to use mental arithmetic to carry out calculations is essential in many aspects of life. Improving these skills will prove useful for daily life, now, during working life, and beyond. In addition, mental arithmetic skills will improve students' ability at maths in general, as will be demonstrated in this chapter.

Curriculum references
KS4 Programme of Study references
1.1a Applying suitable mathematics accurately within the classroom and beyond.
1.3b Understanding that mathematics is used as a tool in a wide range of contexts.
2.1d Select mathematical information, methods, tools and models to use.
2.2p Estimate, approximate and check working.
3.1a Real numbers, their properties and their different representations.
3.1b Rules of arithmetic applied to calculations and manipulations with real numbers.
4a Develop confidence in an increasing range of methods and techniques.

Linear specification
N1.2 Add, subtract, multiply and divide any number.
N1.6 The concepts and vocabulary of factor (divisor), multiple, common factor, highest common factor, least common multiple, prime number and prime factor decomposition.
N1.7 The terms square, positive and negative square root, cube and cube root.
N1.8 Index notation for squares, cubes and powers of 10.

Functional Skills standards
R2 Identify the situation or problems and identify the mathematical methods needed to solve them.
R3 Select a range of mathematics to find solutions.
A1 Apply a range of mathematics to find solutions.
I1 Interpret and communicate solutions to multistage practical problems in familiar and unfamiliar contexts and situations.
I2 Draw conclusions and provide mathematical justifications.

PLTS
Independent enquirers identify questions to answer and problems to resolve. **Creative thinkers** ask questions to extend their thinking. **Reflective learners** assess themselves and others, identifying opportunities and achievements. **Team workers** collaborate with others to work towards common goals. **Self-managers** organise time and resources, prioritising actions. **Effective participators** discuss issues of concern, seeking resolution where needed.

Chapter 2 Number: More number

APP
Numbers and the number system L4.2 Recognise and describe number relationships including multiple, factor and square. **L5.3** Recognise and use number patterns and relationships. **Using and applying mathematics L8.3** Select and combine known facts and problem-solving strategies to solve problems of increasing complexity.

Route mapping

Exercise	Grades				
	D	C	B	A	A*
A	1–12	13	14–15		
B		all			
C		1–7	8–9		
D	all				
E	1–7	8–9			10

Overview test
Question 2 is grade G, question 4 is grade F and question 5 is grade E.
Answer the following questions **without** using a calculator.
1 Write down the following.
a 3 multiples of 7
b 3 square numbers less than 90
c 3 factors of 30
2 Work out the following.
a £21.74 × 6 b £52.50 ÷ 6 c 42 ×268
d 837 ÷ 31 e (2 ÷ 5)2
3 Write down three prime numbers.

Answers to overview test
1 a Any three from: 7, 14, 21, 28, 35, 42 ...
b Any three from: 1, 4, 9, 16, 25, 36, 49, 64, 81
c Any three from: 1, 2, 3, 5, 6, 10, 15, 30
a £130.44 b £8.75 c 11 256
d 27 e 49
3 Any three from: 2, 3, 5, 7, 11, 13, 17, 19 ...

Why this chapter matters
Cross-curricular: The work in this chapter links to the sciences, Design and Technology and any subject that requires 'numbers' knowledge.
Introduction: Numbers are the framework of everyday life, so being able to work with multiples, prime numbers, powers and roots, including prime factors, LCM and HCF is advantageous. A grasp of negative numbers is also a useful skill in relation to temperatures, finance and many other areas.
Discussion points: How will knowledge of numbers be an asset in careers such as weather reporting, geology or finance? Could a marketing executive take on the job without good number skills?
Plenary: Elicit from students, some number-related problems that relate to the discussion points. Ask them to include problems that deal with negative numbers. Work through the problems as a class.

Answers to the quick check test can be found at the end of the chapter.

© HarperCollins*Publishers* 2012

2.1 Multiples, factors, prime numbers, powers and roots

Curriculum references

Functional Skills standards
R3 Select a range of mathematics to find solutions.

PLTS
Independent enquirers, Creative thinkers, Reflective learners, Team workers, Self-managers, Effective participators

APP
Numbers and the number system L4.2; L5.3

Collins references
- Student Book pages 28–30

Learning objectives
- Find multiples and factors. [Grade D]
- Identify square numbers and triangular numbers. [Grade D–C]
- Find square roots. [Grade D–B]
- Identify prime numbers. [Grade D]
- Identify cubes and cube roots. [Grade D–C]

Learning outcomes
- Students **must** be able to identify multiples, factors and prime numbers. [Grade D]
- Students **should** be able to find triangular numbers and cube numbers. [Grade D]
- Students **could** be able to use knowledge of square numbers to find square roots of decimals. [Grade B]

Key words
- cube roots
- square roots
- cubes
- squares
- factor
- triangle number
- multiple
- triangular number
- prime number

Prior knowledge
Most of the learning objectives have been covered previously. This lesson is a reminder of the skills needed.

Common mistakes and remediation
Students often forget that 1 and the number itself are always factors and will omit them from the full list of factors. Some students will think that 1 is a prime number, but forget that it is a square number and a cube number. Regular reminders should help to avoid these errors. As these topics come up within many other topics, e.g. in patterns and sequences, it is always useful to make a point of revisiting or revising these at every opportunity.

Useful tips
Remind students that a prime number always has exactly two factors: itself and 1. This effectively excludes 1, which only has one factor.

Functional maths and problem-solving help
In Exercise 2A, question 2 falls into the category of functional maths (FM) because in real life someone might need an answer to this type of question. Students need to realise that it is a question about multiples. It is a lead-in to the next section on least common multiples. In the plenary it would be useful to discuss this question with students to establish how they set out their working, e.g. 10, 20, 30, 40, 50 ... and 8, 16, 24, 32, 40, 48 ... or whether they used a less systematic approach. Encourage students to present work clearly, so that their method is easy to follow. Stress that students need to read problem-solving (PS) questions 5, 11 and 13 carefully, to find the information they need.

© HarperCollins*Publishers* 2012

Lesson 2.1 Multiples, factors, prime numbers, powers and roots

Starter
- Ask students to explain what multiples, factors, prime numbers, square numbers, triangle numbers, square roots, cubes and cube roots are. If they can't explain, ask them to give an example or two.
- Now ask students to pick out, from the whole numbers 1 to 20, all the multiples of 3 (3, 6, 9, 12, 15 and 18), the factors of 20 (1, 2, 4, 5, 10 and 20), the prime numbers (2, 3, 5, 7, 11, 13, 17 and 19), the squares (1, 4, 9 and 16), the triangular numbers (1, 3, 6, 10 and 15) and finally the cubes (1 and 8).
- Check that students know the square root of 9 (3 or –3) and the cube root of 8 (2) and –8 (–2).

Main lesson activity
- Write the words 'Prime number', 'Square' and 'Cube' as column headings on the board.
- Under the appropriate headings, list the numbers 1 to 20 from the starter. Ask students to complete the lists for all numbers up to 100 to find out how many of each type of number there are. (Prime numbers 25; Squares 10; Cubes 4)
- Extend the lists to 200 for **more able** students. (Prime numbers 46; Squares 14; Cubes 5)
- Highlight numbers that are in two or more columns. (1 and 64 are both squares and cubes.) Point out that square numbers and cube numbers can never be prime numbers.
- Explain that to find a number that is not prime, all that is needed is a factor that is not itself or 1, e.g. 432 is not prime as it is even and the only even prime number is 2.
- Students can now do Exercise 2A, in which question 2 is FM, 5, 11 and 13 are PS and 10 is an assessing understanding (AU) question. **Less able** students may need support for questions 13–15.

Plenary
- Discuss question 2 of Exercise 2A, as described under 'Functional maths and problem-solving help' in order to look at strategies for solving the question.
- Finish by having a counting game using any of the words prime, square, cube or none. The first student says '1' and 'square' (or 'cube'), the second student says '2' and 'prime', the third student says '3' and 'prime', the fourth student says '4' and 'square', and so on. When a mistake is made that person is out and the game starts again from 1.
- **More able** students can include answers such as 5 and square root of 25 or 6 and cube root of 216.

Answers
Answers to Exercise 2A can be found at the end of this chapter.

© HarperCollins*Publishers* 2012

2.2 Prime factors, LCM and HCF

Curriculum references

Functional Skills standards
I2 Draw conclusions and provide mathematical justifications.

PLTS
Independent enquirers, Creative thinkers, Reflective learners, Team workers, Self-managers, Effective participators

APP
Numbers and the number system L5.3

Collins references
- Student Book pages 31–36
- Interactive Book: Common Misconceptions

Learning objectives
- Identify prime factors. [Grade C]
- Identify the least common multiple (LCM) of two numbers. [Grade C]
- Identify the highest common factor (HCF) of two numbers. [Grade C]

Learning outcomes
- Students **must** be able to express a whole number as a product of prime factors. [Grade C]
- Students **should** be able to find the HCF or LCM of two numbers by listing factors or multiples. [Grade C]
- Students **could** be able to find the HCF and LCM of two numbers from their prime factor decompositions. [Grade C]

Key words
- highest common factor (HCF)
- index notation
- least common multiple (LCM)
- prime factor
- prime factor tree
- product
- product of prime factors

Prior knowledge
Students should understand what a factor is and they should know what a prime number is.

Common mistakes and remediation
A common mistake, when students are using the division method, is to leave out the last factor when rewriting the answer in index notation. Encourage students to check their answer by multiplying together the factors and checking that the answer is the original number. When finding prime factors, a common mistake is not using prime factors, e.g. 36 = 2 × 2 × 9. Stress the importance of checking that they have answered the question.

Useful tips
If appropriate, display these facts on a wallchart. A factor is a number that divides exactly into a given number, e.g. 9 is a factor of 27. A prime number is divisible by two numbers, itself and 1, so 1 is not a prime number.

Functional maths and problem-solving help
In functional maths (FM) question 6 of Exercise 2C, students need to realise that it is testing least common multiples, but that they must then interpret the question. **Less able** students may say that 24 is their final answer, rather than 3 packs of cheese slices and 4 packs of bread rolls.
Note: Students need whiteboards and should work in pairs for the starter. (The extended starter deals with prime factors and products of prime factors while the main lesson activity covers LCM and HCF. It may be preferable to split the lesson into two or more sessions so that this starter becomes a starter and a main lesson activity for Exercise 2B and the main lesson activity is split into starter and main lesson activity for LCM and HCF.)

© HarperCollins*Publishers* 2012

Lesson 2.2 Prime factors, LCM and HCF

Starter
- Give students a number. On one whiteboard, each pair must write all the factors of the number and on the other they should have written 'prime'. For every prime number they must hold up both boards. If it is not a prime number, they should simply hold up the board with the factors written on it. For example, if they are given the number 5, then they would write 1, 5 on one board and hold that up alongside the 'prime' board. If they are given the number 6, they would hold up only one board with the numbers 1, 2, 3 and 6 on it.
- Remind students that prime factors are factors that are also prime numbers. Give students a number and ask them to write the prime factors onto the plain boards.
- Work through Examples 1, 2, 3 and 4. Then students can do Exercise 2B, in which question 6 is problem-solving (PS) and 5 is an assessing understanding (AU) question. **Less able** students could work on questions 5–7 with a partner.

Main lesson activity
Least common multiple
- Ask students to give multiples of 2. (2, 4, 6, 8, 10, 12 ...)
  Now ask for multiples of 5. (5, 10, 15, 20 ...)
  Now ask for multiples that are common to both lists. (10, 20, 30 ...)
  Ask students to say which common multiple is the least (10). Explain that is called the least common multiple (LCM). So 10 is the LCM of 2 and 5.
- Work through Example 5. Most students will prefer this method and will have more success with it. Advise **less able** students to concentrate on this method and to complete questions 1–6 of Exercise 2C at this stage.
- Show **more able** students Example 6, in readiness for completing the exercise later.

Highest common factor
- Ask students to work in pairs, writing factors of numbers on their whiteboards. Give one student the number 12 and ask for the factors. (1, 2, 3, 4, 6, 12) Give the second student the number 15 and ask for the factors. (1, 3, 5, 15) Now ask them for the highest number they have in common on their whiteboards. (3) Explain that this is called the highest common factor. Repeat for other numbers.
- Now work through Example 7. Most students will prefer this method and will have more success with it. Advise **less able** students to concentrate on this method and to complete questions 7 and 9 of Exercise 2C, before question 8. Show **more able** students Example 8 and ask them to complete the exercise, in which question 6 is FM; 4, 8 and 9 are PS; and 9 is an AU question.

Plenary
- The Student Book shows two ways to find the LCM, which are likely to suit kinaesthetic learners. A third way, which may suit visual learners more, is to use Venn diagrams with prime factors. For example, draw this Venn diagram for the numbers 42 and 63.

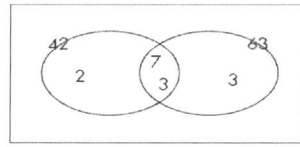

- To find the LCM, multiply all the numbers from all the different regions in the diagram: 2 × 3 × 3 × 7 = 126
- Also, the HCF of the two numbers is the biggest factor shared by both of them. From the diagram, this is the product of all the numbers in the intersection of the sets. (3 × 7 = 21)
- Now let students use their preferred method to find the LCM and HCF of 28 and 40. (LCM = 280, HCF = 4)

Answers
Answers to Exercises 2B and 2C can be found at the end of this chapter.

© HarperCollins*Publishers* 2012

2.3 Negative numbers

Curriculum references

Functional Skills standards
A1 Apply a range of mathematics to find solutions.

PLTS
Independent enquirers, Creative thinkers, Reflective learners, Team workers, Self-managers, Effective participators

APP
Not in assessment criteria.

Collins references
- Student Book pages 37–41
- Interactive Book: Paper Animation

Learning objectives
- Multiply and divide positive and negative numbers. [Grade D]

Learning outcomes
- Students **must** be able to multiply or divide two whole numbers that can be positive or negative. [Grade D]
- Students **should** be able to evaluate calculations involving negative numbers and brackets.
- [Grade D]
- Students **could** be able to solve word problems involving negative numbers. [Grade C]

Key words
- negative
- positive
- order

Prior knowledge
Students must be confident with the order of operations (BIDMAS/BODMAS).

Common mistakes and remediation
The common mistake is that students use these rules when they are set addition or subtraction questions. For example, −3 − 5 will be given as +8, as students interpret the rule that two negatives make a positive incorrectly.
To avoid this, it is important to stress that the rule is: two negative numbers multiplied or divided make a positive number.
Question 2 of Exercise 2D has some parts that will test whether students fully understand this.

Useful tips
To help the class, especially **less able** students, display this information on a wallchart.
- Multiplying or dividing two negative numbers gives a positive number.
- Multiplying or dividing a negative number by a positive, or vice versa, gives a negative number.

$$+ \times + = +$$
$$+ \times - = -$$
$$- \times + = -$$
$$- \times - = +$$

$$+ \div + = +$$
$$+ \div - = -$$
$$- \div + = -$$
$$- \div - = +$$

Functional maths and problem-solving help
Students need to know for question 8 of Exercise 2D: a drop of 2 °C for 12 days requires a calculation of 12 × −2. This question attempts to make the connection. The problem-solving (PS) questions are straightforward.
There are no functional maths (FM) questions in Exercises 2D or 2E.

© HarperCollins*Publishers* 2012

Lesson 2.3 Negative numbers

Starter

×	−3	−2	−1	0	+1	+2	+3
−3				0			
−2				0			
−1				0			
0	0	0	0	0	0	0	0
+1				0			
+2				0			
+3				0			

- Ask students to copy the multiplication table on the left and fill in the top left-hand corner. Explain that you have already filled in the zeros as any number multiplied by zero is zero.

- Check that students' tables appear like this.

×	−3	−2	−1	0	+1	+2	+3
−3	9	6	3	0			
−2	6	4	2	0			
−1	3	2	1	0			
0	0	0	0	0	0	0	0
+1				0			
+2				0			
+3				0			

- Now ask students to read the top row from left to right and continue the pattern: 9, 6, 3, 0, −3, −6, −9
- Ask students to use the pattern to complete the top row.
- Do the same for the second and third rows. It should now appear as shown below.

×	−3	−2	−1	0	+1	+2	+3
−3	9	6	3	0	−3	−6	−9
−2	6	4	2	0	−2	−4	−6
−1	3	2	1	0	−1	−2	−3
0	0	0	0	0	0	0	0
+1				0			
+2				0			
+3				0			

×	−3	−2	−1	0	+1	+2	+3
−3	9	6	3	0	−3	−6	−9
−2	6	4	2	0	−2	−4	−6
−1	3	2	1	0	−1	−2	−3
0	0	0	0	0	0	0	0
+1	−3	−2	−1	0	1	2	3
+2	−6	−4	−2	0	2	4	6
+3	−9	−6	−3	0	3	6	9

Main lesson activity

- Ask students to complete the table, using patterns horizontally and vertically (above).

- Now summarise their results with this table (right).

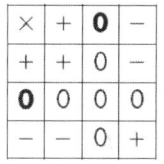 or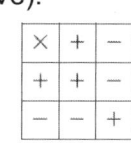

- Explain that, working backwards, they can show exactly the same results for division, e.g. −3 × +2 = −6, so −6 ÷ −3 = +2 and so on. Modify the table to show this.

× or ÷	+	−
+	+	−
−	−	+

- Stress that these results apply for multiplication and division only. Encourage students to learn this fact: two negative numbers multiplied or divided make a positive number. Students can now complete Exercise 2D. **Less able** students should write down the answers so that they can include the necessary working in stages.
- Revise the order of mathematical operations with students, using BIDMAS/BODMAS.
- Students can now complete Exercise 2E.

Plenary

- Ask each student a simple mental question (no writing allowed), e.g. 4 × −3, 8 ÷ −2, −5 × −9.
- Now ask some additions or subtractions to check that the common error, as above, is not being made. If it is, then ask, "You are in a lift; you go down 3 floors, then you go down another 5 floors; how many have you gone down altogether?" Relate this to −3 −5 = +8. Repeat for other numbers.

Answers

Answers to Exercise 2D and E can be found at the end of this chapter.

© HarperCollins*Publishers* 2012

Functional maths
Creating your own allotment

Curriculum references

Functional Skills standards
R2 Identify the situation or problems and identify the mathematical methods needed to solve them.
I1 Interpret and communicate solutions to multistage practical problems in familiar and unfamiliar contexts and situations.

PLTS
Reflective learners, Self-managers, Creative thinkers

APP
Using and applying mathematics L8.3

Collins references
- Student Book pages 44–45

Learning objectives
Use the idea of multiples being evenly spaced and apply this to a practical context.

Key words
- distance
- multiples
- scale

Functional maths help
Students will need to use a simple scale to produce the diagrams of their allotments. Hold a discussion on potential scales and allow students to decide as a class on a suitable scale. For the benefit of **less able** students, you could provide them with a scale, for example: 2 cm : 1 m.

Lesson plans
Build
Have a brief class discussion about allotments and their usefulness for people living in towns and cities. Ask students if they know anyone who rents an allotment. If so, what do these people grow, and what do they do with the produce? Then briefly discuss how, and why, students would divide a 3 m by 5 m allotment if they had one.

Activity 1
Ask students first to look at how many of each plant they can fit in a row and then how many rows they can fit in the allotment. This will help them develop problem-solving skills.

Activity 2
Students are required to work out how to store their gardening tools and equipment. In this functional task, they should make an accurate scale drawing of the shed wall before beginning to add the boxes, as in the table in the Student Book. Encourage them to be creative.

Apply
Refer students to the text in the 'Facts' and 'Handy hints' boxes in the Student Book.
Ask students to solve the functional problems in Activities 1 and 2, working in groups, with some whole-class discussion to guide their progress. **More able** students can work on their own to complete the tasks.
You could use the following questions to act as prompts to encourage students to think about the key elements needed to solve the problem. In particular, these will help **less able** students.
- How many plants of each kind should be grown in the allotment?
- Is there any benefit to restricting the varieties used?
- What combination of plants would yield the most crops?
- How do you balance the number of plants planted with the work required?
- Does the use of different scales help with this problem?
- How much space should be left around the edge of the allotment?

© HarperCollins*Publishers* 2012

Functional maths Creating your own allotment

Less able students may benefit if you define the task in Activity 1 further, by asking them to fill the allotment with only one type of vegetable. This will restrict the number of parameters that they need to negotiate.

Activity 2 will challenge students, and shelving arrangements will vary. Check students' arrangements and accept all valid suggestions. Encourage students to think about how far a box will project. They should try to imagine standing in the shed, and moving among boxes to try to reach a particular one.

Master

In order to demonstrate mastery of the learning objectives, students should be able to:
- Make assumptions about designs.
- Use measurements as the basis of designs.
- Consider various layout plans.
- Present the final designs with scale drawings.
- Clearly communicate the choices made and give justifications for each choice.
- Evaluate each design and decide why one design may be better than another design.

Plenary

Students should present their allotment designs to the class, giving some justification of the choices made.

More able students could bring together their work as a full report to present to an 'allotment committee'.

Then students should present their shed-wall designs, justifying why they placed the boxes where they did. If there is time, invite students to provide feedback of one good thing about each presentation and one thing that could be improved. If there could be cause for offence, divert these comments; instil in students the need to respect one another's viewpoints.

Answers Lessons 2.1 – 2.3

Quick check
1 **a** 4427 **b** 36 **c** 36
2 Answers will vary.
3 **a** 64 **b** 144 **c** 13

2.1 Multiples, factors, prime numbers, powers and roots
Exercise 2A
1 **a** 12 **b** 9 **c** 6 **d** 13 **e** 15 **f** 14 **g** 16 **h** 10
 i 18 **j** 17 **k** 8 (or 16) **l** 21
2 4 packs of sausages and 5 packs of buns (or multiples of these)
3 24 seconds
4 30 seconds
5 12 minutes; Debbie will have run 4 laps; Fred will have run 3 laps.
6 1 + 3 + 5 + 7 + 9 = 25, 1 + 3 + 5 + 7 + 9 + 11 = 36, 1 + 3 + 5 + 7 + 9 + 11 + 13 = 49, 1 + 3 + 5 + 7 + 9 + 11 + 13 + 15 = 64
7 **a** –2 **b** –5 **c** –7 **d** –1 **e** –9 **f** –11 **g** –12 **h** –20 **i** –30 **j** –13
8 **a** 1 **b** 3 **c** 4 **d** 2 **e** 10 **f** –2 **g** –1 **h** 20 **i** 40 **j** –4
9 **a** 1, 3, 6, 10, 15, 21, 28, 36, 45, 55, 66, 78, 91, 105
 b Adding consecutive pairs gives you square numbers.
10

	Square number	Factor of 56
Cube number	64	8
Multiple of 7	49	28

11 **a** These numbers of dots can be arranged in a triangle pattern.
 b 21, 28, 36, 45, 55
12 2, 3 and 12
13 **a** 1, 64, 729, 4096, 15 625
 b 1, 8, 27, 64, 125
 c $\sqrt{a^3} = a \times \sqrt{a}$
 d Square numbers
14 **a** 0.2 **b** 0.5 **c** 0.6 **d** 0.9 **e** 1.2 **f** 0.8 **g** 1.1 **h** 1.5
15 The answers will depend on the approximations made. Your answers should be to the same order as these.
 a 60 **b** 1500 **c** 180

2.2 Prime factors, LCM and HCF
Exercise 2B
1 **a** 84 = 2 × 2 × 3 × 7
 b 100 = 2 × 2 × 5 × 5
 c 180 = 2 × 2 × 3 × 3 × 5
 d 220 = 2 × 2 × 5 × 11
 e 280 = 2 × 2 × 2 × 5 × 7
 f 128 = 2 × 2 × 2 × 2 × 2 × 2 × 2
 g 50 = 2 × 5 × 5
2 **a** 84 = $2^2 \times 3 \times 7$ **b** 100 = $2^2 \times 5^2$
 c 180 = $2^2 \times 3^2 \times 5$
 d 220 = $2^2 \times 5 \times 11$
 e 280 = $2^3 \times 5 \times 7$
 f 128 = 2^7 **g** 50 = 2×5^2
3 1, 2, 3, 2^2, 5, 2 × 3, 7, 2^3, 3^2, 2 × 5, 11, $2^2 \times 3$, 13, 2 × 7, 3 × 5, 2^4, 17, 2 × 3^2, 19, $2^2 \times 5$, 3 × 7, 2 × 11, 23, $2^3 \times 3$, 5^2, 2 × 13, 3^3, $2^2 \times 7$, 29, 2 × 3 × 5, 31, 2^5, 33, 2 × 17, 5 × 7, $2^2 \times 3^2$, 37, 2 × 19, 3 × 13, $2^3 \times 5$, 41, 2 × 3 × 7, 43, $2^2 \times 11$, $3^2 \times 5$, 2 × 23, 47, $2^4 \times 3$, 7^2, 2 × 5^2
4 **a** 2 is always the only prime factor.
 b 64, 128 **c** 81, 243, 729
 d 256, 1024, 4096
 e $2^1, 2^2, 2^3, 2^4, 2^5, 2^6; 3^1, 3^2, 3^3, 3^4, 3^5, 3^6; 4^1, 4^2, 4^3, 4^4, 4^5, 4^6$
5 **a** 2 × 2 × 3 × 5 **b** $2^2 \times 3 \times 5$
 c 120 = $2^3 \times 3 \times 5$, 240 = $2^4 \times 3 \times 5$, 480 = $2^5 \times 3 \times 5$
6 **a** $7^2 \times 11^2 \times 13^2$ **b** $7^3 \times 11^3 \times 13^3$
 c $7^{10} \times 11^{10} \times 13^{10}$
7 Because 3 is not a factor of 40 so it does not divide exactly.

Exercise 2C
1. a 20 b 56 c 6 d 28 e 10 f 15 g 24 h 30
2. They are the two numbers multiplied together.
3. a 8 b 18 c 12 d 30
4. No. The numbers have a common factor. Multiplying them together would mean using this factor twice, thus increasing the size of the common multiple. It would not be the least common multiple.
5. a 168 b 105 c 84 d 84 e 96 f 54 g 75 h 144
6. 3 packs of cheese slices and 4 packs of bread rolls
7. a 8 b 7 c 4 d 14 e 4 f 9 g 5 h 4 i 3 j 16 k 5 l 18
8. a ii and iii b iii
9. 18 and 24

2.3 Negative numbers
Exercise 2D
1. a −15 b −14 c −24 d 6 e 14 f 2 g −2 h −8 i −4 j 3 k −24 l −10 m −18 n 16 o 36
2. a −9 b 16 c −3 d −32 e 18 f 18 g 6 h −4 i 20 j 16 k 8 l −48 m 13 n −13 o −8
3. a −2 b 30 c 15 d −27 e −7
4. a −9 b 3 c 1
5. a 16 b −2 c −12 d −32 e 18
6. −1 × 12, 1 × −12, −2 × 6, 2 × −6, −3 × 4, 3 × −4
7. Any appropriate divisions
8. a −24 b 24 degrees c 3 × −6
9. 13 × −6, −15 × 4, −72 ÷ 4, −56 ÷ −8

Exercise 2E
1. a −4 b −6 c 4 d 45 e 6 f 6
2. a 38 b 24 c −3 d −6 e −1 f 2 g −25 h 25 i 0 j −20 k 4 l 0
3. a (3 × −4) + 1 = −11
   b −6 ÷ (−2 + 1) = 6
   c (−6 ÷ −2) + 1 = 4
   d 4 + (−4 ÷ 4) = 3
   e (4 + −4) ÷ 4 = 0
   f (16 − −4) ÷ 2 = 10
4. a 49 b −1 c −5 d −12
5. a 40 b 1 c 78 d 4
6. Possible answer: 3 × −4 ÷ 2
7. Possible answer: (2 − 4) × (7 − 3)
8. (5 + 6) − (7 ÷ 8) × 9
9. a −15 b Check students' answers.
10. a −16 b −36 c ±10i d ± 12i e 1 f −1 g −l h −1 i −25 j 16

Examination questions
1. a 2 × 3 × 7 b 84
2. a 90 b 240 c 6
3. $2^3 \times 5$
4. Possible answer: 18 and 36 or 4 and 9
5. 5 times
6. 120
7. a x = 5 b 3 × 2 × 5 × 5
8. a p = 2, q = 5 b 10
9. a a = 2, b = 5 (or vice versa)
   b There are two possible solutions, 1 and 6 or 2 and 3.
10. a (3, 8)
    b Any pair from (12, 15) or (15, 18)
11. a $7^2 − 5^2$ = 24. Many other answers.
    b $7^2 − 6^2$ = 13. Many other answers.
12. −2
13. a Could be either.
    b n = 31, p = 5. Many other answers.
14. −4
15. 8^2 = 64, 9^2 = 81
    64 < 72 < 81
    $\sqrt{64} < \sqrt{72} < \sqrt{81}$
    8 < $\sqrt{72}$ < 9
16. −10
17. a 4 b 24

Chapter 3 Number: Fractions, percentages and ratios

Overview

3.1	One quantity as a fraction of another	3.5	Compound interest and repeated percentage change
3.2	Adding and subtracting fractions		
3.3	Increasing and decreasing quantities by a percentage	3.6	Reverse percentages (working out the original quantity)
3.4	Expressing one quantity as a percentage of another	3.7	Ratios

The chapter starts with fraction work. Later lessons include working with percentages and ratios.

Context
The ability to use fractions and percentages is a skill required for much of the mathematics used in everyday life – VAT, profit and loss, interest rates and special offers in shops. Students need to understand the techniques to be used. This chapter covers basic practical uses of ratio and proportion.

Curriculum references
KS4 Programme of Study references
1.1a Applying suitable mathematics accurately within the classroom and beyond.
1.1c Selecting appropriate mathematical tools and methods, including ICT.
2.1a Identify the mathematical aspects of the situation or problem.
2.2b Use knowledge of related problems.
2.2l Calculate accurately, using mental methods or calculating devices as appropriate.
3.1c Proportional reasoning, direct and inverse proportion, proportional change and exponential growth.
4a Develop confidence in an increasing range of methods and techniques.

Linear specification
N1.1 Understand integers and place value to deal with arbitrarily large positive numbers.
N2.1 Understand equivalent fractions, simplifying a fraction by cancelling all common factors.
N2.2 Add and subtract fractions.
N2.5 Understand that 'percentage' means 'number of parts per 100'. Use this to compare proportions.
N2.6 Interpret fractions, decimals, percentages as operators.
N2.7h Calculate with fractions, decimals and percentages, including reverse percentage calculations.

Functional Skills standards
R1 Understand routine and non-routine problems in familiar and unfamiliar contexts and situations.
R2 Identify the situation or problems and identify the mathematical methods needed to solve them.
A1 Apply a range of mathematics to find solutions.
A2 Use appropriate checking procedures and evaluate their effectiveness at each stage.
I1 Interpret and communicate solutions to multistage practical problems in different contexts.
I2 Draw conclusions and provide mathematical justifications.

PLTS
Independent enquirers plan and carry out research, appreciating the consequences of decisions. **Creative thinkers** ask questions to extend their thinking. **Reflective learners** invite feedback and deal positively with praise, setbacks and criticism. **Team workers** reach agreements, managing discussions to achieve results. **Self-managers** seek out challenges or new responsibilities and show flexibility when priorities change. **Effective participators** identify improvements that would benefit others.

APP
Numbers and the number system L5.6 Understand simple ratio. **L7** Understand and use proportionality. **L6.1** Use the equivalence of fractions, decimals and percentages to compare proportions. **Calculating L6.1** Calculate percentages and find the outcome of a given percentage increase or decrease. **L6.3** Use proportional reasoning to solve a problem, choosing the correct numbers to take as 100%, or as a whole. **L6.4** Add and subtract fractions by writing them with a common denominator, calculate fractions of quantities (fraction answers), multiply and divide an integer by a fraction. **L7.3** Add, subtract, multiply and divide

fractions. **L8.1** Use fractions or percentages to solve problems involving repeated proportional changes **Using and applying mathematics L7.1** Solve demanding problems and evaluate solutions.

Route mapping

Exercise	Grades			
	D	C	B	A
A	1–6	7–12		
B	1	2–10	11	
C	1–12	13–15	16	
D	1–10	11–14	15–16	
E	1–4	5–11		
F		1–7	8–15	
G		1	2–11	12–20
H	1–10	11–13		
I		all		
J		1–18	19	

Overview test
Q1 is grade F, Q2 is grade F–E and Q3 is grade C.
1 Cancel these fractions to their simplest form.
  a $\frac{5}{20} =$ b $\frac{4}{14} =$ c $\frac{40}{120} =$
2 Complete the table of equivalent fractions, percentages and decimals, opposite.
3 Find the lowest common multiple (LCM) of 3 and 7.

	Fraction	Percentage	Decimal
a	$\frac{1}{2}$		
b	$\frac{3}{10}$		
c	$\frac{24}{25}$		
d		20%	
e		75%	
f		8%	
g			0.9
h			0.16
i			0.175

Answers to overview test

1 a $\frac{1}{4}$ b $\frac{2}{7}$ c $\frac{1}{3}$

2 a 50%, 0.5   b 30%, 0.3   c 96%, 0.96   d $\frac{1}{5}$, 0.2   e $\frac{3}{4}$, 0.75
   f $\frac{2}{25}$, 0.08   g $\frac{9}{10}$, 90%   h $\frac{4}{25}$, 16%   i $\frac{7}{40}$, 17.5%

3 21

Why this chapter matters
Cross-curricular: This work links to the sciences, Design and Technology, Physical Education, Geography and all subjects where fractions and percentages play a part.
Introduction: One important area that requires an understanding of fractions and percentages is finance. The retail sector also uses percentages when advertising sale bargains.
Discussion points: Why do shops offer 15% off during sales, rather than stating the price? Mia cuts an apple in half and her brother asks for the bigger half. What is wrong with this statement?
Plenary: A bank offers an interest rate of 1.5%, another offers 1%. Which is better? VAT is charged at 17.5%. How much is that on a purchase of £200? Only four-sevenths of club members voted for a candidate to head the committee. The rules say that 75% must be in favour. How can you check if the candidate was elected?

Answers to the quick check test can be found at the end of the chapter.

© HarperCollins*Publishers* 2012

3.1 One quantity as a fraction of another

Curriculum references

Functional Skills standards
R2 Identify the situation or problems and identify the mathematical methods needed to solve them.

PLTS
Independent enquirers, Creative thinkers, Reflective learners, Team workers, Self-managers, Effective participators

APP
Numbers and the number system L6.1

Collins references
- Student Book pages 48–49
- Interactive Book: Multiple Choice Questions

Learning objectives
- Find one quantity as a fraction of another. [Grade D–C]

Learning outcomes
- Students **must** be able to express one quantity as a fraction of another. [Grade D]
- Students **should** be able to compare different proportions using fractions. [Grade D]
- Students **could** be able to solve word problems involving fractions. [Grade C]

Key words
- fraction
- quantity

Prior knowledge
Students should know basic conversions including kilograms to grams and simple everyday facts such as the number of days in March (31).

Common mistakes and remediation
The most common mistake is that students do not match the units before comparing, e.g. 25p as a fraction of £2 may be given as $\frac{25}{2}$ instead of $\frac{25}{200}$. Pointing out this error should help students to avoid it.

Useful tips
Remind students that, to compare two quantities, they must make sure they have common units. Recommend students always to substitute numbers into an expression, using brackets, before working it out. This way, they will get some method marks at least in an examination. If they substitute and work out at the same time and get it wrong, the examiners will not be able to award any method marks as there will be no evidence of working.
When doing simple calculations with direct numbers, it is sometimes helpful to 'say the problem to yourself'. For example, + 2 – – 3 = 5 or plus 5. If you write this on the board, there will be many wrong answers. If you say it to the class as 'plus two minus minus three', there will be fewer wrong answers. But do not encourage students to speak out loud in the examination!

Functional maths and problem-solving help
Functional maths (FM) fits into this topic extremely well. Although question 4 of Exercise 3A is specifically identified as such, many of the other questions have an element of FM. At this level, students simply need to form fractions and apply them.
The problem-solving (PS) aspect is also in evidence in question 12, which requires students to use the information provided to solve them.
Some questions may require students to compare proportions. They will need to interpret this as 'comparing fractions', e.g. asking if 2 out of 7 is more or less than 3 out of 8 should be interpreted as comparing $\frac{2}{7}$ and $\frac{3}{8}$.

Lesson 3.1 One quantity as a fraction of another

Starter
- Use a metre ruler or draw a metre ruler on the board. Show 20 cm on the ruler and ask what fraction this is of the whole ruler. Can this be simplified?
- Repeat with other numbers of centimetres.
- Now ask what fraction 20 cm is of 2 m, etc.
- Look at other measures such as money or time.
- For more able students use fractions. Ask what $12\frac{1}{2}$ cm is as a fraction of the whole ruler ($\frac{1}{8}$) or what $22\frac{1}{2}$ minutes is as a fraction of an hour ($\frac{3}{8}$). Ask students how they work out their answers. Explain that 25 cm is $\frac{1}{4}$ of the whole ruler and $12\frac{1}{2}$ cm is half of that. Similarly for time, explain that $7\frac{1}{2}$ minutes is half of 15 minutes and $22\frac{1}{2}$ minutes is 15 minutes plus $7\frac{1}{2}$ minutes.

Main lesson activity
- Give students three sets of jumbled cards prepared and cut from a grid as shown.
- Ask students to organise the cards so that: a card from Set 1 as a fraction of a card from set 2 = a card from set 3. For example:
  30 cm as a fraction of 1 m = $\frac{3}{10}$

Set 1	Set 2	Set 3
30 cm	1 m	$\frac{3}{10}$
25p	£1	$\frac{1}{4}$
3 hours	1 day	$\frac{1}{8}$
2 days	3 weeks	$\frac{2}{21}$
4 grams	1 kg	$\frac{1}{250}$
5 cl	$\frac{3}{4}$ litre	$\frac{1}{15}$
6 cm²	9 cm²	$\frac{2}{3}$

- Students may spot the correct answers but ultimately, they need to know the conversions.
- Show the class a formal method, e.g. 30 cm as a fraction of 1m is the same as 30 cm as a fraction of 100 cm. So 30 as a fraction of 100 is $\frac{30}{100} = \frac{3}{10}$.

- Work through Example 1 with students. Then ask them to do Exercise 3A, in which question 4 is FM, 12 is PS and 5, 6 and 11 are assessing understanding (AU) questions. The exercise is structured and students should be able to work through the questions. Allow **less able** students to discuss the later questions with a partner, if they have any difficulty.

Plenary
- Explain that later in the course students will need to know fractions of 360° for drawing pie charts.
- Use a circle and ask questions about angles of sectors, such as, "What fraction of a whole circle is 36°?" ($\frac{1}{10}$)

Answers
Answers to Exercise 3A can be found at the end of this chapter.

© HarperCollins*Publishers* 2012

3.2 Adding and subtracting fractions

Curriculum references

Functional Skills standards
A1 Apply a range of mathematics to find solutions.

PLTS
Independent enquirers, Creative thinkers, Reflective learners, Team workers, Self-managers, Effective participators

APP
Calculating L6.4

Collins references
- Student Book pages 49–50
- Interactive Book: Paper Animation

Learning objectives
- Add and subtract fractions with different denominators. [Grade D–C]

Learning outcomes
- Students **must** be able to add or subtract two fractions less than one. [Grade D]
- Students **should** be able to add or subtract two mixed numbers. [Grade D]
- Students **could** be able to solve word problems involving adding or subtracting fractions. [Grade C]

Key words
- denominator
- equivalent fraction
- lowest common denominator

Prior knowledge
Students need to be able to find the least common multiple (LCM) of two or three numbers, and they need to be able to convert mixed numbers into improper fractions and vice versa.

Common mistakes and remediation
Students often fail to convert to equivalent fractions to make the denominators the same – they simply add or subtract the numerators and the denominators. They may change the denominators, but forget to change the numerators. Point out these mistakes at every opportunity and encourage students to take note of the useful tip, below.

Useful tips
To avoid the most common mistake when adding fractions, draw a number line and show that adding numerators and adding denominators always gives a fraction in between the given fractions, and so this method must be incorrect, e.g. $\frac{2}{5}$ is between $\frac{1}{2}$ and $\frac{1}{3}$.
(Correct answer: $\frac{1}{2} + \frac{1}{3} = \frac{3}{6} + \frac{2}{6} = \frac{5}{6}$)

Functional maths and problem-solving help
Functional maths (FM) is again in evidence in this topic. However, questions involving arithmetic with fractions are often poorly answered, simply because standard fraction questions tend to cause problems. For example, in Exercise 3B, question 6, students need to recognise that the calculation is $\frac{2}{5} \times 650$. If students have difficulty interpreting what is required, go through a similar question with simpler numbers. For example: There are 20 students in a school. One-quarter are absent. What is the calculation required? Answer: $\frac{1}{4}$ of 20 or $\frac{1}{4} \times 20 = 5$
More able students can do the question using the more difficult values.
Question 8 highlights the common mistake listed above. Students should spot the mistake immediately. Problem-solving (PS) question 11 is more demanding. Discuss students' methods with them.
Note: Depending on the ability of the group decide whether to show all examples using diagrams or simply to demonstrate the formal written method.

© HarperCollins*Publishers* 2012

Lesson 3.2 Adding and subtracting fractions

Starter
- Write the numbers 2, 3, 4, 5, 6, 7, 8, 9 and 10 on the board. Ask students for the LCM of 2 and 3 (6), 2 and 4 (4), 2 and 5 (10). For more able students use more challenging numbers, e.g. 6 and 8 (24).
- Point out that when fractions are added the lowest common denominator is the LCM of the denominators.

Main lesson activity

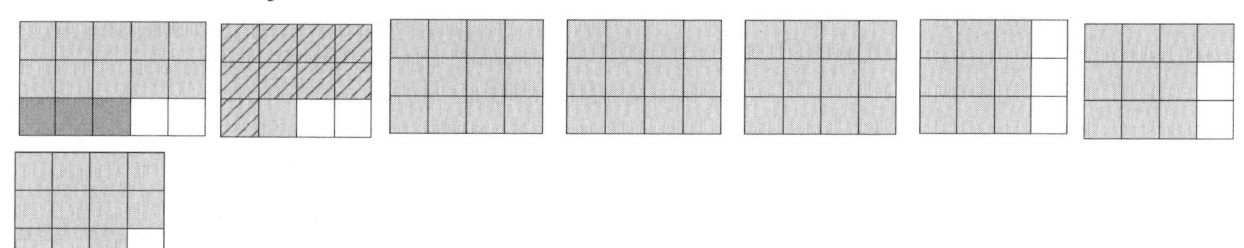

- Write $\frac{2}{3} + \frac{1}{5}$ on the board. Ask students to change each fraction into fifteenths: $\frac{10}{15} + \frac{3}{15}$
- Now draw a 3 by 5 rectangle. Shade 10 squares and then shade 3 squares to show that $\frac{10}{15} + \frac{3}{15} = \frac{13}{15}$
- Now write $\frac{5}{6} - \frac{3}{4}$. Ask students to change each fraction into twelfths: $\frac{10}{12} - \frac{9}{12}$
- Now draw a 3 by 4 rectangle. Shade 10 squares and then shade out 9 of these squares to show that $\frac{10}{12} - \frac{9}{12} = \frac{1}{12}$
- Now write $2\frac{3}{4} - 1\frac{5}{6}$ on the board. Ask students to change the fractions into twelfths: $2\frac{9}{12} - 1\frac{10}{12}$ or $\frac{33}{12} - \frac{22}{12}$
- Draw three 4 by 3 rectangles and shade $2\frac{3}{4}$ as shown. 34 (33 small squares shaded= $\frac{33}{12}$)
- Now draw two 4 by 3 rectangles and shade $1\frac{5}{6}$. (22 small squares shaded = $\frac{22}{12}$)
- Taking 22 small squares away from 33 small squares leaves 11 small squares = $\frac{11}{12}$. So $2\frac{3}{4} - 1\frac{5}{6} = \frac{11}{12}$
- Explain that they can also do these without using diagrams.
- Now show students Examples 2 and 3, which deal with whole numbers and fractions separately.
- Students can now do Exercise 3B.

$$2\frac{3}{4} - 1\frac{5}{6} = 2 + \frac{3}{4} - 1 - \frac{5}{6}$$
$$= 1 + \frac{3}{4} - \frac{5}{6}$$
$$= 1 + \frac{9}{12} - \frac{10}{12}$$
$$= 1 - \frac{1}{12}$$
$$= \frac{11}{12}$$

Plenary
- Write this calculation on the board: $\frac{1}{2} + \frac{1}{3} - \frac{1}{4} - \frac{1}{6}$ ($\frac{6}{12} + \frac{4}{12} - \frac{3}{12} - \frac{2}{12} = \frac{5}{12}$)
- Discuss strategies of what to look for and how to make sure the answer is found most efficiently.
- Ask students to use the fraction button on their calculators to check their answers and correct any errors.
- **Less able** students who have not completed the exercise could use a calculator to do so.

Answers
Answers to Exercise 3B can be found at the end of this chapter.

© HarperCollins*Publishers* 2012

3.3 Increasing and decreasing quantities by a percentage

Curriculum references

Functional Skills standards
A1 Apply a range of mathematics to find solutions.

PLTS
Independent enquirers, Creative thinkers, Reflective learners, Team workers, Self-managers, Effective participators

APP
Calculating L6.1

Collins references
- Student Book pages 51–55
- Interactive Book: Real Life Video

Learning objectives
- Increase and decrease quantities by a percentage. [Grade D–C]

Learning outcomes
- Students **must** be able to increase or decrease a quantity by a given percentage. [Grade D]
- Students **should** be able to answer word problems involving percentage changes. [Grade D]
- Students **could** be able to answer problems involving repeated percentage increases or decreases. [Grade C]

Key words
- multiplier
- percentage

Prior knowledge
Students should be able to work out simple percentages such as $5\% = \frac{5}{100}$
Students need to know how to work out a percentage of a quantity.

Common mistakes and remediation
Errors occur when students are asked to increase or decrease a quantity by a percentage, as they may misread the question and simply work out the percentage. For example, when set the problem of increasing 30 by 5%, students may work out 5% of 30 but not add it on to reach the required answer. Encourage students to read the question again to check that they have answered it correctly.

Useful tips
Make sure students realise that increasing by a percentage will give a larger answer and decreasing will give a smaller answer. Once students have used both methods, as described, suggest that they choose the one they prefer and stick with it.
Remind students that when a percentage is given as, e.g. 5%, it is equivalent to 0.05.

Functional maths and problem-solving help
Many questions in Exercises 3C and 3D come into the category of functional maths (FM) but the most obvious contexts involve money in some form, such as bank interest rates or taxes. Make opportunities to discuss how banks operate, borrowing money from savers and lending it borrowers, usually at a higher rate of interest. Students should also discuss VAT. Point out that VAT may have different rates, depending on what the goods or services are. For example, food is zero-rated. Question 12 of Exercise 3C is a problem-solving (PS) question, which relates to a financial transaction that should be familiar to students.

© HarperCollins*Publishers* 2012

Lesson 3.3 Increasing and decreasing quantities by a percentage

Starter
- Revise the use of percentage multipliers by asking students questions such as, "What is two per cent as a multiplier?" (0.02) Include some of the more difficult questions such as, "What is 0.8% as a multiplier?" (0.008); "What is 3.4% as a multiplier?" (0.034)
- To check prior knowledge, ask students questions about working out a percentage of a quantity, e.g. how would they work out 13% of 40? Lead on to questions involving increasing by a percentage. How would they increase an amount by 13%? Students may suggest working out 13% and adding it on. Repeat for other values.
- Now move on to decreasing by a percentage. Ask students how they would work out 6% of 70. Then ask them how they would decrease 70 by 6%.
- Ask students if they can think what multiplier they would use to increase an amount by 13%. (1.13)

Main lesson activity
- Display this table of multipliers.

1.05	0.05	0.95	1.5	0.5
0.85	0.15	1.15	0.25	1.25
0.75	0.55	0.45	1.55	1.45

- Ask students questions about the multipliers, starting with simple percentages, "What percentage is equivalent to the multiplier 0.05? (5%), 0.45? (45%), 0.5? (50%)" and so on.
- Now ask what percentage is equivalent to the multiplier 1.25 (125%). Check if students understand that this represents a 25% increase. Repeat for 1.5 (50% increase), 1.15 (15% increase), 1.05 (5% increase), 1.55 (55% increase).
- Work through Example 4, pointing out that students can use the 'adding on' or 'subtracting' methods but they need to understand the 'multiplier' method as well.
- Work through Example 5 with students.
- **Less able** students can do Exercise 3C now. They should choose which method to use, ensuring that if they choose to use the multiplier method they can explain why it works.
- Now ask students to identify the multiplier from the table that represents a 5% decrease. (0.95) Repeat for other decreases, 0.85 (15% decrease), 0.75 (25% decrease).
- Advise **more able** students to look at Examples 6 and 7. Go through them with **less able** students. The only difference between the exercises is that Exercise 3C deals with increases and Exercise 3D deals with decreases. Students can now do Exercises 3C and 3D.
- In Exercise 3C, questions 3, 11 and 16 are FM, 12 is PS and 5b and 13–15 are assessing understanding (AU) questions. Be prepared to offer **less able** students extra help, if they need it. Suggest that they work with a partner for any questions they may find difficult.
- In Exercise 3D, questions 5, 10 and 11a are FM, 13, 15 and 16 are PS and 11b, 13 and 14 are AU.
- **Less able** students may find the idea of percentage decrease difficult at first. If so, go through some of the questions with them.

Plenary
- Tell students that they have £100.
- Ask half of the class to increase the £100 by 10% and then decrease the answer by 10%. (£99)
- Ask the other half to decrease the £100 by 10% and then increase the answer by 10%. (£99)
- Then ask all students to show their answer on a whiteboard at the same time. Establish that the result is always £99 but ask students why it is not £100. (The decrease is from a larger amount each time.)

Answers
Answers to Exercises 3C and 3D can be found at the end of this chapter.

© HarperCollins*Publishers* 2012

3.4 Expressing one quantity as a percentage of another

Curriculum references

Functional Skills standards
A2 Use appropriate checking procedures and evaluate their effectiveness at each stage.

PLTS
Independent enquirers, Creative thinkers, Reflective learners, Team workers, Self-managers, Effective participators

APP
Calculating L6.1

Collins references
- Student Book pages 56–58
- Interactive Book: 10 Quick Questions

Learning objectives
- Express one quantity as a percentage of another. [Grade D]
- Work out percentage change. [Grade C]

Learning outcomes
- Students **must** be able to express one quantity as a percentage of another. [Grade D]
- Students **should** be able to express a change as a percentage increase or decrease. [Grade C]
- Students **could** be able to answer word problems involving calculating percentage changes. [Grade C]

Key words
- percentage change
- percentage loss
- percentage increase/decrease
- percentage profit

Prior knowledge
Students need to know how to use a percentage multiplier. They should also know how to write one quantity as a fraction of another quantity.

Common mistakes and remediation
Students frequently forget to make sure the units are the same. Encourage them always to look at units carefully. Often, students do not use the original value as the denominator, but the new value after the increase or decrease. Frequently repeat the formula for percentage change.

Useful tips
Students will mix up increasing a quantity by a percentage and working out a percentage increase. Be careful to use to the correct language to describe each one.

Functional maths and problem-solving help
Students need to be quite clear what information to use when working on functional maths (FM) questions, e.g. question 11. Sometimes questions provide the original and final amounts, as in questions 6 and 7, whereas question 2 simply gives the original amount and the amount he spent, not the amount left after spending. Encourage students, when attempting questions giving original and final amounts, to write down the change before starting to work out the percentage change. Marks are often given for doing this in examinations. Question 9 assesses understanding (AU) and involves very large numbers, but is otherwise structured and straightforward. **More able** students may choose to ignore the final four zeros, but if they do they must take care to be consistent. The outcome is that taking the original amounts, as percentages, gives a final total of 100%.

© HarperCollins*Publishers* 2012

Lesson 3.4 Expressing one quantity as a percentage of another

Starter
- Ask mental questions involving increasing or decreasing quantities by a percentage. For example: Increase £50 by 10% (£55); Increase 100 cm by 2% (102 cm); Decrease 40 kg by 25% (30 kg); Decrease 70 grams by 50% (35 grams)
- Write a table on the board, using the first three columns of the table in the main lesson activity.
- Repeat for other values. Increase the level of difficulty by including decimals or fractions, e.g. increase £18.50 by 10%. (percentage change = £1.85, new amount = £20.35)

Main lesson activity
- Talk about percentage change. If the change is an increase, it is called a percentage increase.
- Ask students what they think each of the following percentage changes is called: Increase 100 cm by 2% (102 cm) Change is a decrease. (percentage decrease); Increase 100 cm by 2% (102 cm) Change is a profit. (percentage profit); Increase 100 cm by 2% (102 cm) Change is a loss. (percentage loss).
- Now ask students to look at the table and see if they know how to get from £50 and £55 to 10%. **Less able** students may need an extra column showing the change. Ask them how to get from £5 and £50 to 10%. (£5 is 10% of £50).
- Repeat for the other rows in the table.

Original amount	Percentage change	New amount	Change
£50	+10%	£55	£5
40 kg	−25%	30 kg	10 kg
100 cm	+2%	102 cm	2 cm
70 grams	−50%	35 grams	35 grams

- Say that to express a quantity as a percentage of another, you write the first quantity as a fraction of the second quantity and then convert to a percentage. Using Examples 8 and 9, explain that this is the same as working out a percentage change; the change in Example 8 is £6 and in Example 9 it is 75 cm.
- Write the formula on the board: $\text{percentage change} = \dfrac{\text{change}}{\text{original amount}} \times 100$
- Show students that the formula works for each row in the table, e.g. for £50 to £55 (change = £5): $\text{percentage change} = \dfrac{5}{50} \times 100 = 10\%$
- Go through Example 10. Students who want to use the multiplier method can also go through Example 11, either now or at the same time as Examples 8 and 9.
- Students can now do Exercise 3E.

Plenary
- Ask the following questions: Express 40p as a percentage of £4 (ask what the obvious mistakes would be); Sam buys a painting for £1500 and sells it for £2000. What is Sam's percentage profit?; If Sam had sold the painting for £1200, what would be the percentage loss?
- Now set a comparison question. Two students take a test scored out of 20. The first student scores 5 and the second scores 10. They retake the test. Both improve their scores by 5. Who has made better progress?
- Discuss the different interpretations of the answers. Some may say both made the same progress as both know the answer to five more questions. Point out that the first student had a percentage increase of 100%, and the second student had a percentage increase of 50%.

Answers
Answers to Exercise 3E can be found at the end of this chapter.

© HarperCollins*Publishers* 2012

3.5 Compound interest and repeated percentage change

Curriculum references

Functional Skills standards
I2 Draw conclusions and provide mathematical justifications.

PLTS
Independent enquirers, Creative thinkers, Reflective learners, Team workers, Self-managers, Effective participators

APP
Calculating L8.1

Collins references
- Student Book pages 58–61
- Interactive Book: 10 Quick Questions

Learning objectives
- Calculate compound interest. [Grade C–B]
- Solve problems involving repeated percentage change. [Grade C–B]

Learning outcomes
- Students **must** be able to calculate the result of a repeated percentage change. [Grade C]
- Students **should** be able to calculate the results of compound interest. [Grade B]
- Students **could** be able to solve practical problems involving repeated percentage change. [Grade B]

Key words
- annual rate
- principal
- multiplier
- compound interest

Prior knowledge
Students should be able to increase an amount by a percentage, preferably using the multiplier method.

Common mistakes and remediation
When doing the year-by-year method, premature rounding or incorrect rounding often leads to incorrect answers. When using the compound interest formula for depreciation, the students often forget that the '+ r%' becomes a '– r%'. Another common mistake is not reading the question properly – does the question ask for the total amount in the bank after *n* years, or for the total interest received after *n* years?

Useful tips
Students often think compound interest only applies to savings account problems. At every opportunity, relate questions to other topics, such as population growth or car depreciation. Encourage the use of a multiplier as the most efficient method. **Less able** students may prefer to do their working one step at a time.

Functional maths and problem-solving help
Functional maths (FM) questions on this topic tend to ask for a prediction, based on current trends. They can take a long time to solve, if students work in single steps. It may be advisable for them to use a trial and improvement method to find the solution, closing in on the correct answer. In Exercise 3F, question 6b, the date is clearly beyond 8 August, so it would be pointless starting again from 1 August. However, as the amount of water on 8 August is nearing 1 million, it is probably easier just to work from that figure in steps, although **less able** students may prefer to use the longer approach, starting from 1 August. In question 12, ensure that students realise that the interest offered by the Bradley bank is paid only once, not each year.

© HarperCollins*Publishers* 2012

Lesson 3.5 Compound interest and repeated percentage change

The method of trial and improvement may also be appropriate for problem-solving (PS) questions on this topic, such as question 3b.
Note: Students should use calculators throughout this lesson.

Starter
- Give the class the following problem:
  Craig invests £100 in a savings account, which earns him 5% interest per annum. How much does Craig have in the account at the end of the first year? (£105), second year? (£110.25) and third year? (£115.76)
- Discuss a quicker method to find out how much Craig had at the end of the third year. (100×1.05^3)
- Students should check the calculation on their calculators.
- Now discuss the advantage of using the multiplier for an investment of 10 years instead of calculating each year separately. (There is just one calculation, so it is much quicker to work out.)

Main lesson activity
- Now check that students can do the calculations using a calculator since the method is already established from the starter. Write on the board: £3000 investment, 6%, 5 years
- Ask students to give the multiplier for the calculation. (1.06) Now ask students for the calculation to work out the final value of the investment. (3000×1.06^5)
- Ask students to do the calculation on their calculators (£4014.68). Ensure that they give the answer in correct money notation.
- Give a few more calculations to work out on their calculators. Use different contexts. The population of a village is 10 000. Each year for three years it falls by 2% of the amount at the end of the year. What is the population after three years? (9411 or 9412)
- Students might prefer to understand the method rather than learn a formula.
- Ask **more able** students to put the method into a formula. They may derive one of the two formulae shown in the Student Book. Refer all students to these formulae as an alternative way of working.
- Work through Example 12. Using calculators, ask students to check that the answer is correct.
- Students can now do Exercise 3F. Remind students to use the multiplier method.

Plenary
- Set these problems.
  - Mr Cash invests £1 000 000 for 5 years at an annual compound interest rate of 5%. By how much does his money grow? (£276281.56)
  - Miss Cash buys a car for £100 000 which depreciates by 5% per year. By how much has the car depreciated after 5 years? (£22 621.91)
  - Mrs Cash invests £1 000 000 at a rate of 5% per annum, compound interest. How many years will it take for this money to reach at least £1.5 million? (9 years)
- The next task is quite challenging and may only be suitable for the **more able** students.
  - Ask students to try to work out what interest rate (nearest whole number) would be needed to double the value of an investment in 5 years. (15%)

Answers
Answers to Exercise 3F can be found in at the end of this chapter.

© HarperCollins*Publishers* 2012

3.6 Reverse percentage (working out the original quantity)

Curriculum references

Functional Skills standards
I2 Draw conclusions and provide mathematical justifications.

PLTS
Independent enquirers, Creative thinkers, Reflective learners, Team workers, Self-managers, Effective participators

APP
Calculating L8.1

Collins references
- Student Book pages 62–65
- Interactive Book: Matching Paper Animations

Learning objectives
- Calculate the original amount, given the final amount, after a known percentage increase or decrease. [Grade C–A]

Learning outcomes
- Students **must** be able to calculate the original amount when the result of a percentage change is known. [Grade B]
- Students **should** be able to solve word problems involving reverse percentage calculations. [Grade A]
- Students **could** be able to demonstrate and explain different methods of calculating reverse percentages. [Grade A]

Key words
- final amount
- original amount
- multiplier
- unitary method

Prior knowledge
Students should be able to work out a multiplier. They also need to know that the whole amount is always equivalent to 100%.

Common mistakes and remediation
A common mistake is to work out 'the percentage of', such as '10% of', rather than working out 10% off. Stress that students should always read the text carefully. The main lesson activity provides an example that could be misconstrued in this way.

Useful tips
Emphasise that this topic is not about working out a percentage, but about working out a quantity by scaling.

Functional maths and problem-solving help
The functional maths (FM) elements of these questions involve interpreting the context to obtain the percentage given, or the multiplier, depending on the method used. In question 10 of Exercise 3G this involves working backwards twice, resulting in the question being a higher grade.
Students can always check answers by working back through the question from their answer to the amount in the question. For example, in question 13 the answer to part **a** is £1600, so 90% of £1600 = £1440 and 90% of £1440 is £1296.
Less able students should work out one year at a time. For example, 90% is equivalent to £1296, so 10% is equivalent to £144 and 100% is equivalent to £1440. Then repeat for the first year, 90% is equivalent to £1440, so 10% is equivalent to £160, so 100% is equivalent to £1600.
The problem-solving (PS) questions (15–19) require students to read them carefully and be systematic.

© HarperCollins*Publishers* 2012

Lesson 3.6 Reverse percentage (working out the original quantity)

Starter
- Tell students that for each statement they must work out 100%. Start with simple percentages: 50% is £3 (100% is £6), 25% is 4 kg (100% is 16 kg), 20% is 5 cm (100% is 25 cm), 10% is 60p (100% is £6).
- Move on to more difficult percentages, using fractions and decimals for the **more able**. Encourage students to give answers in two or more steps: 90% is £18 (10% is £2, so 100% is £20), 75% is 15 kg (25% is 5 kg, so 100% is 20 kg), 37.5% is 30 cm (12.5 % is 10 cm, 25% is 20 cm, 100% is 80 cm).

Main lesson activity
- Develop the questions to give a real context. Ask students to think about this problem: A sale offers 10% off all prices. The sale price of a dress is £18. What was the original price of the dress?
- Point out that '10% off' does not mean '10% of', so they must not work out '10% of'' at any time in this question. On the board, write: 10% off not 10% of
- Now ask students how they would solve the problem. There are two methods. Students should be shown both methods but they should then choose the approach they prefer and stick with it.

The unitary method
- This involves scaling to obtain 1% and then 100%, as used in the starter.
- Show students the three steps for the unitary method in the Student Book then work through the example above.
10% off means the sale price is 90%.
90% is equivalent to £18.
1% is equivalent to 20p, or 10% is equivalent to £2 (dividing by 90 or 9).
So 100% (the original price) = £20.
- Now work through Examples 13 and 14.

The multiplier method
- Show students the two steps of the multiplier method in the Student Book.
- 10% off means a multiplier of 0.9 (writing down the multiplier).
- Working backwards to find the original amount:
£18 ÷ 0.9 gives an original price of £20 (dividing by the multiplier).
Now work through Example 15. Then students can do Exercise 3G.

Plenary
Ask students to help you with a few questions.
- VAT: An MP3 player costs £176.25, including VAT at 17.5%; how much would it have cost without VAT? (£150)
- Salary: Pam now earns £21 840 after a 4% pay rise. How much did she earn before her pay rise? (£21 000)
- Weight: Paul keeps a diary of his weight. He lost 6% of his body weight in one month, then gained 6% of his new body weight during the following month. Did he gain or lose weight overall? What was the percentage change? If he weighs 75 kg now, what was his weight to begin with? (Lose by 0.36%; 75.3 kg)

Answers
Answers to Exercise 3G can be found at the end of this chapter.

© HarperCollins*Publishers* 2012

3.7 Ratio

Curriculum references

Functional Skills standards
A1 Apply a range of mathematics to find solutions.

PLTS
Independent enquirers, Creative thinkers, Reflective learners, Team workers, Self-managers, Effective participators

APP
Numbers and the number system L5.6; Calculating L6.3

Collins references
- Student Book pages 66–74
- Interactive Book: Student Demo

Learning objectives
- Simplify a ratio. [Grade D–C]
- Express a ratio as a fraction. [Grade D]
- Divide amounts into given ratios. [Grade C]
- Complete calculations from a given ratio and partial information. [Grade C–B]

Learning outcomes
- Students **must** be able to express a ratio as a fraction. [Grade D]
- Students **should** be able to divide an amount in a given ratio. [Grade C]
- Students **could** be able to solve a variety of problems involving ratios expressed in words. [Grade C]

Key words
- cancel
- ratio
- common units
- simplest form

Prior knowledge
Students should know the multiplication tables up to 10×10, how to cancel fractions, how to find a fraction of a quantity and how to multiply and divide, both with and without a calculator.

Common mistakes and remediation
Students often forget to express all the quantities in the same units. Keep reminding them to sort out the units first. For example, when expressing 3 : 5 as a fraction, students may simply write it as $\frac{3}{5}$. Ask them, frequently, to tell you the number of parts in the ratio. (8) So in fact, it should be written as $\frac{3}{8}$.

Useful tips
Avoid relating ratios to probability or chance as this will lead to errors later when students think a ratio of 1 : 3 is equal to $\frac{1}{3}$ instead of $\frac{1}{4}$ and $\frac{3}{4}$.

Functional maths and problem-solving help
Most questions in Exercises 3H, 3I and 3J could be considered to be functional maths (FM) as they involve real-life contexts. However, some questions are clearly quite contrived to reinforce the skills required. The more purposeful the question, the more students who will relate to the topic, and the more functional the question will be. For example, in Exercise 3I, questions 6–9, students will readily see the purpose of scales on maps as they can relate it to other subjects. Point out that many maps now simply say, e.g. 1 cm = 3 kilometres.

Starter
- Use quick-fire questions to provide practice in converting between metric units. Cover all units in Exercise 3H, e.g. for money, time, measurement and weight. Write down the conversions for **less able** students. Add other conversions for **more able** students, such as litres and centilitres.

© HarperCollinsPublishers 2012

Lesson 3.7 Ratio

- Write pairs of fractions on the board with either the numerator or denominator from one fraction missing. Ask students to provide the missing number to make the fractions equivalent. Challenge **more able** students with improper fractions using large numbers: $\frac{100}{40} = \frac{?}{5}$ or $\frac{720}{9} = \frac{?}{3} = \frac{?}{1}$. ($\frac{100}{40} = \frac{12.5}{5}$ or $\frac{720}{9} = \frac{240}{3} = \frac{80}{1}$)

Main lesson activity

Common units and ratios as fractions

- Explain that a ratio is a way of comparing the sizes of two or more quantities and that to use ratios any quantities must be given in the same units, e.g. comparing centimetres to centimetres. Say that a ratio can also be expressed as a fraction. The denominator of the fraction is the sum of the numbers in the ratio.
- Ask how many times bigger 1 kilometre is than 1 metre. (1000) Write: 1 metre to 1 kilometre = 1 m : 1 km = 1 : 1000 on the board, and explain that if the units of the quantities are different then one must be converted before the ratio can be formed and simplified. This is because a ratio does not have units.
- Write 1 : 2, 2 : 4, 3 : 6 on the board and explain that these are all equivalent.
- Now write 4 : 12 on the board and ask for some equivalent ratios. (1 : 3, 2 : 6, etc.) Then ask students to draw four squares and shade one of them. □ □ □ ■
- Ask "What is the ratio of shaded to unshaded? (1 : 3) . What fraction are unshaded? ($\frac{3}{4}$) What is the ratio of unshaded to shaded? (3 : 1)".
- Now ask students to draw 8 pint glasses, and shade in three of them (with drinks). Write: 8 pints = 1 gallon.
- Ask "What is the ratio of shaded (drinks) to unshaded (empty)? (3 : 5) What fraction are empty? ($\frac{5}{8}$) What fraction is three pints of a gallon? ($\frac{3}{8}$)". This will reinforce the need to work in one unit (pints).
- Work through Examples 16 and 17, which combine the work on units with the work on ratios. Students can now do Exercise 3H.

Dividing amounts in a given ratio

- Give students a sheet of A4 paper. Ask them to tear it into 16 pieces. Write the ratio 1 : 3 on the board. Tell students to put the pieces into two piles with three in the second pile for every one in the first pile. They have shared the 16 pieces in the ratio 1 : 3. Ask what fraction of the pieces are in the first pile ($\frac{1}{4}$), the second pile ($\frac{3}{4}$)
Say that $\frac{1}{4}$ of 16 pieces = 4 pieces and $\frac{3}{4}$ of 16 pieces = 12 pieces. Work through Examples 18 and 19. Let students do Exercise 3I.

Calculating with ratios when only part of the information is known

- Explain that sometimes, instead of giving the value of the whole amount, students will be told the value of one part of a ratio. For example: I share some money with my brother in the ratio 2 : 3, I get £2. How much does my brother get? (£3) Say that for Exercise 3I the same question would have been: Divide £5 in the ratio 2 : 3.
- Ask **more able** students to rewrite the questions in Exercise 3I in the same way. For **less able** students, talk through a few questions from Exercise 3J and discuss what information is given. Do not work out any answers. Go through Examples 20 and 21. Allow students to choose their favoured method. Let them do Exercise 3J.

Plenary

- For each section, check understanding by going over the FM, PS and AU questions in Exercises 3H–3J.

Answers

Answers to Exercises 3H, 3I and 3J can be found at the end of this chapter.

© HarperCollins*Publishers* 2012

Functional maths
Understanding VAT

Curriculum references

Functional Skills standards
A1 Apply a range of mathematics to find solutions.
I1 Interpret and communicate solutions to multistage practical problems in familiar and unfamiliar contexts and situations.
R1 Understand routine and non-routine problems in familiar and unfamiliar contexts and situations.

PLTS
Reflective learners, Effective participators

APP
Using and applying mathematics L7.1;
Calculating L8.1

Collins references
- Student Book pages 78–79
- Interactive Book: Real Life Video

Learning objectives
- Find percentage increases and reductions.
- Understand that to remove a percentage increase you have to reduce the new price by a different amount.

Key words
- increase
- percentage
- reduction
- VAT

Functional maths help
Point out to students that VAT is applied to many purchases in daily life. Every time they go to the shop to buy a bar of chocolate or soft drink, VAT is added. Suggest to students that they ask their parents or guardians for some till slips, or that they ask for and keep the till slips the next couple of times they go to the shops. Most till slips have a row for VAT.
Ask them to calculate the amounts to see if they are correct.
Less able students will benefit from being told that calculating VAT can be likened to finding percentage increases. For example, the original cost plus the percentage rate equals the total cost.

Let's look at the simple example of an item that costs £10: £10 + (20% × £10) = £12.00

$$\frac{20}{100} \times £10 = £2.00$$

so £10 + £2.00 = £12.00

Explain to students that shop items are usually marked as £9.99, for example. They should remember that in this case, they can still work out VAT by rounding the amount to £10.

Lesson plans
Build
Hold a class discussion to find out what students know about working out VAT. Discuss real-life situations where VAT applies, and when these types of calculations are needed.

Activity 1
Refer students to the first question in 'Getting started' in the Student Book. They should be able to answer this question with little guidance. (• £483 • £493.50 • £504)
Less able students may need help in completing the second question, which involves reversing the percentage calculations. (• £391.30 • 382.98 • £375)

© HarperCollins*Publishers* 2012

Functional maths Understanding VAT

Activity 2
While students work on the first four points of the text in 'Your task', you may find that **less able** students need help in understanding that reducing VAT by 2.5% is not the same as reducing the final price by 2.5%. For example: £100 + 17.5% VAT is £117.50. £117.50 – 2.5% is 117.5 × 0.975 = £114.56 not £115, as £100 + 15% VAT would be, so the shop would make a loss on every sale.
To structure this further, provide some prices inclusive of 17.5% VAT. Ask what price the shop should sell them for after the reduction in VAT. Encourage students to work on several prices before they draw conclusions.

Activity 3
Have a class discussion about what shopkeepers do now to come into line with the increased VAT rate. List all ideas on the board.
Then ask students, in pairs or groups, to find case studies about this topic. They can research this information on the internet, but it would be better if they could find some shop or supermarket managers to interview.
Make sure students understand that they must identify the mathematics the shops used, and evaluate the approach that was taken by the shops.

Apply
More able students should be able to do the second activity with little guidance. They may also reach the final conclusion $\frac{115}{117.5} = \frac{1150}{1175} = \frac{230}{235} = \frac{46}{47}$ of the old price, will result in the new price. **Less able** students may need some help with this last fractional calculation.
They may also be able to conclude that when VAT returns to 17.5% all shopkeepers need to do is to reverse the last calculation and then multiply their prices by $\frac{47}{46}$.
Before students work on their case studies, have a class discussion about how they will go about it. **Less able** students may need help with questions they should ask in order to get the results they need.

Master
If you intend that students should show mastery in this task, allow discussion about the different multipliers and different prices, but try not to guide students through examples.
In order to demonstrate mastery of the learning objectives, students should be able to:
- Calculate percentage increases and reductions.
- Understand how to remove a percentage increase.
- Clearly state the multipliers of $\frac{46}{47}$ and $\frac{47}{46}$, and understand that these give exact conversions.

Plenary
Look at different ways that shops advertise goods and the way some goods need VAT added to arrive at the full price. Discuss net and gross prices and 'earners'.

© HarperCollins*Publishers* 2012

Answers Lessons 3.1 – 3.7

Quick check

1 a $\frac{2}{5}$ b $\frac{3}{8}$ c $\frac{3}{7}$

2
Fraction	Percentage	Decimal
$\frac{3}{4}$	75%	0.75
$\frac{2}{5}$	40%	0.4
$\frac{11}{20}$	55%	0.55

3 a £23 b £4.60 c 23p

3.1 One quantity as a fraction of another
Exercise 3A

1 a $\frac{1}{3}$ b $\frac{1}{5}$ c $\frac{2}{5}$ d $\frac{5}{24}$ e $\frac{2}{5}$
  f $\frac{1}{6}$ g $\frac{2}{7}$ h $\frac{1}{3}$

2 $\frac{3}{5}$

3 $\frac{12}{31}$

4 20 weeks

5 Jon saves $\frac{30}{90} = \frac{1}{3}$

  Matt saves $\frac{35}{100}$ which is greater than $\frac{1}{3}$, so Matt saves the greater proportion of his earnings.

6 $\frac{13}{20} = \frac{65}{100}, \frac{16}{25} = \frac{64}{100}$, so the first mark is better.

7 $\frac{1}{8}$

8 $\frac{5}{12}$

9 $\frac{1}{5}$

10 $\frac{3}{20}$

11 $\frac{3}{10}$

12 32 or 26

3.2 Adding and subtracting fractions
Exercise 3B

1 a $\frac{8}{15}$ b $\frac{7}{12}$ c $\frac{3}{10}$ d $\frac{11}{12}$ e $\frac{1}{10}$ f $\frac{1}{8}$
  g $\frac{1}{12}$ h $\frac{1}{3}$ i $\frac{7}{9}$ j $\frac{5}{8}$ k $\frac{3}{8}$ l $1\frac{1}{15}$

2 a $3\frac{31}{45}$ b $4\frac{47}{60}$ c $\frac{41}{72}$ d $\frac{29}{48}$ e $1\frac{43}{48}$ f $1\frac{109}{120}$
  g $1\frac{23}{30}$ h $1\frac{31}{84}$

3 $\frac{1}{20}$

4 a $\frac{1}{6}$ b 30

5 No, one eighth is left, which is 12.5 cl, so enough for one cup but not two cups.

6 260

7 Three-quarters of 68

8 He has added the numerators and added the denominators instead of using a common denominator.
Correct answer is $3\frac{7}{12}$.

9 Possible answer: The denominators are 4 and 5. I first find a common denominator. The lowest common denominator is 20 because 4 and 5 are both factors of 20. So I am changing the fractions to twentieths. One-quarter is the same as five-twentieths (multiplying numerator and denominator by 5). Two-fifths is the same as eight-twentieths (multiplying numerator and denominator by 4). Five-twentieths plus eight-twentieths = thirteen-twentieths.

10 £51

11 10 minutes

© HarperCollins*Publishers* 2012

3.3 Increasing and decreasing quantities by a percentage
Exercise 3C
1. **a** 1.1 **b** 1.03 **c** 1.2 **d** 1.07 **e** 1.12
2. **a** £62.40 **b** 12.96 kg **c** 472.5 g **d** 599.5 m **e** £38.08 **f** £90 **g** 391 kg **h** 824.1 cm **i** 253.5 g **j** £143.50 **k** 736 m **l** £30.24
3. £29 425 – 7% pay rise
4. 1 690 200
5. **a** Bob: £17 325, Anne: £18 165, Jean: £20 475, Brian: £26 565 **b** 5% of different amounts is not a fixed amount. The more pay to start with, the more the increase (5%) will be.
6. £411.95
7. 193 800
8. 575 g
9. 918
10. 60
11. TV: £294, microwave: £86.40, CD: £138, stereo: £35.40
12. £10
13. **c** Both same as 1.05 × 1.03 = 1.03 × 1.05
14. **d** Shop A as 1.04 × 1.04 = 1.0816, so an 8.16% increase.
15. £540.96
16. Calculate the VAT on certain amounts, and $\frac{1}{6}$ of that amount. Show the error grows as the amount increases. After €600 the error is greater than €10, so the method works to within €10 with prices up to €600.

Exercise 3D
1. **a** 0.92 **b** 0.85 **c** 0.75 **d** 0.91 **e** 0.88
2. **a** £9.40 **b** 23 kg **c** 212.4 g **d** 339.5 m **e** £4.90 **f** 39.6 m **g** 731 m **h** 83.52 g **i** 360 cm **j** 117 min **k** 81.7 kg **l** £37.70
3. £5525
4. **a** 52.8 kg **b** 66 kg **c** 45.76 kg
5. Mr Speed: £176, Mrs Speed: £297.50, James: £341, John: £562.50
6. 448
7. 705
8. £18 975
9. 66.5 mph
10. No, as the total is £101. She will save £20.20, which is less than the £25 it would cost to join the club.
11. **a** 524.8 units **b** Less gas since 18% of the smaller amount of 524.8 units (94.464 units) is less than 18% of 640 units (115.2 units). I used 619.264 units.
12. TV £227.04, DVD player £172.80
13. 10% off £50 is £45; 10% off £45 is £40.50; 20% off £50 is £40
14. £765
15. 1.10 × 0.9 = 0.99 (99%)
16. Offer A gives 360 grams for £1.40, i.e. 0.388 pence per gram. Offer B gives 300 grams for £1.12, i.e 0.373 pence per gram, so offer B is the better offer. Or offer A is 360 for 1.40 = 2.6 g/p, offer B is 300 for 1.12 = 2.7 g/p, so offer B is better.

3.4 Expressing one quantity as a percentage of another
Exercise 3E
1. **a** 25% **b** 60.6% **c** 46.3% **d** 12.5% **e** 41.7% **f** 60% **g** 20.8% **h** 10% **i** 1.9% **j** 8.3% **k** 45.5% **l** 10.5%
2. 32%
3. 6.5%
4. 33.7%
5. **a** 49.2% **b** 64.5% **c** 10.6%
6. 17.9%
7. 4.9%
8. 90.5%
9. **a** Brit Com: 20.9%, USA: 26.5%, France: 10.3%, Other 42.3% **b** Total 100%, all imports
10. Stacey had the greater percentage increase.
    Stacey: (20 – 14) × 100 ÷ 14 = 42.9%
    Calum: (17 – 12) × 100 ÷ 12 = 41.7%
11. Yes, as 38 out of 46 is over 80% (82.6%)

Answers Lessons 3.1 – 3.7

3.5 Compound interest and repeated percentage change
Exercise 3F
1 a i 10.5 g ii 11.03 g iii 12.16 g iv 14.07 g
  b 9 days
2 12 years
3 a £14 272.27 b 20 years
4 a i 2550 ii 2168 iii 1331
  b 7 years
5 a £6800 b £5440 c £3481.60
6 a i 1.9 million litres ii 1.6 million litres iii 1.2 million litres
  b 10 August
7 a i 51 980 ii 84 752 iii 138 186
  b 2021
8 a 21 years b 21 years
9 3 years
10 30 years
11 1.1 × 1.1 = 1.21 (21% increase)
12 Bradley Bank account is worth £1032, Monastery Building Society account is worth £1031.30, so Bradley Bank by 70p
13 4 months: fish weighs 3×1.1^4 = 4.3923 kg; crab weighs 6×0.9^4 = 3.9366 kg
14 4 weeks
15 20

3.6 Reverse percentage (working out the original quantity)
Exercise 3G
1 a 800 g b 250 m c 60 cm d £3075 e £200 f £400
2 80
3 T shirt £8.40, Tights £1.20, Shorts £5.20, Sweater £10.75, Trainers £24.80, Boots £32.40
4 £833.33
5 £300
6 240
7 £350
8 4750 blue bottles
9 £23.10
10 300 cm^3
11 8 cm
12 5 cm

13 a £1600
  b With 10% cut each year he earns £1440 × 12 + £1296 × 12 = £17 280 + £15 552 = £32 832.
  With immediate 14% cut he earns £1376 × 24 = £33 024, so correct decision.
14 a 30% b 15%
15 Less, by 0.25%
16 £900
17 Calculate the pre-VAT price for certain amounts, and $\frac{5}{6}$ of that amount. Show the error grows as the amount increases. Up to €280 the error is less than €5.
18 £1250
19 £1250
20 Baz has assumed that 291.2 is 100% instead of 112%. He rounded his wrong answer to the correct answer of £260.

3.7 Ratio
Exercise 3H
1 $\frac{7}{10}$
2 $\frac{2}{5}$
3 a $\frac{2}{5}$ b $\frac{3}{5}$
4 a $\frac{7}{10}$ b $\frac{3}{10}$
5 Amy $\frac{3}{5}$, Katie $\frac{2}{5}$
6 a Fruit crush $\frac{5}{32}$, lemonade $\frac{27}{32}$
  b The second recipe
7 13.5 litres
8 a $\frac{1}{2}$ b $\frac{7}{20}$ c $\frac{3}{20}$
9 James $\frac{1}{2}$, John $\frac{3}{10}$, Joseph $\frac{1}{5}$
10 Sugar $\frac{5}{22}$, flour $\frac{3}{11}$, margarine $\frac{2}{11}$, fruit $\frac{7}{22}$
11 3 : 1
12 $\frac{1}{7}$
13 1 : 1 : 1

Answers Lessons 3.1 – 3.7

Exercise 3I
1 **a** 160 g : 240 g **b** 80 kg : 200 kg **c** 150 : 350 **d** 950 m : 50 m **e** 175 min : 125 min **f** £20 : £30 : £50 **g** £36 : £60 : £144 **h** 50 g : 250 g : 300 g
2 **a** 175 **b** 30%
3 **a** 40% **b** 300 kg
4 21 horses
5 **a** No, Yes, No, No, Yes **b** Possible answers: W26, H30; W31, H38; W33, H37
6 **a** 1 : 400 000 **b** 1 : 125 000 **c** 1 : 250 000 **d** 1 : 25 000 **e** 1 : 20 000 **f** 1 : 40 000
7 **a** 1: 1 000 000 **b** 47 km **c** 0.8 cm
8 **a** 1 : 250 000 **b** 2 km **c** 4.8 cm
9 **a** 1 : 20 000 **b** 0.54 km **c** 40 cm
10 **a** 4 : 3
  b 90 miles
  c Both arrive at the same time.
11 0.4 metres
12 **a** 1 : 1.6 **b** 1 : 3.25 **c** 1 : 1.125 **d** 1 : 1.44 **e** 1 : 5.4 **f** 1 : 1.5 **g** 1 : 4.8 **h** 1 : 42 **i** 1 : 1.25

Exercise 3J
1 **a** 3 : 2 **b** 32 **c** 80
2 **a** 100 **b** 160
3 0.4 litres
4 102
5 1000 g
6 Jamie has 1.7 pints, so he has enough.
7 8100
8 5.5 litres
9 **a** 14 in **b** 75 min ($1\frac{1}{4}$ h)
10 **a** 11 pages **b** 32%
11 Kevin £2040, John £2720
12 C, F, T, T
13 51
14 100
15 40 cc
16 **a** 160 cans **b** 48 cans
17 **a** Lemonade 20 litres, ginger 0.5 litres
  b This one, in part **a** there are 50 parts in the ratio 40 : 9 : 1, so ginger is $\frac{1}{50}$ of total amount; in part **b** there are 13 parts in the ratio 10 : 2 : 1, so ginger is $\frac{1}{13}$ of total amount. $\frac{1}{13} > \frac{1}{50}$
18 225 kg
19 54

Examination questions
1 £332.80
2 **a** 48.1 seconds
  b i 44.1 seconds
    ii Di (40.23 seconds) **iii** Di
3 £141
4 £2200 per month
5 £220
6 $\frac{9}{40}$
7 $4\frac{1}{12}$ pints
8 2 tins
9 **a** $\frac{312}{77}, \frac{54}{17}, \frac{22}{7}, \frac{221}{71}$
  b $\frac{22}{7}$
10 Yes, investment will be worth £4008.46
11 8% decrease
12 Estimate: 80% (78.4%)
13 **a** No, only enough for 6 days: $5 \div \frac{4}{5} = 6\frac{1}{4}$ or $5 \div \frac{2}{5} = 12.5$, so $12\frac{1}{2}$ meals
  b $2\frac{2}{3}$
14 £7375.53
15 194.6%
16 **a** 90% **b** £152 000
17 **a** 18 adults, 108 children **b** 1 : 4
18 4 more red balls
19 Yes, $100 \times 0.96^9 = 69.3$ kg
20 392 500 square kilometres
21 Jill is correct $0.4 \times 0.75 = 0.3$, so 30% of the original price is equal to 70% off.
22 Not correct, since $0.64^5 = 0.107$, lost 100% − 10.7% = 89.3% of its original contents.
23 20%
24 60 men (and 50 women)

Chapter 4 Number: Proportions

Overview

| 4.1 Speed, time and distance | 4.3 Best buys |
| 4.2 Direct proportion problems | 4.4 Density |

In the core section of Chapter 3, Lesson 3.7, students learned about ratio – what it is, how to simplify a ratio and express it as a fraction, how to divide amounts into given ratios and solve ratio problems. In this chapter, students will look at speed–distance–time problems and then study direct proportion, using the unitary method, which leads to best buys. The chapter concludes with a lesson on density, mass and volume problems.

Context

There are many practical uses of proportion. This chapter covers those most commonly used, including recipes, comparing petrol consumption, looking at speed over time, density and, especially, looking at best buys.

Curriculum references
KS4 Programme of Study references
1.1a Applying suitable mathematics accurately within the classroom and beyond.
1.3b Understanding that mathematics is used as a tool in a wide range of contexts.
2.2b Use knowledge of related problems.
2.2l Calculate accurately, using mental methods or calculating devices as appropriate.
3.1c Proportional reasoning, direct proportion.
4d Work on problems that arise in other subjects and in contexts beyond the school.

Linear specification
N3.1 Use ratio notation, including reduction to its simplest form and its various links to fraction notation.
N3.3h Solve problems involving ratio and proportion, including the unitary method of solution and repeated proportional change.
G3.7 Understand and use compound measures including density.

Functional Skills standards
R2 Identify the situation or problems and identify the mathematical methods needed to solve them.
A1 Apply a range of mathematics to find solutions.
A2 Use appropriate checking procedures and evaluate their effectiveness at each stage.
I2 Draw conclusions and provide mathematical justifications.

PLTS
Independent enquirers explore issues, events or problems from different perspectives; analyse and evaluate information, judging its relevance and value. **Creative thinkers** generate ideas and explore possibilities; question their own and others' assumptions. **Reflective learners** invite feedback and deal positively with praise, setbacks and criticism. **Team workers** collaborate with others to work towards common goals. **Self-managers** work towards goals, showing initiative, commitment and perseverance. **Effective participators** identify improvements that would benefit others as well as themselves.

APP
Numbers and the numbers system L7.1 Understand and use proportionality. **Space, shape and measure L7.6** Understand and use measures of speed (and other compound measures such as density or pressure) to solve problems. **Using and applying mathematics L7.2** Give reasons for choice of presentation, explaining selected features and showing insight into the problems structure.

© HarperCollins*Publishers* 2012

Chapter 4 Number: Proportions

Route mapping

Exercise	Grades D	C	B
A	1–9	10–18	
B	1–10	11–13	
C	all		
D			all

Overview test
Questions 1 and 2 are grade F, question 3 is grade C.
1 Simplify these fractions by cancelling.
  a $\dfrac{15}{30}$ b $\dfrac{18}{72}$ c $\dfrac{44}{66}$
2 Find the following quantities.
  a $\dfrac{3}{4}$ of £48 b $\dfrac{1}{8}$ of 20 cm c $\dfrac{2}{3}$ of 336 litres
3 If I am in a car, travelling at 60 mph, how long will it take me to drive 150 miles?

Answers to overview test
1 a $\dfrac{1}{2}$ b $\dfrac{1}{4}$ c $\dfrac{2}{3}$
2 a £36 b 2.5 cm c 224 litres
3 2.5 hours

Why this chapter matters
Cross-curricular: This work has links with Physical Education, Geography and any subject that requires an understanding of speed or proportion.
Introduction: This chapter covers ground with which many students will already be familiar, through sport, cooking and making purchases at supermarkets. The link with sport will find ready acceptance with many students, especially in relation to record holders. The introductory page in the Student Book offers several lead-ins to the topics in the chapter.
Discussion points: How fast can a domestic cat run? Is a dog faster or slower? How do the speeds achieved by animals compare with those achieved by machines? How many students in the class own mobile phones? What is this number, as a proportion of the number of students in the class? As a proportion of the number of students in the school? How many people live in the UK? What is the average population per square kilometre? How does this compare to other countries?
Plenary: Draw the discussion together to summarise how to make comparisons. Encourage students to compare 'like with like'. Discuss how to compare value for money when shopping for food in a supermarket. Ask students how they would find the best buy from competing brands of rice or vanilla ice-cream.

Answers to the quick check test can be found at the end of the chapter.

© HarperCollins*Publishers* 2012

4.1 Speed, time and distance

Curriculum references

Functional Skills standards
A1 Apply a range of mathematics to find solutions.

PLTS
Independent enquirers, Creative thinkers, Reflective learners, Team workers, Self-managers, Effective participators

APP
Shape, space and measure L7.6

Collins references
- Student Book pages 82–85
- Interactive Book: 10 Quick Questions

Key words
- average
- distance
- speed
- time

Learning objectives
- Recognise the relationship between speed, distance and time. [Grade D–C]
- Calculate average speed from distance and time. [Grade D–C]
- Calculate distance travelled from the speed and the time. [Grade D–C]
- Calculate the time taken on a journey from the speed and the distance. [Grade D]

Learning outcomes
- Students **must** be able to calculate an average speed. [Grade D]
- Students **should** be able to solve practical problems involving speed, distance and time. [Grade C]
- Students **could** be able to convert units of speed. [Grade C]

Prior knowledge
Students should know how to multiply and divide, and they should also be familiar with common units for speed.

Common mistakes and remediation
Students often fail to realise that 30 minutes is not 0.3 of an hour or 15 minutes is not 0.15 of an hour. Pointing out these common errors before they are made can often prevent them.

Useful tips
Students may not relate units (mph) to formulae.
Explain that 'per' means 'over' in Latin and is related to division, so miles per hour means 'miles over hours', i.e. distance ÷ time.
If appropriate, display the distance–speed–time triangle.
Ensure that students know that average speed $= \dfrac{\text{total distance}}{\text{total time}}$.

Functional maths and problem-solving help
Speed, time and distance offer excellent contexts for functional maths (FM) in situations that are familiar and relevant for students. They should relate the questions to real-life experience and decide if their answers sound reasonable or sensible. For example, in question 11 of Exercise 4A, they should ask themselves if the average speed for the whole journey is between the two speeds given. If not, it is an indication that they have made a mistake. **More able** students should also realise that as the train is travelling for a longer time at the higher speed the answer will be closer to this speed than the lower one. Problem-solving questions 17 and 18b are also in familiar contexts. Students need to take care to read the questions carefully, especially 18b, to find all the information they need.

© HarperCollins*Publishers* 2012

Lesson 4.1 Speed, time and distance

Starter
- Ask students simple quick-fire questions about speed. They could display answers on small whiteboards.
    - Speed 30 mph. How far in 30 minutes? (15 miles)
    - Speed 60 mph. How far in two hours? (120 miles)
    - Speed 20 mph. How far in one hour 30 minutes? (30 miles)
    - Distance 40 miles; time one hour. What is the average speed? (40 mph)
    - Distance 60 miles; time two hours. What is the average speed? (30 mph)
    - Distance 10 miles; time 30 minutes. What is the average speed? (20 mph)
    - Speed 50 mph; distance 50 miles. How long? (one hour)
    - Speed 60 mph; distance 30 miles. How long? (30 minutes)
    - Speed 30 mph; distance 90 miles. How long? (3 hours)

Main lesson activity
- Remind students of a question from the starter and ask them how they worked out the answer. Speed 60 mph, How far in two hours? Ask, "How did you get 120 miles?" Prompt students to give a formula, e.g. 60 miles in one hour, so in two hours 60 × 2 = 120. (speed × time = distance) Repeat for each type of question.
    - Distance 60 miles; time two hours. Ask, "How did you get 30 mph?" Prompt students to give a formula, e.g. 60 miles in two hours, 60 ÷ 2 = 30 miles in one hour, so 30 mph. (distance ÷ time = speed)
    - Speed 30 mph; distance 90 miles. How long? (one hour) Again prompt them to give a formula, e.g. 30 mph is 30 miles per hour, so 30 miles in one hour, 90 ÷ 30 = 3, so 3 hours. (distance ÷ speed = time)
- Draw the triangle to show the three formulae.

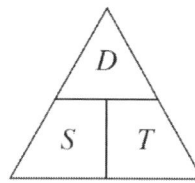

- Remind students how to convert hours and minutes into hours and vice versa: 20 minutes = 20 ÷ 60 hour = $\frac{1}{3}$ hour, 0.8 of an hour = 0.8 × 60 minutes = 48 minutes.
- Go through Examples 1, 2 and 3. Remind **less able** students to change hours and minutes into hours; reinforce this point when looking at Examples 2 and 3.
- Students can now do Exercise 4A, in which questions 10–12 and 18a are FM, 17 and 18b are PS and 15 is an assessing understanding (AU) question. It would be a good idea for **less able** students to work with a partner to complete questions 15–18.

Plenary
- Describe a two-part journey: 100 miles at an average speed of 40 mph followed by 90 miles at an average speed of 60 mph. Go through the process of calculating the average speed for the whole journey. (190 miles in four hours = 47.5 mph)
- Discuss the checking methods outlined above to test whether answers are sensible. Point out that 100 miles at 40 mph followed by 100 miles at 60 mph would give an average of 50 mph, so the answer to the problem will be less than 50 mph.

Answers
Answers to Exercise 4A can be found at the end of this chapter.

4.2 Direct proportion problems

Curriculum references

Functional Skills standards
A2 Use appropriate checking procedures and evaluate their effectiveness at each stage.

PLTS
Independent enquirers, Creative thinkers, Reflective learners, Team workers, Self-managers, Effective participators

APP
Numbers and the numbers system L7.1

Collins references
- Student Book pages 85–87
- Interactive Book: Multiple Choice Questions

Learning objectives
- Recognise and solve problems, using direct proportion. [Grade D–C]

Learning outcomes
- Students **must** be able to answer straightforward problems involving direct proportion. [Grade D]
- Students **should** be able to recognise and use direct proportion in practical situations. [Grade D]
- Students **could** be able to solve more complex problems that involve direct proportion. [Grade C]

Key words
- direct proportion
- unit cost
- unitary method

Prior knowledge
Students should know how to multiply and divide without using a calculator.

Common mistakes and remediation
Errors will occur when some information in the question leads to an answer that is not sensible. For example, when adapting a recipe, students may give answers of 1.5 eggs (see main lesson activity). Point out that in practical situations answers must always be sensible.

Functional maths and problem-solving help
The functional maths (FM) aspect of Exercise 4B is fairly self-evident and leads into the next section on best buys.
Students should be comfortable with the unitary method as a safe, reliable method that can always be used but **more able** students will be expected to adapt questions to use more efficient methods. For example, in question 7b, recognising that 150 kg is three-quarters of 200 kg and using mental methods to work out three-quarters of 12 weeks = 9 weeks.
Students should read problem-solving (PS) questions 9b and 11 carefully to find the information they need. In question 11, for example, the machine needs to pause between shredding sessions. The pause after all the paper is shredded is not included.

Lesson 4.2 Direct proportion problems

Starter
- Put the basic diagram below on the board and ask the students to give an answer.

  [6] cost [36p]. How much do [5] cost?

  Students will have an intuitive idea of this and will answer 30p.
- Discuss the idea of finding the cost of one item and then scaling up to the required number. Repeat with other values, e.g. eight cost 56p, how much will six cost?
- For **less able** students, use whole-number answers.
- For **more able** students, use decimals, e.g. seven cost £10.50, how much do six cost?

Main lesson activity
- Explain that the method they have used in the starter, finding the cost of one item and then scaling, is called the *unitary method*.
- Write this recipe on the board.
  For four people:
  100 g butter
  300 g flour
  2 eggs
  1 cup of milk
- Ask students how they could change this for three people – don't worry about the eggs at this stage. (Divide by four and multiply by three.)
- Now say that you only have 300 g of butter and 600 g of flour but plenty of eggs and milk and ask how many you could serve and why. (eight, only enough flour for eight)
- Go through Examples 4 and 5.
- Now students can do Exercise 4B, in which questions 6, 7, 12 and 13 are FM, 9b and 11 are PS and 10 is an assessing understanding (AU) question. **Less able** students may work with a partner to complete questions 11–13 if you feel it is necessary.

Plenary
- On the board, write a simple recipe for six people. Ask students how much of each ingredient would be needed for one person, two people, 12 people, and so on. Include an item, such as two eggs, which cannot be easily divided. Such a problem would not occur in an examination, but could come up in real life.
- As a lead-in to the next section on best buys, ask students which is better value: three chocolate bars for 90p or five chocolate bars for £1.50? (Neither. Both work out at 30p for each bar.)
- Now discuss with students how they would compare values.

Answers
Answers to Exercise 4B can be found at the end of this chapter.

© HarperCollins*Publishers* 2012

4.3 Best buys

Curriculum references

Functional Skills standards
R2 Identify the situation or problems and identify the mathematical methods needed to solve them.
I2 Draw conclusions and provide mathematical justifications.

PLTS
Independent enquirers, Creative thinkers, Reflective learners, Team workers, Self-managers, Effective participators

APP
Numbers and the numbers system L7.1

Collins references
- Student Book pages 88–90
- Interactive Book: Common Misconceptions

Learning objectives
- Find the cost per unit weight. [Grade D]
- Find the weight per unit cost. [Grade D]
- Use the above to find which product is the cheaper. [Grade D]

Learning outcomes
- Students **must** be able to decide on the best buy in practical situations. [Grade D]
- Students **should** be able to explain their reasoning when identifying a best buy. [Grade D]
- Students **could** be able to demonstrate a range of methods for finding a best buy. [Grade D]

Key words
- best buy
- better value
- value for money

Prior knowledge
Students need to know how to multiply and divide, with and without a calculator.

Common mistakes and remediation
Students often make mistakes by not making the units the same for each item. Stress that they must compare like with like. Students may not realise which is the best buy once they have completed the calculations. Careful thought and practice should help to minimise this problem.

Useful tips
People often assume that larger items are always the cheapest. This will not always be true.

Functional maths and problem-solving help
Students will realise this topic is extremely attuned to functional maths (FM).
Some may say that supermarkets often display the unit cost on their price labels. Point out that these are not always accurate; when goods have extra quantities for free, these are sometimes not included on the label. For example, a pack of nine toilet rolls might normally sell for £1.80 making them 20p per roll. A pack of nine with three extra free should also sell for £1.80, making the unit cost 15p but the label may still be marked as 20p per roll. Encourage students to check for this when they visit a supermarket. Without a calculator it is easier to use the label and multiply 20p by 12 to see if it gives the sale price.
Question 4 of Exercise 4C assumes that students know basic conversions: such as 1 litre = 100 cl. You may need to provide **less able** students with this information.
Problem-solving (PS) questions 5, 6 and 7 are set in familiar contexts. Again, students need to read them carefully to be able to answer them correctly.

© HarperCollins*Publishers* 2012

Lesson 4.3 Best buys

Starter
- Draw two jars of different sizes on the board. Label both jars 'jam'. Write 250 g and £0.55 on one and 600 g and £1.20 on the other. Ask students if they can tell which jar gives the better value for money.
- Explore various methods. For example:
  - Find the cost of 50 g (11p and 10p).
  - Find a common multiple of 250 and 600 (3000), giving 12 × 55 = 660 and 5 × 120 = 600.
- Students may still be confused about which is better value. Emphasise the phrase 'More jam per penny'.
- Point out that this is a clue as to how to work out the problem as 'More jam per penny' can be worked out as jam ÷ money, with the word 'per' being replaced with ÷.

Main lesson activity
- Continue with the example used in the starter. Make a list of possible answers for students to write down. Talk about each one and how to obtain it. Explain that you have written them both in pence to make it easier to do the calculations.

	250 g for 55p	**600 g for 120p**	
Dividing by 5	50 g for 11p	50 g for 10p	Dividing by 12
Dividing by 250	1 g for 0.22p	1 g for 0.2p	Dividing by 600
Multiplying by 12	3000 g for £6.60	3000 g for £6	Multiplying by 5
Dividing by 55	Number of grams per penny = 250 ÷ 55 = 4.54	Number of grams per penny = 600 ÷ 120 = 5	Dividing by 120

- Now work through Examples 6, 7, 8 and 9. Explain that different students will prefer different methods and that the more ideas they are comfortable with the easier they will find this topic.
- Students can now do Exercise 4C, in which questions 2–4 are FM, 5–7 are PS and 2 and 7 are assessing understanding (AU) questions.

Plenary
- Ask students if they think the larger quantities will be proportionally cheaper than the smaller quantities. Why might it not always be the best option to buy the biggest?
- Discuss the practicalities of buying large quantities to save money. For example, is it worth buying two loaves on a special offer when one will be thrown away as it does not get eaten?

Answers
Answers to Exercise 4C can be found at the end of this chapter.

4.4 Density

Curriculum references

Functional Skills standards
A1 Apply a range of mathematics to find solutions.

PLTS
Independent enquirers, Creative thinkers, Reflective learners, Team workers, Self-managers, Effective participators

APP
Shape, space and measure L7.6

Collins references
- Student Book pages 91–92
- Interactive Book: Multiple Choice Questions

Learning objectives
- Solve problems involving density. [Grade B]

Learning outcomes
- Students **must** be able to calculate density when given mass and volume. [Grade B]
- Students **should** be able to carry out a variety of calculations involving density. [Grade B]
- Students **could** be able to solve practical problems that involve density. [Grade B]

Key words
- density
- mass
- volume

Prior knowledge
Students should understand that mass and weight are different. They should also know the common units used for mass (grams and kilograms) and the common units used for volume (cm^3 and m^3).

Common mistakes and remediation
Students often use the wrong formula, e.g. $D = M \times V$. Encourage them to learn either the formula or the triangle with M in the top space.
They may also misunderstand the difference between mass and density, saying wood floats because it is light.

Useful tips
Density is mass per unit volume, or density = mass divided by volume.
The units for density are g/cm^3 or kg/m^3.
Students can use the units to work out the formula. For example:
grams → mass   / → divide cm^3 → volume

Functional maths and problem-solving help
The functional maths (FM) in this topic is revealed in the use of density, as a property of a substance, to solve problems. Question 11 of Exercise 4D, a problem-solving (PS) question, is an excellent example of the type of dilemma that someone, such as a jeweller or an antiques merchant, may face. Such a person would need to find out the density of various metals. To do this, they must check the mass and the volume accurately. This question can be linked to science, where volumes are found by displacement of water from a container.
Challenge students to estimate the mass of various objects in the room, if they were made from gold; a gold board cleaner (approximately 15 cm by 5 cm by 4 cm) would have a volume of 300 cm_3 and therefore a mass of 6000 grams or 6 kg.
More able students could work out the figures for larger objects, which are more difficult to measure.

© HarperCollins*Publishers* 2012

Lesson 4.4 Density

Starter
- Ask: Which is heavier, a tonne of feathers or a tonne of bricks? (both the same – both one tonne)
- Now ask: Which is heavier, a box filled with feathers or the same box filled with bricks? (bricks)
- Ask them to explain why the box of bricks is heavier. (density is higher) Try to introduce the word 'density' into their answers.
- Put some standard (equal-sized) tins of food on display (or ask students to imagine them). Ask students, without lifting them, to try to arrange the tins in order of mass. For example, for a tin of tomatoes, a tin of baked beans and a tin of fruit, although they are not much different in mass, generally the tomatoes are lightest (400 grams), then fruit (410 grams), then baked beans (420 grams).

Main lesson activity
- Explain that the mass of an object is determined by its volume and also its density.
- Ask students to name materials that they would consider to be heavy, such as lead and some other metals.
- Now ask students to suggest materials that they think are light. Some may say paper or water.
- Point out that water is quite heavy, as lifting a full bucket of water can be difficult. Also point out that 1 m^3 of water weighs one tonne. So the density of water is one tonne/m^3.
- Write the formula for density on the board.

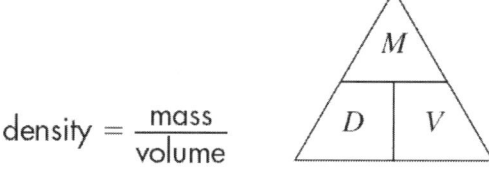

$$\text{density} = \frac{\text{mass}}{\text{volume}}$$

Show students the formula, using a triangle.
- Work through Examples 10 and 11.
- Before they start Exercise 4D, warn students that in some questions the units are mixed up; in question 6, mass is in kilograms but density is in g/cm³. Students can now complete the exercise, in which question 13 is FM, 11 is a PS question and 12 is an assessing understanding (AU) question. **Less able** students may need extra support for some of the questions.

Plenary
- Ask a student to draw the mass–density–volume triangle on the board.
- Ask quick-fire, simple questions, using all versions of the formula.
  - Find the density, if mass = 2 kg and volume = 10 cm³ (0.2 kg/cm³).
  - Find the mass, if density = 2 g/cm³ and volume = 25 cm³ (50 g).
  - Find the volume, if mass = 50 g and density = 5 g/m³ (10 m³).
- Now challenge **more able** students with a PS question linked to ratio, which they learned about in the core section: Two objects have volumes in the ratio 2 : 3. The mass of the objects is also in the ratio 2 : 3. Which object has the greater density? (They are both the same.) Students can check by using actual values:
20 cm³ and 30 cm³, 40 kg and 60 kg → density = 2 kg/cm³

Answers
Answers to Exercise 4D can be found at the end of this chapter.

Functional maths
Organising your birthday dinner

Curriculum references

Functional Skills standards
R2 Identify the situation or problems and identify the mathematical methods needed to solve them.
I2 Draw conclusions and provide mathematical justifications.

PLTS
Independent enquirers, Reflective learners, Self-managers, Team workers

APP
Using and applying mathematics L7.2

Collins references
- Student Book pages 96–97
- Interactive Book: Real Life Video

Learning objectives
- Consolidate number work on proportion.
- Convert between metric units of measure.
- Break down a complex task into smaller manageable tasks.

Key words
- Ratio

Functional maths help
This task builds on the skills students used in 'Planning a dinner party' (Chapter 1). Encourage students to link the ingredients they need to the list of 'Prices of ingredients'. Students will need calculators, and they will need to think about the practical aspect of multiplying or dividing recipe ingredients in order to vary the number of servings. Remind students:

- When planning, **more able** students should add themselves to the number of guests.
  (If **less able** students find it difficult to work out the ingredients according to the numbers plus themselves, you could allow these students to assume they are already included in the numbers: 8, 10, 15 rather than 9, 11, 16)
- Students may need to round the amounts of ingredients they buy, as not all the ingredients can be bought in exact quantities needed.
- They may have ingredients left over, and will need to think about what to do with these, e.g. add them to the dish when making it, freeze them, or cook them for another meal.

Lesson plans
Build
Discuss the last two points above. Would it matter if a few extra grams of a main ingredient, e.g. chicken livers, were added to the dish? (probably not) How could adding extra of, e.g. garlic or peppers affect a dish, e.g. by adding too much flavour or digestive upsets? Are any ingredients already available at home, e.g. salt or pepper?

Activity 1
Ask students to answer the four 'Getting started' questions. (• 175 g pasta = 25p • three sausages = 94p • four portions and 100 g over • 350 g for £2.50 are cheaper (± 0.71p per gram; 200 g for £1.90 = ± 0.95p per gram).

Activity 2
Refer students to question 1 of 'Your task' and ask them to work through it. Here are suggestions for working out ingredients and costs.
- Chilli pasta: to serve nine people, multiply ingredients by five; to serve eight, multiply by four.
- Ragu Bolognese: to serve nine, make full recipe (leftovers can be frozen); to serve eight, halve ingredients.
- Wild mushroom tart: to serve nine, triple ingredients (three tarts, with leftovers); to serve eight, double ingredients.
- Steak and kidney pudding: to serve eight or nine, make 1 ½ times the recipe, or double ingredients to make two puddings and have leftovers.

© HarperCollins*Publishers* 2012

Functional maths Organising your birthday dinner

Activity 3
Now refer students to question 2 of 'Your task', in which students must work out the ingredients and costs for each recipe for 10 (or 11) people and 15 (or 16) people. Suggestions for recipe calculations:
- Chilli pasta: to serve 10, multiply by five; to serve 11, multiply by six; to serve 15–16, multiply by eight.
- Ragu Bolognese: to serve 10–11 people, make two-thirds (though difficult to divide) or three-quarters (easier), or make full recipe (freeze leftovers); to serve 10–11, make 900 g–1 kg pasta; to serve 15–16, 1.4 kg.
- Wild mushroom tart: to serve 10–11, triple ingredients; to serve 15–16, quadruple ingredients.
- Steak and kidney pudding: to serve 10–11, double ingredients (and have leftovers); to serve 15–16, triple ingredients (and have leftovers).

Activity 4
Refer students to question 3 of 'Your task'. Discuss what they need to do: set a budget for the ingredients for the main course with £75 as a starting point. They should evaluate how realistic this budget is, taking into consideration the number of guests that could attend and choices offered, and the fact that some will eat meat and some will be vegetarians. It would be a good idea for students to work in pairs or small groups for this activity.

Activity 5
Refer students to the 'Extension'. First, they must work out who will arrive first. They could use, e.g. 40 or 50 mph. (At 40 mph, Sam will arrive at 5.22 pm; Charlie will arrive at 5.37 pm. At 50 mph, Sam will arrive at 5.12 pm. Charlie will arrive at 5.18 pm. So Sam will get there first.) Secondly, the speed they will need to travel to arrive at the same time (which will be 5.05 pm) is 60 mph.
(Sam: $\frac{35}{60} \times 60 = 35$ minutes; Charlie: $\frac{65}{60} \times 60 = 65$ minutes)

Apply
Once students have completed the first activity, check that their calculations are correct. Encourage them to ask their own questions before continuing with the second activity.
Provide **less able** students with these suggestions to calculate the ingredients to get them started. (Make a list of the ingredients for each recipe; Work out what they need to buy according to the price ingredients list, e.g. they need 225 g chicken livers, so they will have to buy the whole tub and have leftover chicken livers; Collate the ingredients in another list.
Once students have completed Activity 2, go through the calculations as a class. Activity 4 is best done in pairs or groups.

Master
Students should be able to:
- Accurately calculate the cost of ingredients, in total and per portion, for each recipe.
- Work out the ingredients and costs of recipes to serve varying numbers of people.
- Evaluate the budget for a main course, considering the number of guests.

© HarperCollins*Publishers* 2012

Answers Lessons 4.1 – 4.4

Quick check

1 a $\frac{3}{5}$ **b** $\frac{1}{5}$ **c** $\frac{1}{3}$ **d** $\frac{16}{25}$
e $\frac{2}{5}$ **f** $\frac{3}{4}$ **g** $\frac{1}{3}$

2 a £12 **b** £33 **c** 175 litres **d** 15 kg **e** 40 m
f £35 **g** 135 g **h** 1.05 litres

4.1 Speed, time and distance
Exercise 4A
1. 18 mph
2. 280 miles
3. 52.5 mph
4. 11.50 am
5. 500 seconds
6. **a** 75 mph **b** 6.5 hours
   c 175 miles **d** 240 km
   e 64 km/h **f** 325 km
   g 4.3 h (4 h 18 min)
7. **a** 7.75 h **b** 52.9 mph
8. **a** 2.25 h **b** 99 miles
9. **a** 1.25 h **b** 1 h 15 min
10. **a** 48 mph **b** 6 h 40 min
11. **a** 120 km **b** 48 km/h
12. **a** 30 min **b** 6 mph
13. **a** 10 m/s **b** 3.3 m/s
    c 16.7 m/s **d** 41.7 m/s
    e 20.8 m/s
14. **a** 90 km/h **b** 43.2 km/h
    c 14.4 km/h **d** 108 km/h
    e 1.8 km/h
15. **a** 64.8 km/h **b** 28 s
    c 8.07
16. **a** 6.7 m/s **b** 66 km
    c 5 minutes **d** 133.3
17. 7 minutes
18. **a** 20 mph **b** 07 30

4.2 Direct proportion problems
Exercise 4B
1. 60 g
2. £5.22
3. 45
4. £6.72
5. **a** £312.50 **b** 8
6. **a** 56 litres **b** 350 miles
7. **a** 300 kg **b** 9 weeks
8. 40 seconds
9. **a i** 100 g, 200 g, 250 g, 150 g
   ii 150 g, 300 g, 375 g, 225 g
   iii 250 g, 500 g, 625 g, 375 g
   b 24
10. Peter: £2.30 ÷ 6 = 38.33p each; I can buy four packs (24 sausages) from him (£9.20).
    Paul: £3.50 ÷ 10 = 35p each; I can only buy two packs (20 sausages) from him (£7).
    I should use Peter's shop to get the most sausages for £10.
11. 11 minutes 40 seconds + 12 minutes = 23 minutes 40 seconds
12. Possible answer:
    30 g plain flour (rounding to nearest 10 g)
    60 ml whole milk (rounding to nearest 10 ml)
    1 egg (need an egg)
    1 g salt (nearest whole number)
    10 ml beef dripping or lard (rounding to nearest 10 ml)
13. 30 litres

4.3 Best buys
Exercise 4C
1. a £4.50 for a 10-pack
   b £1.08 for 6
   c £2.45 for 1 litre
   d Same value
   e 29p for 250 g
   f £1.39 for a pack of 6
   g £4 for 3
2. a Large jar as more g per £
   b 600 g tin as more g per p
   c 5 kg bag as more kg per £
   d 75 ml tube as more ml per £
   e Large box as more g per £
   f Large box as more g per £
   g 400 ml bottle as more ml per £
3. a £5.11
   b Large tin (small £5.11/l, medium £4.80/l, large £4.47/l)
4. a 95p b Family size
5. Bashir's
6. Mary
7. Kelly

4.4 Density
Exercise 4D
1. 0.75 g/cm^3
2. 8 g/cm^3
3. 32 g
4. 120 cm^3
5. 156.8 g
6. 3200 cm^3
7. 2.72 g/cm^3
8. 36 800 kg
9. 1.79 g/cm^3 (3 sf)
10. 1.6 g/cm^3
11. First statue is the fake as density is approximately 26 g/cm^3
12. Second piece by 1 cm^3
13. 0.339 m

Examination questions
1. 4 minutes
2. a 75%
   b 36 000 litres
3. 8 mph
4. a £105
   b 70%
5. 4.17 kg
6. a 140 km
   b 100 km/h
7. 11 : 8
8. 0.16 km^2
9. Small 3.33p per ml, large 3.125p per ml, so large is better value.

Chapter 5 Geometry: Shapes

Overview

5.1 Circumference and area of a circle	5.3 Sectors
5.2 Area of a trapezium	5.4 Volume of a prism
	5.5 Cylinders

This chapter covers the surface area and volume of prisms and solids, as well as the area and perimeter of circles, sectors and trapezia.

Context
Students will be familiar with the methods of calculating areas and volumes of simple shapes such as rectangles, parallelograms, triangles and cuboids. This chapter develops skills for dealing with more complex plane and 3D shapes, including compound shapes that require separate calculations for different parts.

Curriculum references
KS4 Programme of Study references
1.1a Applying suitable mathematics accurately within the classroom and beyond.
1.1c Selecting appropriate mathematical tools and methods, including ICT.
2.1a Identify the mathematical aspects of the situation or problem.
2.2b Use of knowledge of related problems.
2.2l Calculate accurately, using mental methods or calculating devices as appropriate.
3.2a Properties and mensuration of 2D and 3D shapes.
4a Develop confidence in an increasing range of methods and techniques.

Linear references
G4.1h Calculate perimeters and areas of shapes made from triangles and rectangles. Extend to other compound shapes.
G4.3 Calculate circumferences and areas of circles.
G4.3h Calculate lengths of arcs and areas of sectors.
G4.4 Calculate volumes of right prisms and of shapes made from cubes and cuboids.
G4.5h Solve mensuration problems involving more complex shapes and solids.

Functional Skills standards
R1 Understand routine and non-routine problems in different contexts.
R2 Identify the problem and identify the mathematical methods needed to solve it.
R3 Select a range of mathematics to find solutions.
A1 Apply a range of mathematics to find solutions.
A2 Use appropriate checking procedures and evaluate their effectiveness at each stage.
I1 Interpret and communicate solutions to multistage practical problems in different contexts.
I2 Draw conclusions and provide mathematical justifications.

PLTS
Independent enquirers identify questions to answer and problems to resolve; analyse and evaluate information, judging its relevance and value. **Reflective learners** evaluate experiences and learning to inform future progress. **Self-managers** work towards goals, showing initiative, commitment and perseverance. **Creative thinkers** ask questions to extend their thinking. **Effective participators** discuss issues of concern, seeking resolution where needed.

APP
Shape, space and measure L6.10 Know and use the formulae for the circumference and area of a circle. **L7.2** Calculate lengths, areas and volumes in plane shapes and right prisms. **Using and applying mathematics L8.3** Select and combine known facts and problem-solving strategies to solve complex problems.

Chapter 5 Geometry: Shapes

Route mapping

Exercise	\multicolumn{5}{c}{Grades}
	D	C	B	A	A*
A	1–6	7–14			
B	1–4	5–10			
C				1–6	7–10
D		1–5	6–10		
E			1–10		11–14

Overview test
Question 1 is grade F, questions 2 and 3 are grade E and questions 4 and 5 are grade D.

1 Work out the perimeter and area of this rectangle.

   9 cm
   3 cm

   a Perimeter = _____ cm   **b** Area = _____ cm^2

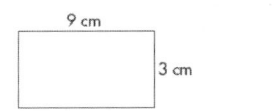

2 Work out the volume and total surface area of this cuboid.

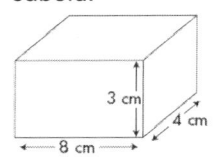

   3 cm, 4 cm, 8 cm

   a Volume = _____ cm^3 **b** Area = _____ cm^2

3 Work out the perimeter and area of this triangle.

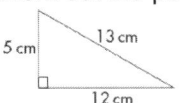

   5 cm, 13 cm, 12 cm

   a Perimeter = _____ cm   **b** Area = _____ cm^2

4 Work out the perimeter and area of this parallelogram.

   5 cm, 3 cm, 8 cm

   a Perimeter = _____ cm   **b** Area = _____ cm^2

5 Work out the circumference and area of this circle.

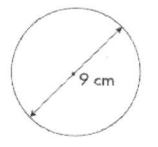

   9 cm

   a Perimeter = _____ cm   **b** Area = _____ cm^2

Answers to overview test
1 **a** 24 cm **b** 27 cm^2
2 **a** 96 cm^3 **b** 136 cm^2
3 **a** 30 cm **b** 30 cm^2
4 **a** 26 cm **b** 24 cm^2
5 **a** 28.3 cm **b** 63.6 cm^2

Why this chapter matters
Cross-curricular: This chapter links to all branches of the sciences, Design and Technology.
Introduction: People have always needed to measure areas and volumes, so have developed methods and units of measurement. The text describes everyday examples of when measurement of area and volume is important and useful. Ask students to think of other everyday examples.
Discussion points: The text mentions acres. An acre is about half the size of a football pitch. What other units of area do students know? Make sure they know the metric units. You could mention the hectare (10 000 m^2), which is the metric version of the acre. Make sure students know the metric units of volume. There is a distinction between liquid measure (capacity) and spatial volume. Litres are used for the former. Students need to know the connection between litres and cm^3.
Plenary: Ask: For what shapes can you find the area? The volume? What formulae do you know?

Answers to the quick check test can be found at the end of the chapter.

© HarperCollins*Publishers* 2012

5.1 Circumference and area of a circle

Curriculum references

Functional Skills standards
A1 Apply a range of mathematics to find solutions.

PLTS
Independent enquirers, Creative thinkers, Effective participators

APP
Shape, space and measure L6.10

Collins references
- Student Book pages 101–102
- Interactive Book: Multiple Choice Questions

Learning objectives
- Calculate the circumference and area of a circle. [Grade D–C]

Learning outcomes
- Students **must** be able to calculate the circumference and area of a circle. [Grade D]
- Students **should** be able to calculate the perimeter and area of shapes that include parts of circles. [Grade C]
- Students **could** be able to answer practical problems involving circumference and area of circles. [Grade C]

Key words
- π
- area
- circumference

Prior knowledge
Students should have covered the area and circumference of a circle previously. The content of this lesson is mainly for revision purposes.

Common mistakes and remediation
Less able students may be confused by πr^2 and $(\pi r)^2$. They may find it helpful to write the formula for the area of the circle in the form $A = \pi \times r \times r$ until their confidence in using the formula (and their calculator) improves.
Some students do not round numerical answers to an appropriate degree of freedom. Remind them of this from time to time.

Useful tips
This lesson should be revision. Find out what students already know and decide how much of the exercise is necessary. Tell students that in this and future lessons they may need to give an answer in terms of π rather than a numerical answer.

Functional maths and problem-solving help
In Exercise 5A, question 9 is functional maths (FM), and it is worth pointing out that although the diameter is only 25% larger, the pizza is over 50% larger. In question 13, which is not FM, there is an implicit assumption that the trunk is circular. Encourage students to consider whether this is a reasonable assumption.

© HarperCollins*Publishers* 2012

Lesson 5.1 Circumference and area of a circle

Starter
- Take a circular object, such as a Frisbee or a plate (paper or ceramic), into the classroom and ask students to discuss in pairs how they would find the circumference. What measurements are needed? Take suggestions, which could include using a tape measure.
- Try to elicit the formula $C = \pi d$ or $C = 2\pi r$. Students need to be familiar with both.
- Repeat with the area when the formula is $A = \pi r^2$.

Main lesson activity
- You could offer students a sketch proof of $A = \pi r^2$. Ask them to imagine a circle cut into slices.

- Now rearrange the slices in the form of an approximate parallelogram.
- The top and bottom of the parallelogram must equal the circumference of the circle. As the circumference of a circle is given by $C = 2\pi r$, each side must be πr. The length of each triangular section is approximately r. Therefore, the area of a circle is approximately $\pi r \times r = \pi r^2$. **More able** students should be able to see how the diagrammatical approximation improves as the circle is divided into more slices.
- Work through Examples 1 and 2 in the Student Book if necessary.
- Now students can complete Exercise 5A, in which the functional maths (FM) question is 9, the problem-solving (PS) question is 5 and the assessing understanding (AU) questions are 4, 8, 11 and 14.
- Students need not complete all the questions if they show a good understanding in the starter. Note that question 7 requires students to work backwards from the circumference, which will be more demanding. It might be worth having a plenary during the lesson to set a similar problem such as: If you want to draw a circle with a circumference of 20 cm, to what radius should you open the compasses? (3.2 cm)

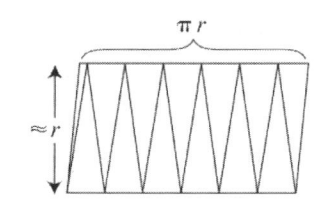

Plenary
- The bull's-eye illusion consists of five concentric circles. Draw it on the board and ask students if they think the shaded inner area is larger than the shaded outer area. Most will tend to think that it is. Say: Suppose the innermost ring has a radius of 1 cm, the next 2 cm and so on.

- Work out the two areas. (Students should find they are both the same area.)

Answers
Answers to Exercise 5A can be found at the end of this chapter.

© HarperCollins*Publishers* 2012

5.2 Area of a trapezium

Curriculum references

Functional Skills standards
R3 Select a range of mathematics to find solutions.

PLTS
Independent enquirers, Creative thinkers, Effective participators

APP
Shape, space and measure L7.2

Collins references
- Student Book pages 102–105
- Interactive Book: Multiple Choice Questions

Learning objectives
- Find the area of a trapezium. [Grade D–C]

Learning outcomes
- Students **must** be able to calculate the area of a trapezium. [Grade D]
- Students **should** be able to calculate the areas of more complex shapes that involve trapeziums. [Grade D]
- Students **could** be able to identify trapeziums in practical problems and use this to calculate areas. [Grade C]

Key words
- trapezium

Prior knowledge
Students should know how to find the areas of triangles, rectangles and parallelograms.

Common mistakes and remediation
The most common error is not identifying the height correctly. Give students lots of examples in different orientations and make sure they can identify the height.
Students often think that a square or rectangle is not a trapezium. Emphasise that these are special cases. In the main lesson activity, show students that the formula for the area still works for these shapes.

Useful tips
Point out that $\frac{(a + b)}{2}$ is just the mean of the two parallel sides; average length × width is a generalisation of the formula for the area of a rectangle. This may help them to remember the formula for the area of a trapezium.

Functional maths and problem-solving help
The main lesson activity shows one way of demonstrating the formula. In Exercise 5B, question 3, which assesses understanding (AU), gives another way. Yet another way is to compare the trapezium to rectangles with height h, one of length a and the other of length b. Question 7 is functional maths (FM).

Starter
- Ask students to draw a trapezium on their mini whiteboards, if they have them. Then ask them to draw a second trapezium that is very different. They should then draw a third trapezium that is different again.
- Share examples with the whole class and discuss what the examples have in common.

© HarperCollins*Publishers* 2012

Lesson 5.2 Area of a trapezium

Main lesson activity
- Explain that the area of a trapezium is given by A = half the sum of the parallel sides × height.
  This can be seen to be the case by splitting a trapezium into two triangles.
  Area of triangle 1 = $\frac{1}{2}$ base × height = $\frac{1}{2}ah$
  Area of triangle 2 = $\frac{1}{2}$ base × height $\frac{1}{2}bh$
  Therefore, total area = $\frac{1}{2}ah \times \frac{1}{2}bh = \frac{1}{2}h(a+b) = \frac{1}{2}(a+b)h$

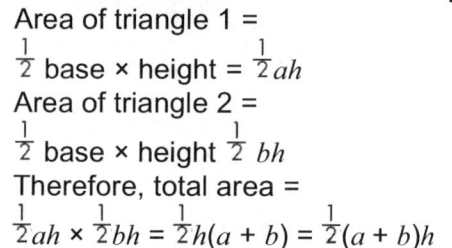

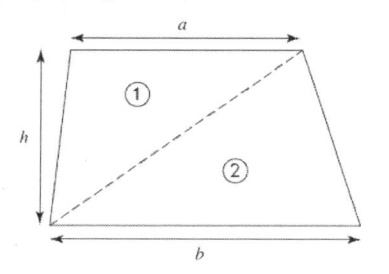

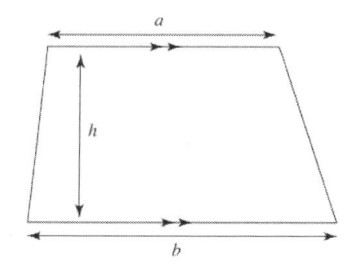

- Emphasise in this context that height means perpendicular height, and that we add the two parallel sides. Ensure that students give final answers to a suitable degree of accuracy, usually three significant figures.
- Discuss what distances should be measured to find the area in the examples students drew in the starter. In some examples, the height may not be obvious (when there is no perpendicular line joining the two parallel sides). Challenge **more able** students to do these examples. Then use them as a teaching point.
- Ask students to draw accurately a trapezium with a given area, e.g. 20 cm². Suggest that each student checks his or her neighbour's drawing.
- If necessary, look at Example 3 with students. Then students can complete Exercise 5B, in which question 7 is FM, the problem-solving (PS) question is 4 and the AU questions are 3, 6, 8 and 9. **Less able** students can complete difficult questions with a partner.

Plenary
- Draw the following diagrams on the board:
- Ask: Which has the largest area? (all the same except the triangle, which is half the size)

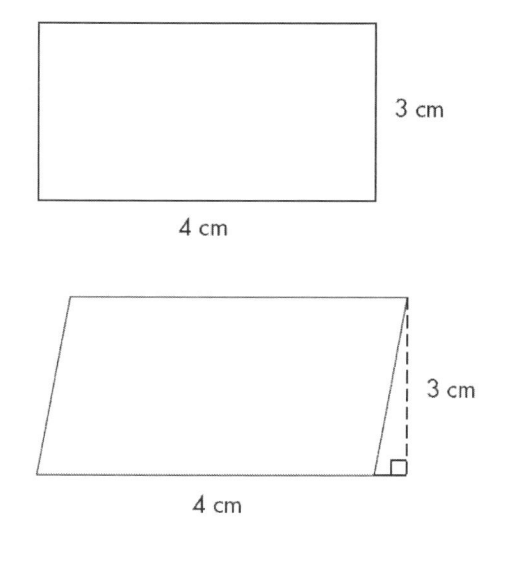

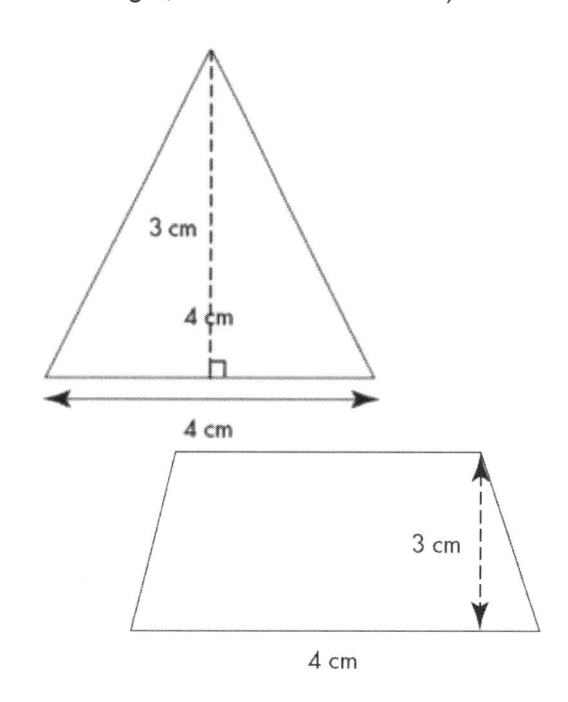

Answers
Answers to Exercise 5B can be found at the end of this chapter.

5.3 Sectors

Curriculum references

Functional Skills standards
I2 Draw conclusions and provide mathematical justifications.

PLTS
Independent enquirers, Creative thinkers, Effective participators

APP
Shape, space and measure L7.2

Collins references
- Student Book pages 105–107
- Interactive Book: 10 Quick Questions

Learning objectives
- Calculate the length of an arc and the area of a sector. [Grade A–A*]

Learning outcomes
- Students **must** be able to calculate the area of a sector. [Grade A]
- Students **should** be able to calculate the perimeter of a sector. [Grade A]
- Students **could** be able to calculate the areas of segments of a circle. [Grade A*]

Key words
- arc
- sector
- subtend

Prior knowledge
Students need to be able to calculate the area and circumference of a circle, as covered in Lesson 5.1.

Common mistakes and remediation
Students often forget that the perimeter of a sector will be the sum of the arc length plus two radii, not just the arc length alone. Emphasise this point.

Useful tips
Encourage students to use the correct terms 'arc', 'sector' and 'subtend'.

Functional maths and problem-solving help
A pizza slice is the most common everyday example of a sector and it can be used to illustrate a sector.
Question 9 is the functional maths (FM) question in Exercise 5C.

Starter
- Recap on students' knowledge of circles.
- Draw circles of radii 5 cm and 10 cm on the board. Ask students what they can tell you about the circumferences. (They may tell you how to work them out.) Try to elicit the fact that one circumference is double the other.
- Ask if you can generalise. (If the radius of a circle is doubled, so is the circumference.)
- Now ask about the areas. Is the same thing true? (This time one area is four times the other.)

© HarperCollinsPublishers 2012

Lesson 5.3 Sectors

Main lesson activity
- Following their work on the area and circumference of circles in Lesson 5.1, most students should be in a position to begin to discover the formulae for the arc perimeter and area of a sector.
- Ask students to write down the formula for the circumference of a circle. Then ask them to write a formula for the curved part of a semi-circle, and again for a quarter circle. Can they make the leap to finding the formula for an arc sector, given a sector angle?
- Assist if necessary by, for example, drawing a sector with a given sector angle and asking what fraction of the circle (or circumference) has been drawn. The area of a sector could be found in a similar way.
- Explain that if an arc ab subtends an angle $\theta°$ at the centre of a circle, then the arc length ab is given by:

$$\text{arc length} = \frac{\theta}{360} \times 2\pi r$$

The area of the sector is given by:

$$\text{area} = \frac{\theta}{360} \times 2\pi r^2$$

- Emphasise that the angle tells you the fraction of a whole circle you are considering. The formula is less important.
- Work through Example 4 briefly.
- Students can now complete Exercise 5C, in which question 9 is FM, 6 and 8 are problem-solving (PS) questions and the assessing understanding (AU) question is 7. If **less able** students are struggling to complete the exercise, you could advise them to work with a partner who is **more able**.

Plenary
- Draw two sectors: one with an angle of 90° and radius 4 cm, the other with an angle of 45° and radius 8 cm.
- Ask: Which arc has the bigger arc length? Which sector has the bigger area?
- Students should think about this in pairs for a couple of minutes and then take their comments. (They have the same arc length. The second is double the area of the first.)
- Ask students for more examples of sectors with the same arc length as the first two. (Other examples are 180° and 2 cm; 60° and 6 cm; 120° and 3 cm; 22.5° and 16 cm.)

Answers
Answers to Exercise 5C can be found at the end of this chapter.

© HarperCollins*Publishers* 2012

5.4 Volume of a prism

Curriculum references

Functional Skills standards
R2 Identify the situation or problems and identify the mathematical methods needed to solve them.

PLTS
Independent enquirers, Creative thinkers, Effective participators

APP
Shape, space and measure L7.2

Collins references
- Student Book pages 108–110
- Interactive Book: Common Misconceptions

Learning objectives
- Calculate the volume of a prism. [Grade C–B]

Learning outcomes
- Students **must** be able to calculate the volume of a prism. [Grade C]
- Students **should** be able to answer straightforward practical problems involving prisms. [Grade C]
- Students **could** be able to answer more complex practical problems involving prisms. [Grade B]

Key words
- cross-section
- prism

Prior knowledge
Students need to be able to calculate the areas of rectangles, triangles, trapezia and compound shapes.

Common mistakes and remediation
Students will sometimes mistake the length of the prism for a length occurring in the common cross-section. They should be able to avoid this mistake by drawing a good diagram and solving the problem in two distinct stages.

Useful tips
Use the image of a loaf of bread, which gives similar slices wherever you cut it, as long as you cut it correctly.

Functional maths and problem-solving help
In common usage, the word prism refers to a shape with a triangular cross-section. Students may be familiar with the use of transparent prisms to split light into different colours. Swimming pools are often in the shape of prisms.
Question 3 of Exercise 5D is functional maths (FM) and deals with swimming pool measurements. Question 9 is also FM.
Note: Try to find pictures of Anish Kapoor's sculpture (referred to in question 7 of Exercise 4D) on the internet.

© HarperCollins*Publishers* 2012

Lesson 5.4 Volume of a prism

Starter
- Ask students if they have heard the word 'prism' before and what they understand by it.
- Show students as many examples of prisms as you can find (e.g. packaging boxes or pictures).
- Discuss what they have in common (i.e. a constant cross-section parallel to one plane). Include some examples that are not prisms to make the point.

Main lesson activity
- Draw a rectangle, triangle, circle and trapezium on the board. Ask students to recall the areas of these shapes.
  Extend the diagrams into 3D prisms by adding depth. Label them: rectangular prism, triangular prism, circular prism, trapezoidal prism.
- Ask what the rectangular prisms and circular prisms are more commonly called (cuboid, cylinder). Why is it that the other two prisms do not have special names?
- Explain that the volume of a prism is given by the product of the area of the cross-section of the prism and its length. This might more usefully be written as $V = A \times$ length.
- For most problems, it is therefore important to stress that students calculate the cross-sectional area of the prism as an essential first step.
- Show students that this is equivalent to the formula for the area of a cuboid. The formula for the volume of a prism is a generalisation of that.
- For most problems, stress that students should begin by calculating the cross-sectional area of the prism.
- If you have some boxes, e.g. cardboard, you could ask students to make appropriate measurements so that they can calculate the volume. Also, work through Example 5.
- Students can now complete Exercise 5D, in which questions 3 and 9 are FM, 4 and 7 are problem-solving (PS) questions and 6, 8 and 10 are assessing understanding (AU) questions. **Less able** students can complete questions they may find difficult with a partner who is **more able**.

Plenary
- Ask students to sketch a prism with a volume of 60 cm^3. This could be done in pairs. Use mini whiteboards if you have them. Ask one or two students who have drawn interesting prisms to show them to the class.

Answers
Answers to Exercise 5D can be found at the end of this chapter.

© HarperCollins*Publishers* 2012

5.5 Cylinders

Curriculum references

Functional Skills standards
A2 Use appropriate checking procedures and evaluate their effectiveness at each stage.

PLTS
Independent enquirers, Creative thinkers, Effective participators

APP
Shape, space and measure L7.2

Collins references
- Student Book pages 111–113
- Interactive Book: Matching Pairs

Learning objectives
- Calculate the volume and surface area of a cylinder. [Grade B–A*]

Learning outcomes
- Students **must** be able to calculate the volume of a cylinder. [Grade B]
- Students **should** be able to calculate the surface area of a cylinder. [Grade B]
- Students **could** be able to answer practical problems involving surface area or volume of a cylinder. [Grade B]

Key words
- cylinder
- surface area
- volume

Prior knowledge
Students should know how to find the volume of a prism, as covered in Lesson 5.4, and be able to calculate the area of a circle as covered in Lesson 5.1.

Common mistakes and remediation
As in Lesson 5.1, **less able** students may be confused between πr^2 and $(\pi r)^2$. They may find it helpful to use a simpler version of the formula in the form $V = \pi \times r \times r \times h$ until their confidence improves.

Useful tips
Students should be able to give numerical answers and answers in terms of π.

Functional maths and problem-solving help
Questions 6 and 8 of Exercise 5E are functional maths (FM), and discuss a food can and a drinks can. It would be useful to have some examples at hand. These questions are less about right answers and more about applying the ideas in this chapter in a real context. Question 12 is also FM, and is about the cylinder of a car engine. More likely than not, only the **more able** students will be able to complete it. The same goes for question 11, which is problem-solving (PS), as these questions are both grade A*.

© HarperCollins*Publishers* 2012

Lesson 5.5 Cylinders

Starter
- Take a trundle wheel into the class or display a picture of one. Explain that a trundle wheel is used to measure distances. It has a wheel with a circumference of 1 metre.
- Ask for the diameter of the wheel (31.8 cm) and the area (796 cm^2). Students can work in pairs, as this is revision of formulae that they will need during the lesson.

Main lesson activity
- Use an A4 sheet of paper to make a hollow cylinder by bending the long edges together. Remind students that a cylinder is a circular prism.
- Tell them that the sheet of paper is 29.7 cm long and 21.0 cm wide.
- Ask how they can find the area of the curved surface of the cylinder. This should be easy: 29.7 × 21.0 cm^2.
  Now set the task of finding the volume of the cylinder. They could work in pairs to complete this. (The circumference of the base is 21.0 cm; radius = 3.34 cm; area = 35.09 cm^2; length = 29.7 cm; volume = 1040 cm^3 to 3 sf)
- You could set an extension task of finding the volume of the cylinder formed by putting the shorter edges together. Is it the same? (No: volume = 1470 cm^3 to 3 sf)
- Now consider a cylinder with two circular ends. The total surface area consists of a rectangle and two circles.
- Demonstrate that the curved surface area is πdh or $2\pi rh$, and the total surface area is $\pi dh + 2\pi r^2$ or $2\pi rh + 2\pi r^2$. Students can use either of these. You could show **more able** students that this is equivalent to $2\pi r(r + h)$.
- Work through Examples 6 and 7 with students.
- Now students can complete Exercise 5E, in which questions 6, 8 and 12 are FM, 11 is a PS question and the assessing understanding (AU) questions are 3, 4 and 10. **Less able** students can complete questions they may find difficult with a partner.

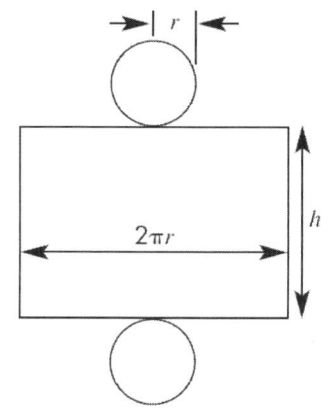

Plenary
- Ask students (in pairs if you like) to find out the amount of metal that is used to make a 2p coin. Discuss their answers. If they use 2 mm for the thickness of the coin and 26 mm for the diameter, they should get an answer of about 1060 mm^3 for the volume. Discuss the accuracy of the coin's 'thickness' measurement. Do the rim and the pattern make a difference?

Answers
Answers to Exercise 5E can be found at the end of this chapter.

© HarperCollins*Publishers* 2012

Functional maths
Fitting a carpet

Curriculum references

Functional Skills standards
R2 Identify and obtain necessary information to tackle the problem.
A1 Apply mathematics in an organised way to find solutions to straightforward practical problems for different purposes.
I1 Interpret and communicate solutions to practical problems, drawing simple conclusions and giving explanations.

PLTS
Independent enquirers, Creative thinkers, Reflective learners

APP
Shape, space and measure L7.2
Using and applying mathematics L8.3

Collins references
- Student Book pages 118–119
- Interactive Book: Real Life Video

Learning objectives
- Find the perimeter and area of rectangles and compound shapes.

Key words
- area
- perimeter
- compound shapes
- rectangle

Functional maths help
The main learning point of this activity is to let students discover that different shapes can have the same area.
Students will need to gain understanding of how to deal with the constraints of the floor coverage. Discuss the various possible solutions, which depend on the room design.
You may choose to restrict students to rectangles only, or you may wish to build skills using this shape and then allow students to apply their skills to different areas or different shapes.

Lesson plans
Build
Have a brief class discussion about buying and fitting, or laying, carpets. You could ask the following questions.
- Have you had personal experience of buying a carpet and having it fitted? (Some students' parents may work in the carpeting industry and may be able to share information.)
- How did your family or friends choose the carpet? (What qualities did they look for in the carpeting?)
- How were the carpets fitted? (Who laid them and how – were they laid wall-to-wall, or did they only partially cover the floor in the room(s)?

Ask students to share their experiences and knowledge of carpeting with the class.

© HarperCollins*Publishers* 2012

Functional maths Fitting a carpet

Activity 1
Refer students to the introduction for the activity in the Student Book, and ask them to consider these three points (a summary of the text in the Student Book):
- Sketch two potential shapes for the new room and plan their dimensions.
- Consider the area of each room to be covered by the carpet.
- Calculate the carpet dimensions for each room.

Draw some of the shapes on the board with the dimensions and work through the calculations with the class.

Activity 2
After students have read the information about Steve and the extension to his house, plus the criteria that he has provided, ask students to work out the size of the carpet he should buy (see 'Your task'). **Less able** students can work in pairs or groups. **More able** students should be able to do the task on their own.

Ask students to draw several rectangles that give an area of 30 m^2. (Possible answers are 1 m × 30 m, 2 m × 15 m, 3 m × 10 m, 5 m × 6 m if we restrict to whole metre answers. If you allow students to work with decimals or fractions the list extends, for example, 4.8 m^2 × 6.2 m^2 = 29.76 m^2. The carpet must cover at least half, or 50%, of the floor area, which measures 15 m^2; three-quarters, or 75%, of the floor area is 22.5 m^2. So the total size Steve wants to cover is between 15 m^2 and 22.5 m^2. Different sizes of carpets that could fit the criterion are: 3 m × 5 m, 4 m × 4 m, 3 m × 6 m, 4 m × 5 m.)

Activity 3
Once students have completed Activity 2, they should refer to the Extension in 'Your task' and work out how much skirting board Steve will need for each room design, remembering that skirting is not needed for doorways. Students can use their measurements for possible room sizes as a basis for their calculation.

Activity 4
Students can independently look up the costs of both carpet and skirting board to extend this activity to include the cost of the various options.

Apply
In Activities 1, 2 and 3, provide minimal guidance, although **less able** students may need some support.

Ask students to design rooms with compound shapes and to exchange examples. Students should be able to show mastery of this example if they are working on a higher scheme of work. This will provide students with an opportunity to look at different ways the carpet can be laid.

At the end of the task students should evaluate their approach to the problem.

Master
In order to demonstrate mastery of the learning objectives, students should be able to:
- Create shapes with the required area.
- Calculate the amount of carpet required.
- Give dimensions of the amount of carpet that can be ordered.
- Make suggestions about which dimensions might be the best design for the room.

Plenary
Ask students to describe the strategies they used to produce their answers.
Then ask them to compare the different-sized rooms they have considered and to try to work out the best way of laying various carpets for the least possible cost.

© HarperCollins*Publishers* 2012

Answers Lessons 5.1 – 5.5

Quick check
1 a 90 mm², b 40 cm², c 21 m²
2 120 cm³

5.1 Circumference and area of a circle
Exercise 5A
1 a 8 cm, 25.1 cm, 50.3 cm²
  b 5.2 m, 16.3 m, 21.2 m²
  c 6 cm, 37.7 cm, 113 cm²
  d 1.6 m, 10.1 m, 8.04 m²
2 a 5π cm b 8π cm c 18π m d 12π cm
3 a 25π cm² b 36π cm²
  c 100π cm² d 0.25π m²
4 8.80 m
5 4 complete revolutions
6 1p : 3.1 cm², 2p : 5.3 cm², 5p : 2.3 cm², 10p : 4.5 cm²
7 0.83 m
8 38.6 cm
9 Claim is correct (ratio of the areas is just over 1.5 : 1)
10 a 18π cm² b 4π cm²
11 9π cm²
12 9π m²
13 Yes, if the classroom is at least 10 metres by 10 metres (diameter 9.9 m).
14 45 complete revolutions

5.2 Area of a trapezium
Exercise 5B
1 a 30 cm² b 77 cm² c 24 cm² d 42 cm²
  e 40 m² f 6 cm g 3 cm h 10 cm
2 a 27.5 cm, 36.25 cm²
  b 33.4 cm, 61.2 cm²
  c 38.5 m, 90 m²
3 The area of the parallelogram is $(a + b)h$.
  This is the same as two trapezia.
4 Two of 20 cm² and two of 16 cm²
5 a 57 m²
  b 702.5 cm²
  c 84 m²
6 47 m²
7 4, because the total area doubled is about 32 m²
8 80.2%
9 1 100 000 km²
10 160 cm²

5.3 Sectors
Exercise 5C
1 a i 5.59 cm ii 22.3 cm²
  b i 8.29 cm ii 20.7 cm²
  c i 16.3 cm ii 98.0 cm²
  d i 15.9 cm ii 55.6 cm²
2 2π cm, 6π cm²
3 a 73.8 cm b 20.3 cm
4 a 107 cm² b 173 cm²
5 43.6 cm
6 a $\dfrac{180}{\pi}$
  b If arc length is 10 cm, distance along chord joining the two points of the sector on the circumference will be less than 10 cm, so angle at centre will be less than 60°.
7 $(36\pi - 72)$ cm²
8 36.5 cm²
9 16 cm (15.7)
10 a 13.9 cm
   b 7.07 cm²

5.4 Volume of a prism
Exercise 5D
1 a i 21 cm² ii 63 cm³
  b i 48 cm² ii 432 cm³
  c i 36 m² ii 324 m³
2 a 432 m³ b 225 m³ c 1332 m³
3 a A cross-section parallel to the side of the pool always has the same shape.
  b About 3 hours
4 7.65 m³
5 a 21 cm³, 210 cm³
  b 54 cm², 270 cm²
6 146 cm³
7 78 m³ (78.3 m³)
8 327 litres
9 1.02 tonnes
10 672 cm²

5.5 Cylinders
Exercise 5E
1 a i 226 cm^3 **ii** 207 cm^2
  b i 14.9 cm^3 **ii** 61.3 cm^2
  c i 346 cm^3 **ii** 275 cm^2
  d i 1060 cm^3 **ii** 636 cm^2
2 a i 72π cm^3 **ii** 48π cm^2
  b i 112π cm^3 **ii** 56π cm^2
  c i 180 cm^3 **ii** 60π cm^2
  d i 600π m^3 **ii** 120π m^2
3 £80
4 1.23 tonnes
5 665 cm^3
6 Label should be less than 10.5 cm wide so that it fits the can and does not overlap the rim and more than 23.3 cm long to allow an overlap.
7 332 litres
8 There is no right answer. Students could start with the dimensions of a real can. Often drinks cans are not exactly cylindrical. One possible answer is height of 6.6 cm and diameter of 8 cm.
9 1.71 g/cm^3
10 7.78 g/cm^3
11 About 127 cm
12 A diameter of 10 cm and a length of 5 cm give a volume close to 400 cm^3 (0.4 litres).
13 B has volume 200π, C has volume 400π, D has volume 337.5π so order is B, D, C.
14 $πr^2h = r^3$ so $πh = r$, $h = r ÷ π ≈ 0.32r$.

Examination questions
1 a 66.5 cm^2
  b 855.5 cm^2
2 a 320π cm^2 **b** 4
3 11 777 cm^3
4 9.08 m
5 480π cm^2
6 $\frac{3}{8}$

Chapter 6 Algebra: Expressions and equations 1

Overview

6.1a	Basic algebra: Substitution	6.3a	Solving linear equations: Fractional equations
6.1b	Basic algebra: Expansion	6.3b	Solving linear equations: Brackets
6.1c	Basic algebra: Simplification	6.3c	Solving linear equations: Variables on both sides
6.2	Factorisation	6.4	Setting up equations
		6.5	Trial and improvement

These lessons are aimed at building upon the students' knowledge of algebraic manipulation.

Context
Solving equations, dealing effectively with algebraic expressions, and rearranging formulae is the basis of many subjects ranging from mathematics, science and economics to architecture.

Curriculum references
KS4 Programme of Study references
2.1c Simplify the situation or problem in order to represent it mathematically, using appropriate variables, symbols, diagrams and models.
2.2j Reason inductively, deduce and prove.
2.2o Record methods, solutions and conclusions.
2.4d look for equivalence in relation to both the different approaches to the problem and different problems with similar structures.
3.1e Linear, quadratic and other expressions and equations.

Linear specification
N4.2 Distinguish in meaning between the words 'equation', 'formula' and 'expression'.
N5.1 Manipulate algebraic expressions by collecting like terms, by multiplying a single term over a bracket and by taking out common factors.
N5.4h Set up and solve simple linear equations including simultaneous equations.
N5.8 Use systematic trial and improvement to find the approximate solutions of equations where there is no simple analytical method of solving them.

Functional Skills standards
R2 Identify the situation or problems and identify the mathematical methods needed to solve them.
A1 Apply a range of mathematics to find solutions.
I2 Draw conclusions and provide mathematical justifications.

PLTS
Independent enquirers explore issues, events or problems from different perspectives. **Creative thinkers** ask questions to extend their thinking. **Reflective learners** assess themselves and others, identifying opportunities and achievements. **Team workers** collaborate with others to work towards common goals. **Self-managers** work towards goals, showing initiative, commitment and perseverance. **Effective participators** discuss issues of concern, seeking resolution where needed.

APP
Algebra L6.2 Construct and solve linear equations with integer coefficients, using an appropriate method. **L7.2** Use algebraic and graphical methods to solve simultaneous linear equations in two variables. **L7.4** Use formulae from mathematics and other subjects; substitute numbers into expressions and formulae; derive a formula and, in simple cases, change its subject. **L8.2** Manipulate algebraic formulae, equations and expressions, finding common factors and multiplying two linear expressions. **L8.3** Derive and use more complex formulae and change the subject of a formula. **Using and applying mathematics L8.3** Select and combine known facts and problem-solving strategies to solve problems of increasing complexity.

© HarperCollins*Publishers* 2012

Chapter 6 Algebra: Expressions and equations 1

Route mapping

Exercise	Grades D	Grades C	Grades B
A	1–10	11–19	
B	1	2–4	
C	1–3	4–11	
D	1	2–5	6
E	1	2–3	
F	all		
G	1–2	3–6	
H	1–8	9–17	
I		all	

Overview test

Questions 1, 11 and 12 are grade G, questions 2–4 and 13–17 are grade F, questions 5–9 and 18–20 are grade E, question 10 is grade D

1. $2 + 3 \times 4 =$
2. $-7 - (-3) =$
3. $4(2 + 3) =$
4. $5(4 - 1) =$
5. $x(4 + 2) =$
6. $3x(6 - 2) =$
7. $10 + (2x + 4x) =$
8. $3 \times 2a =$
9. $2p \times 4p =$
10. $a^3 \times a =$
11. Put brackets in this calculation to make the answer 49.
    $5 \div 2 \times 3 \div 4 = 49$
12. Put brackets in this calculation to make the answer 3.
    $19 - 2 \times 4 + 4 = 3$
13. Solve for x: $5x = 25$
14. Solve for x: $x + 8 = 10$
15. Solve for x: $x - 6 = 12$
16. Solve for x: $12x = 3$
17. Solve for x: $\dfrac{x}{5} = 2$
18. Solve for x: $x + 6 = 3$
19. Solve for x: $\dfrac{2}{x} = 4$
20. Solve for x: $8 - x = 12$

Answers to overview test

1. 14
2. –4
3. 20
4. 15
5. $6x$
6. $12x$
7. $10 + 6x$
8. $6a$
9. $8p^2$
10. a^4
11. $(5 + 2) \times (3 + 4) = 49$
12. $19 - 2 \times (4 + 4) = 3$
13. 5
14. 2
15. 18
16. 0.25
17. 10
18. –3
19. 0.5
20. –4

Why this chapter matters

Cross-curricular: This work has links with the sciences, Art and Crafts, Physical Education and languages.

Introduction: Algebra can seem remote from everyday arithmetic and problems but, in reality, it is part of our everyday world. It allows us to carry out calculations in standard ways and to describe those calculations to others.

Discussion points: Has anyone heard of the 'Golden Ratio'? How does it relate to a sheet of A4 paper? If you cut the largest square that you can, from an A4 sheet, how does the size of the rectangle cut off relate to the size of the original sheet? Where else might students use algebra in everyday life?

Plenary: A woman deposits £100 in a bank account at an interest rate of 2.5%. What will the amount be after five years? (£102.50 after one year, ...) Is there an easier way? How can we find it?

Answers to the quick check test can be found at the end of the chapter.

© HarperCollins*Publishers* 2012

6.1a Basic algebra: Substitution

Curriculum references

Functional Skills standards
A1 Apply a range of mathematics to find solutions.

PLTS
Independent enquirers, Creative thinkers, Reflective learners, Team workers, Self-managers, Effective participators

APP
Algebra L7.4

Collins references
- Student Book pages 122–125
- Interactive Book: Common Misconceptions

Learning objectives
- Recognise expressions, equations, formulae and identities. [Grade D–C]
- Substitute into, manipulate and simplify algebraic expressions. [Grade D–C]

Learning outcomes
- Students **must** be able to substitute numbers into expressions and formulae. [Grade D]
- Students **should** be able to use formulae in realistic contexts. [Grade C]
- Students **could** be able to expand brackets and simplify expressions. [Grade C]

Key words
- brackets
- expression
- simplification
- coefficient
- term
- expand
- formula
- substitute
- equation
- identity
- like terms
- variable

Prior knowledge
Students should be conversant with the basic language of algebra, BIDMAS/BODMAS, and negative numbers.

Common mistakes and remediation
Students often ignore the rules of BIDMAS/BODMAS. And, they may misunderstand algebraic conventions, e.g. substituting the value $t = 4$ into the expression $2t + 7$ to get $24 + 7$ instead of $2 \times 4 + 7$. They may simplify expressions correctly, e.g. to $2t + 7$ then make an error in the next step, such as $2t + 7 = 9t$. Discourage students from trying to do these three things at once: substituting, working out and evaluating. If they get one of these things wrong, they score zero marks.

Useful tips
Encourage students, when substituting numbers into an expression, to use brackets before simplifying and finding the value, to gain some method marks in an examination. If they substitute and work out at the same time, and it is wrong, examiners will award no method marks, as there will be no evidence of working. When doing simple calculations with direct numbers, tell students that it is sometimes helpful to 'say the problem to yourself', e.g. $+ 2 - - 3 = 5$ or plus 5. If you write this on the board, many answers will be wrong. If you say it as 'plus two minus minus three', fewer answers will be wrong.

Functional maths and problem-solving help
Functional maths (FM) question 10 in Exercise 6A is a standard 'formula in words' question but part **c** requires students to solve a practical problem. FM question 11 has a hint which will help students with part **a**. Part **b** requires a strategy that students should be encouraged to write down or explain before starting the problem.

© HarperCollins*Publishers* 2012

Lesson 6.1a Basic algebra: Substitution

Problem-solving (PS) question 15 requires students to find values that make the area of a rectangle and a triangle the same. Rather than just trying out random values, encourage students to think about any connections between the two shapes. They should conclude that a triangle is half of a rectangle. PS question 17 has a straightforward first part followed by two parts that require students to analyse the problem and decide on the mathematics to solve it.

	a		
	2	3	5
2	6	8	12
b 4	8	10	14
6	10	12	1

Starter
- Draw this grid on the board without the answers filled in:
- **More able** students should be able to cope with a rule such as $a2 - 2b$, where values for a and/or b are negative.

> Get the pupils to fill in the values by using a formula, e.g. $2a + b$.
> Explain, if needed, that $2a + b$ is 2 times a, plus b.
> Give a time limit or make it a race.
> For **more able** students, make values fractional and/or negative.

Main lesson activity
Substitution
- Ask students how they worked out the calculations in the starter.
- If there were errors (which is very likely if powers and negative numbers were used), ask what caused the errors to be made. This is almost certainly not seeing the calculation clearly, e.g. 2×-3 or -32 instead of $(2)(-3)$ or $(-3)2$.
- Take an expression such as $3a + 2b - 3c$ and values $a = 2$, $b = -4$ and $c = -5$. Ask students to work it out and note their answer. Collect answers and methods from the class. It is very unlikely that everyone will get the correct answer (+13) and methods will vary. Do not say what the correct answer is at this stage.
- Write $3(2) + 2(-4) -3(-5)$ on the board and ask students to work it out. Ask if anyone now has a different answer to what they had previously, but still do not give the correct answer.
- Write down $6 - 8 + 15$ and ask the students to work this out. Check answers.
- Discuss which was the easier calculation.
- Encourage **less able** students to adopt the procedure of:
    - substituting values into the expression, replacing letters with numbers in brackets.
    - evaluating each term individually
    - working out the final answer.
- **More able** students may feel this is a waste of time, but warn them that any errors that the examiner cannot see could result in zero marks, so showing the method is vital.
- Take time to emphasise the use of BIDMAS/BODMAS once substitution has taken place.
- Go through the introductory text and Examples 1, 2 and 3 in this section.
- Students can now do Exercise 6A

Plenary
- Ask students to work out the value of expressions such as:
    - $2x + 4$, $2(x + 2)$ for $x = 5$
    - $3a + 9b$ and $3(a + 3b)$ for $a = 2$, $b = 3$, etc.
- They should get the same answers.
- Ask if they can see the connection. They should be able to appreciate that $2x + 4 = 2(x + 2)$, etc.

Answers
Answers to Exercise 6A can be found at the end of this chapter.

© HarperCollins*Publishers* 2012

6.1b Basic algebra: Expansion

Curriculum references

Functional Skills standards
A1 Apply a range of mathematics to find solutions.

PLTS
Independent enquirers, Creative thinkers, Reflective learners, Team workers, Self-managers, Effective participators

APP
Algebra L7.4

Collins references
- Student Book pages 126–127
- Interactive Book: Common Misconceptions

Learning objectives
- Recognise expressions, equations, formulae and identities. [Grade D–C]
- Substitute into, manipulate and simplify algebraic expressions. [Grade D–C]

Learning outcomes
- Students **must** be able to substitute numbers into expressions and formulae. [Grade D]
- Students **should** be able to use formulae in realistic contexts. [Grade C]
- Students **could** be able to expand brackets and simplify expressions. [Grade C]

Key words
- brackets
- expression
- simplification
- coefficient
- formula
- substitute
- equation
- identity
- term
- expand
- like terms
- variable

Prior knowledge
Students need to be able to multiply algebraic terms together.

Common mistakes and remediation
Students may multiply the first term properly but then just write down the subsequent terms without multiplying them; for example,
$$5(3a + 2b) = 15a + 2b \text{ (it should be } 15a + 10b\text{)}.$$

Useful tips
Encourage students to work out the number calculations and each different letter calculation separately. Stress that, when they are multiplying one term outside the brackets by one or more terms inside the brackets, students must multiply everything inside the brackets by the number or algebraic term outside the brackets.

Functional maths and problem-solving help
The problem-solving (PS) questions and all the other questions in Exercise 6B are structured and straightforward. The main skill with both questions is to interpret the question and communicate the solution.

© HarperCollins*Publishers* 2012

Lesson 6.1b Basic algebra: Expansion

Starter
- Ask a variety of multiplication questions, e.g. 5×2, $5 \times 2a$, $5 \times 2a^2$, $5a \times 2a$, $5a \times 2b$, $5a \times 2a^2$, $5a^2 \times 2a$, $5a \times 2a^3$.

Main lesson activity
- Write on the board:
  $3 \times 4 + 3 \times 5$ and $3 \times (4 + 5)$
  Ask students to work out these calculations. They should get the same answer. Write
  $3 \times (4 + 5) = 3 \times 4 + 3 \times 5$.

- Now write $2 \times (x + 5)$ and ask students, using the previous example as a clue, to write this without brackets.
- Look for the answer $2 \times x + 2 \times 5$. Simplify this to $2x + 10$.
- Repeat with $3(y - 6)$ to get the answer $3y - 18$.
- Remind **less able** students that there is an invisible multiplication sign between the number outside the brackets and the opening bracket. Show this as $3 \times (y - 6)$ and the answer as $3 \times y - 3 \times 6$ if necessary.
  Emphasise that students must evaluate the multiplications and simplify expressions.
- Go through this section in the Student Book and familiarise the students with the term 'expansion'.
- **Less able** students may need to write in the multiplication signs before the opening brackets, and expand the terms, again with multiplication signs, before simplifying them. They could draw arcs as they multiply the terms, to keep track. For example:
  $5(3a + 2b) = 15a + 10b$
- **Less able** students may also find the 'minus sign outside the bracket' a difficult concept, but this is essential for **more able** students as it is often assessed in proof questions which are grade A or A*.
- Students can now do Exercise 6B, in which question 3 is an assessing understanding (AU) question and 4 is PS. **Less able** students should work carefully through question 1, to gain practice, before moving on to questions 2–4.

Plenary
- Write on the board: $3 \times 4 + 3 \times 5 = 3 \times (4 + 5)$. Students will have seen this from the main lesson activity.
- Now write $2x + 10 = 2(x + 5)$. This was also part of the main lesson activity.
- Explain that this is the reverse process of expanding.
- Now ask the students if they can 'reverse' $3x - 18$, $4x - 20$, $4y + 6x$.
- Tell them that they will come across this later when they learn about factorisation.

Answers
Answers to Exercise 6B can be found at the end of this chapter.

© HarperCollins*Publishers* 2012

6.1c Basic algebra: Simplification

Curriculum references

Functional Skills standards
A1 Apply a range of mathematics to find solutions.

PLTS
Independent enquirers, Creative thinkers, Reflective learners, Team workers, Self-managers, Effective participators

APP
Algebra L7.4

Collins references
- Student Book pages 127–129
- Interactive Book: Common Misconceptions

Learning objectives
- Recognise expressions, equations, formulae and identities. [Grade D–C]
- Substitute into, manipulate and simplify algebraic expressions. [Grade D–C]

Learning outcomes
- Students **must** be able to substitute numbers into expressions and formulae. [Grade D]
- Students **should** be able to use formulae in realistic contexts. [Grade C]
- Students **could** be able to expand brackets and simplify expressions. [Grade C]

Key words
- brackets
- expression
- simplification
- coefficient
- formula
- substitute
- equation
- identity
- term
- expand
- like terms
- variable

Prior knowledge
Students should have completed and understood Section 6.1b Basic algebra: Expansion.

Common mistakes and remediation
Students may try to combine unlike terms, e.g. $4a + 5 = 9a$. (The correct answer is that this cannot be simplified.)

Useful tips
Encourage students to work out the number calculations and each different letter calculation separately. Advise them to take time to identify like terms, and remember they cannot add unlike terms.

Functional maths and problem-solving help
The functional maths (FM) question in Exercise 6C is structured but could easily be set without the structure as a more open-ended question, as below.

    A two-carriage train has f first-class seats and $2s$ standard-class seats.
    A three-carriage train has $2f$ first-class seats and $3s$ standard-class seats.
    $f = 15$ and $s = 80$.
    On a weekday, 5 two-carriage trains and 2 three-carriage trains travel from Hull to Liverpool.
    On average in any day, half of the first-class seats are used at a cost of £60.
    On average in any day, three-quarters of the standard-class seats are used at a cost of £40.
    The rail company pays £30 000 a day to operate this route. How much profit do they make on an average day?

In this case, students will need to work out their own strategy (identifying the situation and mathematics needed).
Refer students to the 'Hints and tips' box next to problem-solving (PS) question 11 of Exercise 6C.

© HarperCollins*Publishers* 2012

Lesson 6.1c Basic algebra: Simplification

Starter
- Ask students if it always true that $a \times b = b \times a$? (Yes) Can this be written in a simpler form? ($ab = ba$).
- **Less able** students may need to see this demonstrated with numbers.
- Repeat with $a + b = b + a$, $a - b = b - a$ and $a \div b = b + a$.
- Ask the following quick fire questions.
  - What is 3 apples + 4 apples? (7 apples)
  - What is $3a + 4a$? ($7a$)
  - What is 3 apples + 4 apples + 2 bananas + 3 bananas? (7 apples + 5 bananas)
  - What is $3a + 4a + 2b + 3b$? ($7a + 5b$)
  - Expand $4(3 + 5a)$. ($12 + 20a$)
  - Expand $3(3a - 2)$. ($9a - 6$)

Main lesson activity
  Simplification
- Ask students if they can simplify $3a + 4a$. ($7a$) Write on the board: $3a + 4a = 7a$
- Ask **less able** students, "What is 3 apples plus 4 apples?" Ask if they can express this mathematically (they may be able to do so, as it was a 'quick-fire' question in the starter). Repeat with:
  - $6b - 2b$ ($4b$) (6 bananas minus 2 bananas).
  - $3ab + 7ba$ ($= 10ab$). If this causes confusion, refer to the starter when it was established that $ab = ba$. Students may ask why the answer is not written as $10ba$. Say that it could be written that way but, as a normal convention, letters in algebra are usually written alphabetically.
- Now ask for the simplification of $3a + 5b + 2a + 4b$ ($5a + 9b$). This may lead to answers such as $14ab$.
  Explain 'like terms'. Say that only like terms can be combined.
- **Less able** students may prefer to rewrtite the calculation as $3a + 2a + 5b + 4b$.
- Point out to all students that once they have identified the like terms, they only need to combine the coefficients.
- Repeat with $6a + 2b - 3a + 5b$. ($3a + 7b$) and $5a - 4b + 9a + 5b$. ($13a + b$)
  This last example is easier to see if written as $5a + 9a - 4b + 5b$, and **less able** students may see the answers more easily as $5 + 9$ and $-4 + 5$.
  Expand and simplify
- Go through this section and Examples 4 and 5 in the Student Book, to familiarise students further with the process of simplification.
- Students can now do Exercise 6C, in which question 9 is FM, 2 and 10 are assessing understanding (AU) questions and 3 and 11 are PS. Encourage **less able** students to do questions 1–3 and then to work with a partner or a **more able** student for the rest of the exercise, if necessary.

Plenary
- Write on the board $7a + 6b$. Ask students to write an expression such as $3a + 2b + 4a + 4b$, for which the answer is $7a = 6b$. Make sure some expressions contain minus signs.
- Collect and compare answers.
- Repeat for a similar expression with a negative term, such as $5a - 3b$.

Answers
Answers to Exercise 6B can be found at the end of this chapter.

6.2 Factorisation

Curriculum references

Functional Skills standards
I2 Draw conclusions and provide mathematical justifications.

PLTS
Independent enquirers, Creative thinkers, Reflective learners, Team workers, Self-managers, Effective participators

APP
Algebra L8.2

Collins references
- Student Book pages 130–131
- Interactive Book: Matching Pairs

Learning objectives
- Factorise an algebraic expression. [Grade D–B]

Learning outcomes
- Students **must** be able to factorise a simple expression. [Grade D]
- Students **should** be able to decide whether expressions can be factorised or not. [Grade C]
- Students **could** be able to explain how expressions can be factorised in different ways. [Grade C]

Key words
- common factor
- factorisation

Prior knowledge
Being able to expand brackets is the first step to being able to factorise, because one is the inverse of the other. Students should be able to find the highest common factor (HCF) of two or more numbers.

Common mistakes and remediation
Students may just find a factor, and not the highest factor.
In a question such as $4xy + 2x$, students may write $2x(2y + 0)$ as they think that there is nothing left when $2x$ is taken from the second term. Stress that they are dividing, not subtracting.

Useful tips
Advise students that they can always check that an expression has been factorised correctly by expanding the answer. When they are factorising, encourage students to look for the HCF and not just any factor.

Functional maths and problem-solving help
In Exercise 6D, functional maths (FM) question 2 and problem-solving (PS) question 6 are structured and straightforward. In this case students will need to explain their answers clearly (communicate solutions).

© HarperCollins*Publishers* 2012

Lesson 6.2 Factorisation

Starter
- Ask students to give the HCF of:
  - 6 and 9 (3)
  - $6a^2$ and $9a$ ($3a$)
  - $6a$ and 9 (3)
  - $6a^2$ and $9a^2$. ($3a^2$)

- **Less able** students may find it easier to see the common factors if the expressions are broken down into products, e.g. 6 and 9 become 3 × 2 and 3 × 3, $6a$ and 9 become 3 × $2a$ and 3 × 3, $6a$ and $9a$ become $3a$ × 2 + $3a$ × 3, and so on.

Main lesson activity
- Write some expansions on the board, such as:
  - $2(x + 5) = 2x + 10$
  - $3(x - 2) = 3x - 6$
  - $5(x + 3) = 5x + 15$.
- Now write $2x + 14 = \ldots$ and ask students if they can write down the expansion that it comes from. Most students will appreciate that the answer is $2(x + 7)$.
- Repeat with other examples such as $5x - 15$, $6x + 18$.
- Explain that what they have been doing is called factorisation and it is the reverse of expanding.
- Now ask students to factorise $4x + 12$. Many of them may say $2(2x + 6)$. Explain that this is a partial factorisation; the correct answer is $4(x + 3)$.
- Explain that they must take the HCF out of each term.
- Now ask for the factorisation of $x^2 + 3x$. This may need some more explanation, such as $x \times x + 3 \times x = x(x + 3)$.
- Work through Example 6, which includes a variety of different factorisations.
- Emphasise to the students that they can multiply out the factored expression to check that the answer is the same as the original expression.
- All but the most **more able** students will most likely find this topic difficult. Initially, restrict **less able** students to examples with positive values and those that do not require a letter to be taken out as a factor.
  Alternatively, encourage these students to write terms as products, as in the starter.
- Students can now do Exercise 6D, in which 2 is FM, 4 and 5 are assessing understanding (AU) questions, and 6 is a PS question. Encourage **less able** students to complete question 1 as far as they can on their own, but be prepared to offer support for questions 2–5.

Plenary
- Ask students to supply expressions with terms that have an HCF of 6, $6a$, $6a2$, $6ab$, $6ab^2$, etc. Ask for several expressions for each HCF. (This is the reverse of the starter.)

Answers
Answers to Exercise 6D can be found at the end of this chapter.

© HarperCollins*Publishers* 2012

6.3a Solving linear equations: Fractional equations

Curriculum references

Functional Skills standards
A1 Apply a range of mathematics to find solutions.

PLTS
Independent enquirers, Creative thinkers, Reflective learners, Team workers, Self-managers, Effective participators

APP
Algebra L6.2

Collins references
- Student Book pages 132–133
- Interactive Book: Common Misconceptions

Learning objectives
- Set up equations from given information and then solve them. [Grade D–C]
- Solve equations in which the variable appears as part of the numerator of a fraction. [Grade D–C]
- Solve equations where you have to expand brackets first. [Grade D–C]
- Solve equations where the variable appears on both sides of the equals sign. [Grade D–C]

Learning outcomes
- Students **must** be able to solve simple linear equations. [Grade C]
- Students **should** be able to solve linear equations that contain brackets. [Grade D]
- Students **could** be able to solve equations where the variable occurs on both sides. [Grade C]

Key words
- brackets
- rearrange
- do the same to both sides
- solution
- solve
- equation

Prior knowledge
Students need to be able to solve simple one- and two-step equations where the variable appears only on one side of the equation, e.g. $2x - 4 = 6$.

Common mistakes and remediation
The main mistake made with fractional equations is that students will try to solve the problem 'out of order'. For example, for $\frac{x+5}{3} = 7$, students may try to subtract 5 rather than multiply by 3 as the first step, as they don't realise that the term $x + 5$ must remain together until the denominator is removed.

Useful tips
Remind students always to check answers in the original equation.
Using brackets around the numerator may help some students to understand and/or remember that they must remove the denominator before dealing with the numerator (if it is more complex than just x or $3y$, etc.).

Functional maths and problem-solving help
There are no functional maths (FM) questions in Exercise 6E. In general, the questions are intended to assess students' understanding of the processes rather than their ability simply to solve equations.

© HarperCollins*Publishers* 2012

Lesson 6.3a Solving linear equations: Fractional equations

Note: This is the first of three sections on solving linear equations. For **more able** students, the sections could be combined, as essentially the process is the same for solving all linear equations.

Starter
- Ask students to solve the following equations.
  - $x - 4 = 6$
  - $2x - 4 = 6$
  - $2x + 5 = 6$
  - $\dfrac{x}{3} = 6$
- Discuss the methods used to solve the equations.
- Make sure students are aware of the need to 'do the same thing to both sides'.

Main lesson activity
- Read out the two 'think of a number problems' below. For each problem, ask students to form the equation and solve it.
  I am thinking of a number. I divide it by 2 and add 6. The answer is 11. What number did I think of?
  I am thinking of another number. I add 6 then divide the result by 2. The answer is 11. What number did I think of?
  The equations are $\dfrac{x}{2} + 6 = 11$ and $\dfrac{x+6}{2} = 11$, with answers of 10 and 16 respectively.
- Discuss the differences between these equations, then work through the algebraic solution. It is important that students appreciate that when rearranging, the first operation is different for both equations.
- **Less able** students may find this hard to see, so encourage them to recognise the expressions that contain x, i.e. $\dfrac{x}{2}$ and $x + 6$. The equations could then be seen as: $\square + 6 = 11$ and $\dfrac{\square}{2} = 11$.
- Work through this section and Examples 7, 8 and 9 to familiarise the students more with how to solve equations with fractions. Make sure they check the answers in the original equation.
- Students can now do Exercise 6E, in which questions 2 and 3 are assessing understanding (AU) questions. Advise **less able** students to complete question 1, for plenty of practice, before moving on to 2 and 3. **More able** students could omit some of the parts of question 1, if they are successfully solving the equations.

Plenary
- Ask the class to help solve these three equations. Ask students to point out things to watch out for. For example:
  How might they be confused? What mistakes might they make?
  - $x + \dfrac{10}{2} = 20$
  - $\dfrac{x}{2} + 10 = 20$
  - $\dfrac{x + 10}{2} = 20$

Answers
Answers to Exercise 6E can be found at the end of this chapter.

© HarperCollins*Publishers* 2012

6.3b Solving linear equations: Brackets

Curriculum references

Functional Skills standards
A1 Apply a range of mathematics to find solutions.

PLTS
Independent enquirers, Creative thinkers, Reflective learners, Team workers, Self-managers, Effective participators

APP
Algebra L6.2

Collins references
- Student Book pages 134–135
- Interactive Book: Common Misconceptions

Learning objectives
- Set up equations from given information and then solve them. [Grade D–C]
- Solve equations where you have to expand brackets first. [Grade D]
- Solve equations where you have to expand brackets first. [Grade D–C]
- Solve equations where the variable appears on both sides of the equals sign. [Grade D–C]

Learning outcomes
- Students **must** be able to solve simple linear equations. [Grade C]
- Students **should** be able to solve linear equations that contain brackets. [Grade D]
- Students **could** be able to solve equations where the variable occurs on both sides. [Grade C]

Key words
- brackets
- rearrange
- do the same to both sides
- solution
- solve
- equation

Prior knowledge
Students must know how to expand brackets.

Common mistakes and remediation
Students lose marks by multiplying only the first part of the terms inside the brackets by the outside. Remind students to watch out for this.

Useful tips
Make sure students understand that they must multiply out brackets before they can solve the equation. They must multiply everything inside the brackets by what is in front of the opening bracket. They must take care when dealing with negative numbers. Once they have removed the brackets, they can solve the equations as before.

Functional maths and problem-solving help
There are no functional maths (FM) questions, as such, in Exercise 6F.
Problem-solving (PS) question 3 has an open-ended element: $x + 5 + 2(x + 5) = 60$, $x = 15$. This doesn't fit the conditions. Eventually, when students put the sum of ages equal to 90, they obtain the correct answer. Students will need to work methodically and break the problem down.
The problem-solving and assessing understanding (AU) questions are intended to establish students' understanding of the processes rather than their ability simply to solve equations.

© HarperCollins*Publishers* 2012

Lesson 6.3b Solving linear equations: Brackets

Starter
- Ask the following quick-fire questions:
  - What is 3 × £4.25? (£12.75)
    What is 3 × £4 + 3 × 25p? (£12 + 75p = £12.75)
    What is 3 × (£4 + 25p)? (£12 + 75p = £12.75)
    What is 3(£4 + 25p)? (£12 + 75p = £12.75)
  - Is 5(2 + 4p) = £10.20? (No = 10 + 20p)
  - What is $5(2x + 3y)$? ($10x + 15y$)

- Discuss the methods used to solve the equations. Make sure students are aware of the need to 'do the same thing to both sides'.

Main lesson activity
- Ask the class to think of a number, add 3 to it and then multiply the result by 5. Ask some students to give their result and ask other students if they can give the original number thought of. Do this for five or six numbers.
- Now ask if students can explain how they worked out the original number. Try to establish that they first divided the result by 5 and then subtracted 3.
- Now show the original problem as an equation: $5(x + 3)$ = result. **Less able** students may wish to see this as a flow diagram. Demonstrate with one of the results from earlier, say $5(x + 3) = 50$, for which the starting number x is 7.
- Now work through the equations in Examples 10 and 11.
- Students can now do Exercise 6F, in which question 2 is an AU question and 3 is PS. Advise **less able** students to complete question 1 first, then move on to 2 and 3.

Plenary
- Write the following on the board:
- $5(x + 3) - 2(x + 6) = 12$
- Ask students to expand, simplify and solve ($x = 3$)
- Repeat with $6(x + 7) + 2(x - 3) = 12$. ($x = -2$)

Answers
Answers to Exercise 6F can be found at the end of this chapter.

© HarperCollins*Publishers* 2012

97

6.3c Solving linear equations: Variables on both sides

Curriculum references

Functional Skills standards
A1 Apply a range of mathematics to find solutions.

PLTS
Independent enquirers, Creative thinkers, Reflective learners, Team workers, Self-managers, Effective participators

APP
Algebra L6.2

Collins references
- Student Book pages 135–136
- Interactive Book: Common Misconceptions

Learning objectives
- Solve equations where the variable appears on both sides of the equals sign. [Grade D–C]
- Set up equations from given information and then solve them. [Grade D–C]
- Solve equations where you have to expand brackets first. [Grade D–C]
- Solve equations where the variable appears on both sides of the equals sign. [Grade D–C]

Learning outcomes
- Students **must** be able to solve simple linear equations. [Grade C]
- Students **should** be able to solve linear equations that contain brackets. [Grade D]
- Students **could** be able to solve equations where the variable occurs on both sides. [Grade C]

Key words
- brackets
- rearrange
- do the same to both sides
- solution
- solve
- equation

Prior knowledge
As in Lesson 6.3b, students should know how to expand brackets.

Common mistakes and remediation
Students often get the wrong answer by not changing signs when they change sides when rearranging equations.
They do not always appreciate that the sign relates to the number that follows it. Encourage students to be logical and to rearrange before simplifying, and not try to do both at the same time. Students may find it helpful to remember 'change side, change sign' when they move a term from one side of the equals sign to the other.

Functional maths and problem-solving help
There are no functional maths (FM) questions in Exercise 6G.
Problem-solving (PS) questions 2 and 5 in Exercise 6G have hints which should make the questions accessible to the majority of students. The process skill is to interpret the question and communicate the solution.

© HarperCollins*Publishers* 2012

Lesson 6.3c Solving linear equations: Variables on both sides

Starter
- Display these algebra pyramids. Explain that the terms in the bricks on the bottom row have been added, to give the middle row. The number in the top row is the sum of the terms in the bottom row. The problem is to solve to find x. In the first one, $5x + 7 = 22$, so $x = 3$.

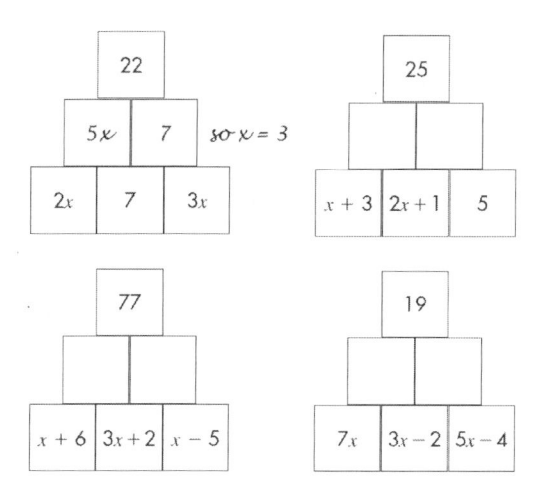

Pyramid 1: Top: 22; Middle: $5x$, 7 (so $x = 3$); Bottom: $2x$, 7, $3x$

Pyramid 2: Top: 25; Bottom: $x + 3$, $2x + 1$, 5

Pyramid 3: Top: 77; Bottom: $x + 6$, $3x + 2$, $x - 5$

Pyramid 4: Top: 19; Bottom: $7x$, $3x - 2$, $5x - 4$

Main lesson activity
- Say that you are thinking of a number. You multiply it by 2 then add 5 to it. The final result is 8 more than the original number you thought of. Ask students if they know the number you started with. This will not be as easy to solve as previous similar questions, as there is no simple method of using inverse operations.
- If no student comes up with the correct answer of 3, then tell them the answer and work through the problem: $2 \times 3 + 5 = 3 + 8$. Then set this up as an equation: $2x + 5 = x + 8$.
- Ask students why this is different from previous problems. They should see that the variable is on both sides. Work through the method of solution:
$$2x - x = 8 - 5$$
$$x = 3$$
- It is important to emphasise that the variable terms need to be collected on one side (normally the left hand side, but this is not always the case), and the numbers on the other side.
- Now work through the equations in Examples 12, 13 and 14.
- Students can now do Exercise 6G, in which questions 2 and 5 are PS and 4 and 6 are assessing understanding (AU) questions. **Less able** students may find questions 3–5 difficult to access. Once they have completed 1 and 2 (grade D questions), they could work with a partner or a **more able** student to attempt the remaining questions, which are grade C.

Plenary
- Write on the board: $5(x + 3) = 2(x + 6)$.
- Ask students to expand, simplify and solve. ($x = -1$)
- Repeat with $6(x - 5) = 2(x + 3)$. ($x = 9$)

Answers
Answers to Exercise 6G can be found at the end of this chapter.

© HarperCollins*Publishers* 2012

6.4 Setting up equations

Curriculum references

Functional Skills standards
R2 Identify the situation or problems and identify the mathematical methods needed to solve them.

PLTS
Independent enquirers, Creative thinkers, Effective participators

APP
Algebra L6.2

Collins references
- Student Book pages 137–140

Learning objectives
- Set up equations from given information, and then solve them. [Grade D–C]

Learning outcomes
- Students **must** be able to set up and solve an equation in a particular context. [Grade D]
- Students **should** be able to use equations when solving real life problems. [Grade D]
- Students **could** be able to interpret the solution of an equation in a real situation. [Grade C]

Key words
- do the same to both sides
- equation
- rearrange
- solve

Prior knowledge
Students must be able to solve simple equations.

Common mistakes and remediation
Students often find that the most difficult aspect is in accessing questions to set up the equations. Encourage them to follow the steps as outlined. This should limit the errors to those for solving any equations (mistakes with rearranging and incorrect calculations) covered in the previous lesson.

Useful tips
Tell students that this lesson is about setting up equations from real-life or practical situations, and that they should define the 'unknown' as a variable, say x; then set up an equation using x.

Functional maths and problem-solving help
In Exercise 6H the functional maths (FM) and problem-solving (PS) questions are 1, 3, 5 to 10, 12, 14, 16 and 17, which involve the process skills as outlined in the main lesson activity. It may be worthwhile asking students to write these down, or you could display them on the board or on a poster. Ensure that students follow the steps in each question. Other questions have hints.

Lesson 6.4 Setting up equations

Starter

- Ask students to formulate equations for simple word problems such as: If you cut m cm from 2 metres of ribbon how much is left? (($200 - m$) cm) the side of a square is c metres, so what is its area? ($c^2 \, m^2$)
- Ask students to name any simple formulae they know. They may provide these in words that, as a class, you can convert into algebra.

Main lesson activity

- Explain to students that they will look at problems from real-life or practical situations and solve them by using a variable such as x to represent the unknown. While the unknown is often defined in the question, sometimes it needs to be identified from the information provided.
- Point out that they should read the question thoroughly and identify key words.
- Give students the following problem: My sister is five years younger than I am. The sum of our ages is 41. How old am I?
- Ask students to identify the unknown. Clearly the unknown is 'my age'. Calling this x (or A), ask students to identify the necessary information and translate it into mathematical symbolism. Sister's age is $x - 5$; sum of ages $x + x - 5 = 41$; $2x - 5 = 41$, $2x = 46$, $x = 23$.
- Having solved the problem (my age is 23), explain that students should check that the solution fits the original problem. In doing so, students should be aware of the process skills:
    - Define the unknown if necessary.
    - Put given information in terms of the unknown.
    - Set up an equation.
    - Collect terms if necessary and solve the equation.
    - Check that the answer fits the original problem.
- **Less able** students may need help in accessing the questions and deciding on the variable. They may also need help in breaking down the problems.
- Work through Examples 14 and 15.
- Now students can complete Exercise 6H, in which the FM questions are 1 and 5–8, 14 and 17, the PS questions are 3, 9, 10, 12 and 16 and the assessing understanding (AU) question is 15. **Less able** students can complete the earlier questions and then work with a partner, if necessary. Provide support as needed.

Plenary

- Ask students to make up simple problems such as: Mary goes to the shop with a £10 note and spends £x. She gets £3 change. How much did she spend?
- Although this is a trivial question, set it up as an equation, e.g. $10 - x = 3$.
- Ask for similar problems that involve multiplication and division.

Answers

Answers to Exercise 6H can be found at the end of this chapter.

6.5 Trial and improvement

Curriculum references

Functional Skills standards
R2 Identify the situation or problems and identify the mathematical methods needed to solve them.

PLTS
Independent enquirers, Creative thinkers, Effective participators

APP
Not in assessment criteria.

Collins references
- Student Book pages 140–142

Learning objectives
- Estimate the answer to some equations that do not have exact solutions, using the method of trial and improvement. [Grade C]

Learning outcomes
- Students **must** be able to use trial and improvement to solve an equation. [Grade C]
- Students **should** be able to state the accuracy of a solution found by trial and improvement. [Grade C]
- Students **could** be able to use trial and improvement to find solutions to real-life problems. [Grade C]

Key words
- comment
- guess
- decimal place
- trial and improvement

Prior knowledge
Students should understand approximation and estimation. They should be able to apply BIDMAS/BODMAS, particularly working out a power before a multiplication or division.

Common mistakes and remediation
Students often do not test the halfway value, as they think the answer is obvious. This would lose a mark in the examination. Remind them of the need to test this point whenever errors arise. Students also tend to keep improving their guesses until the answer is within one decimal place of the target value. Emphasise the need simply to check the bounds to the same degree of accuracy as required, and then the halfway value.

Useful tips
Students will need calculators for this lesson. If their calculators have a cube key, make sure they can use it. If not, then they should know how to use the power key. Students will also need to be able to do a calculation such as $2.7^3 - 2 \times 2.7$. (= 14.283)

Functional maths and problem-solving help
In Exercise 6I the functional maths (FM) and problem-solving (PS) questions are 7 and 9. The process skill is setting up the problem in such a way that the working can be followed. Using a table is vital to keep the working logical.

© HarperCollins*Publishers* 2012

Lesson 6.5 Trial and improvement

Starter
- Write this calculation on the board: $2.7^3 - 2 \times 2.7$, and ask students to work it out. (14.283) It is unlikely that everyone in class will have obtained the correct value.
- Mention the cube button, which up-to-date calculators have, or the power key, although the power key on older calculators often involves '2nd function' keys and proves difficult for **less able** students to use.
- Another way to do the calculation is: $2.7 \times 2.7 \times 2.7 - 2 \times 2.7 = 14.283$. Make sure students understand that 2.7^3 means $2.7 \times 2.7 \times 2.7$.
- Repeat with similar calculations until students can do these.

Main lesson activity
- Write the following on the board: $x^2 + x = 12$.
- Ask students if they can solve it for x. If they need a clue, say that it is a whole number.
- Obtain, or give, the answer of 3 and show that $3 \times 3 + 3 = 12$.
- Below the first equation, write: $x^2 + x = 20$ and repeat the process. (4)
- Now, below the other two equations, write: $x^2 + x = 15$. Ask students if this can be solved. Some may not think so as solutions of 3 and 4 have already been used. Others may say that the answer will be between 3 and 4. Ask how they can be sure of this.
- Explain that (at this level), the equation cannot be solved so the only method is guesswork.
- Ask students to find the answer by guessing a number between 3 and 4 and trying it out. Say that if their first guess doesn't work they should try another.
- You could point out that there is no exact answer, so as soon as they get an answer very close to 15 they can give you the value. Given a reasonably large class, it should not be long before the answer comes up: 3.405... or 3.4.
- Check this in the original equation: 3.4 gives an answer of 14.96.
- Explain that this is called 'trial and improvement' and that it is used to find the answer to certain types of equations that cannot be solved using 'rearrangement'.
- Ask students if they can see one significant difference in this type of equation. Hopefully they will spot that this type of equation has an x term with a power. Explain that to use this process efficiently they must work in a systematic way, using a table.
- Work through Example 16, filling in a table from the start. Students may query why the midpoint of the two 1 dp values needs to be tested. This is because we are dealing with curves and there is no guarantee that an x value that gives an answer closest to the target value is closest to the real answer.
- Now students can complete Exercise 6I. Remind them always to use a table. **Less able** students may need help with using their calculators.
- In this exercise, question 9 is FM and 7 is PS. **Less able** students can work with a partner if they are having difficulty with the work.

Plenary
- Ask students to find the solution to $x^5 = 100$ using trial and improvement. (2.5 to 1dp)
- Then write the following on the board:
  $x^2 = 100, x = 10$ $x^3 = 100, x = 4.6$ $x^4 = 100, x = 3.2$ $x^5 = 100, x = 2.5$
- Ask students to guess the answer to $x^6 = 100$ and then try it out. (2.15 or 2.2)

Answers
Answers to Exercise 6I can be found at the end of this chapter.

Functional maths
Temperature scales

Curriculum references

Functional Skills standards
R3 Select a range of mathematics to find solutions.
A1 Apply a range of mathematics to find solutions.
I2 Draw conclusions and provide mathematical justifications.

PLTS
Creative thinkers, Reflective learners

APP
Using and applying mathematics L8.1; Calculating L6.1; Algebra L8

Collins references
- Student Book pages 150–151

Learning objectives
- Add, subtract, multiply and divide negative numbers
- Substitute numbers into formulae

Key words
- negative numbers
- scales
- formula

Functional maths help
All students should be familiar with temperature scales. Start a discussion, first about the temperature in the classroom. Discuss the temperatures at which students would need to wear a coat, can walk around in a shirt. Ask about the temperature at which water freezes. In these discussions, students will probably use the terms 'Centigrade', 'Celsius' and 'Fahrenheit'. If they give a temperature value without saying the scale, ask them to define the scale they are using. They are more likely to refer to 'Celsius' if they have used it in science.

Lesson plans
Build
After an initial discussion about the basic idea of temperature, focus on the temperature scales. Use the information in the Student Book to discuss the origin and fixed points of the Fahrenheit, Celsius and Kelvin scales. Point out that both the Celsius and Fahrenheit scales measure temperatures in **degrees** but the units on the Kelvin scale are called **kelvin**.

Ask students why the Celsius scale is based on the freezing and boiling point of water. Introduce the concept of absolute zero, the temperature at which no more heat can be removed from a system. **More able** students could carry out more research, by means of the internet.

Activity 1
Ensure that students have read about the three temperature scales. Refer them to the first activity in 'Your task'. Ask students to draw three vertical, parallel lines, all of equal length, one for each temperature scale. They should label the left line °F (Fahrenheit), the second line °C (Celsius) and the third line K (Kelvin) and mark the bottom of the Kelvin line 0 to represent absolute zero.

To complete the activity, students need to determine the highest temperature that they will need to find (327.5°, the melting point of lead) and mark the top of the Celsius line at an appropriately high temperature (such as 400°C.) Ask students if they know what values should go at the bottom of the other two scales (–273.15°, approximated to –273° on the Celsius scale; –459.67° approximated to –460°on the Fahrenheit scale). Explain that the equivalent temperatures, on the three scales, should align horizontally on the three vertical lines. The main values (rounded where appropriate) that the students are asked to mark are given in the following table.

Temperature	Absolute zero	Freezing point mercury	Zero on Fahrenheit scale	Freezing point water	Human body temperature	50°C equivalent points	Boiling point water	Combustion point paper	Melting point of lead	Boiling point mercury	Highest points on the three temperature lines
Fahrenheit	–460°	–37°	0°	32°	98.5°	122°	212°	451°	622°	674°	752°
Celsius	–273°	–39°	–18°	0°	37°	50°	100°	233°	328°	357°	400°
Kelvin	0	234	255	273	310	323	373	506	601	630	673

© HarperCollins*Publishers* 2012

Functional maths Temperature scales

Ask students to mark 0° on the Celsius scale. Then ask them to mark the equivalent values on the other two scales (273 on the Kelvin scale and 32° on the Fahrenheit scale).
Use a similar procedure, starting with 100° on the Celsius scale (373 on the Kelvin scale and 212° on the Fahrenheit scale). It is important to mark the temperatures in the right places on the scales. Now compare and mark some other standard points, such as body temperature, the temperature at which lead melts and the temperature at which paper burns.
Less able students may benefit from marking equal divisions (say, 10) on each scale between freezing and boiling point of water, and marking, say 10, 20, … on the Celsius scale, 50, 68, … on the Fahrenheit scale and 283, 293, … on the Kelvin scale.

Activity 2
Now ask students to mark 50 on the Celsius scale. Ask them to try to work out the equivalent value on the other scales and establish a rule for the relationship between the scales. Start with the Kelvin and Celsius scales. Give a few equivalent values. Students should see that, numerically, the Kelvin temperature is 273 more than the Celsius temperature. Ask how this could be expressed as a rule. Repeat with the Celsius and Fahrenheit scales. This is harder to see, so if no progress is being made, subtract 32 from the F values. This then relates 0 to 0 and 180 to 100. If students do not spot the connection suggest a scale factor approach. Eventually, they should arrive at $F = 1.8C$. Now add the 32 back on so $F = 1.8C + 32$. Depending on their ability, students could try rearranging these formulae. Most students could manage the first but the second is more complicated.
More able students could combine these relationships to find a relationship between F and K.
$$K = \frac{5}{9}F + 255$$
Very able students may be able to solve equations, setting both variables as x, to find the temperatures at which both scales read the same. This is not possible for K and C as they always differ by 273 but for F and C it is −40° and for F and K it is 573°.

Activity 3
Provide internet access or selected newspapers and ask students to find current weather maps of Britain showing maximum and minimum temperatures. Ask students to use the conversions developed in Activity 2 to convert these weather figures to alternative scales.

Apply
Use a class discussion to determine the range of the scales for the first activity. Ask why it is important to know the high and low points to be marked. Emphasise that, although the numbers marked are very different, the temperatures are equivalent across the scales. Discuss the choice of standard temperatures and ask students to think of alternatives that would serve the same purpose. Mercury's freezing and boiling points, for example, are within the chosen temperature range.

Master
In order to demonstrate their mastery of the learning objectives, students should be able to:
- Add, subtract, multiply and divide negative numbers accurately.
- Substitute numbers into formulae and derive solutions.

Plenary
Discuss where temperatures are used in everyday life, for example, on weather maps, car displays, washing machines, room thermostats. Use the internet to look at other scales and their practical uses. Discuss which scale is the most useful for daily life.

© HarperCollins*Publishers* 2012

Answers Lessons 6.1 – 6.5

Quick check
1 a $2x + 12$ b $4x - 12$ c $12x - 6$
2 a $5y$ b $4x - 3$ c $-x - 4$
3 a $6x$ b $8y^2$ c $2c^3$
4 a $x = 1$ b $x = 3$ c $x = 9$ d $x = 8$
  e $x = 24$ f $x = 15$

6.1a Basic algebra: Substitution
Exercise 6A
1 a 13 b –3 c 5
2 a 2 b 8 c –10
3 a 6 b 3 c –2
4 a –7 b –10 c 6.5
5 a –4.8 b 48 c 32
6 a 1.4 b 1.4 c –0.4
7 a 13 b 74 c 17
8 a 75 b 22.5 c –135
9 a 2.5 b –20 c 2.5
10 a £4 b 13 km
   c No, 5 miles is 8 km so fare would be £6.50.
11 a $\dfrac{150}{n}$ b £925
12 a $2 \times 8 + 6 \times 11 - 3 \times 2 = 76$
   b $5 \times 2 - 2 \times 1 + 3 \times 8 = 12$
13 a £477.90
   b £117.90 still owed (debit)
14 a One odd one even value, different from each other.
   b Any valid combination, e.g. $x = 1$, $y = 2$
15 Any values such that $2lw = bh$
16 a i Odd ii Odd iii Even iv Odd
   b Any valid expression such as $xy + z$
17 a £20
   b i –£40 ii Delivery cost will be zero.
   c 40 miles
18 A expression, B formula, C identity, D equation
19 a First term is cost of petrol, each mile is a tenth of £0.98. Second term is the hire cost divided by the miles.
   b 29.8p per mile

6.1b Basic algebra: Expansion
Exercise 6B
1 a $6 + 2m$ b $10 + 5l$
  c $12 - 3y$ d $20 + 8k$
  e $6 - 12f$ f $10 - 6w$
  g $10k + 15m$ h $12d - 8n$
  i $t^2 + 3t$ j $k^2 - 3k$
  k $4t^2 - 4t$ l $8k - 2k^2$
  m $8g^2 + 20g$ n $15h^2 - 10h$
  o $y^3 + 5y$ p $h^4 + 7h$
  q $k^3 - 5k$ r $3t^3 + 12t$
  s $15d^3 - 3d^4$ t $6w^3 + 3tw$
  u $15a^3 - 10ab$ v $12p^4 - 15mp$
  w $12h^3 + 8h^2g$ x $8m^3 + 2m^4$
2 a $5(t - 1)$ and $5t - 5$
  b Yes, when $t = 4.50$, $5(t - 1)$ is $5 \times 3.50 = £17.50$.
3 He has worked out 3×5 as 8 instead of 15 and he has not multiplied the second term by 3. Answer should be $15x - 12$.
4 a $3(2y + 3)$ b $2(6z + 4)$ or $4(3z + 2)$

6.1c Basic algebra: Simplification
Exercise 6C
1 a $7t$ b $9d$ c $3e$ d $2t$ e $5t^2$ f $4y^2$
  g $5ab$ h $3a^2d$
2 a $2x$ and $2y$ b a and $7b$
3 a $3x - 1 - x$ b $10x$ c 25 cm
4 a $22 + 5t$ b $21 + 19k$
  c $22 + 2f$ d $14 + 3g$
5 a $2 + 2h$ b $9g + 5$
  c $17k + 16$ d $6e + 20$
6 a $4m + 3p + 2mp$ b $3k + 4h + 5hk$
  c $12r + 24p + 13pr$
  d $19km + 20k - 6m$
7 a $9t^2 + 13t$ b $13y^2 + 5y$
  c $10e^2 - 6e$ d $14k^2 - 3kp$
8 a $17ab + 12ac + 6bc$
  b $18xy + 6ty - 8tw$
  c $14mn - 15mp - 6np$
  d $8r^3 - 6r^2$

9 **a** $5(f + 2s) + 2(2f + 3s) = 9f + 16s$
  b £$(270f + 480s)$
  c £42 450 − £30 000 = £12 450
10 For x-coefficients, 3 and 1 or 1 and 4; For y-coefficients, 5 and 1 or 3 and 4 or 1 and 7
11 $5(3x + 2) − 3(2x − 1) = 9x + 13$

6.2 Factorisation
Exercise 6D
1 **a** $6(m + 2t)$ **b** $3(3t + p)$ **c** $4(2m + 3k)$
  d $4(r + 2t)$ **e** $m(n + 3)$ **f** $g(5g + 3)$
  g $2(2w − 3t)$ **h** $y(3y + 2)$ **i** $t(4t − 3)$
  j $3m(m − p)$ **k** $3p(2p + 3t)$ **l** $2p(4t + 3m)$
  m $4b(2a − c)$ **n** $5bc(b − 2)$
  o $2b(4ac + 3de)$ **p** $2(2a^2 + 3a + 4)$
  q $3b(2a + 3c + d)$ **r** $t(5t + 4 + a)$
  s $3mt(2t − 1 + 3m)$ **t** $2ab(4b + 1 − 2a)$
  u $5pt(2t + 3 + p)$
2 **a** Mary has taken out a common factor.
  b Because the bracket adds up to £10.
  c £30
3 **a**, **d**, **f** and **h** do not factorise.
  b $m(5 + 2p)$ **c** $t(t − 7)$ **e** $2m(2m − 3p)$
  g $a(4a − 5b)$ **i** $b(5a − 3bc)$
4 **a** Bernice
  b Aidan has not taken out the largest possible common factor. Craig has taken m out of both terms but there isn't an m in the second term.
5 There are no common factors.
6 $4x^3 − 12x, 2x − 6$

6.3a Solving linear equations: Fractional equations
Exercise 6E
1 **a** 30 **b** 21 **c** 72 **d** 12 **e** 6 **f** $10\frac{1}{2}$
  g −10 **h** 7 **i** 11 **j** 2 **k** 7 **l** $2\frac{4}{5}$
  m 1 **n** $11\frac{1}{2}$ **o** $\frac{1}{5}$
2 Accept any valid equations.
3 **a** Amanda
  b First line: Betsy adds 4 instead of multiplying by 5.
  Second line: Betsy adds 5 instead of multiplying by 5.
  Fourth line: Betsy subtracts 2 instead of dividing by 2.

6.3b Solving linear equations: Brackets
Exercise 6F
1 **a** 3 **b** 7 **c** 5 **d** 3 **e** 4 **f** 6
  g 8 **h** 1 **i** $1\frac{1}{2}$ **j** $2\frac{1}{2}$ **k** $\frac{1}{2}$ **l** $1\frac{1}{5}$
  m 2 **n** −2 **o** −1 **p** −2 **q** −2 **r** −1
2 Any values that work, e.g. $a = 2$, $b = 3$ and $c = 30$.
3 55

6.3c Solving linear equations: Variables on both sides
Exercise 6G
1 **a** $x = 2$ **b** $y = 1$ **c** $a = 7$ **d** $t = 4$
  e $p = 2$ **f** $k = −1$ **g** $m = 3$ **h** $s = −2$
2 $3x − 2 = 2x + 5, x = 7$
3 **a** $d = 6$ **b** $x = 11$ **c** $y = 1$ **d** $h = 4$
  e $b = 9$ **f** $c = 6$
4 $6x + 3 = 6x + 10$; $6x − 6x = 10 − 3 = 7$, which is obviously false. Both sides include $6x$, which cancels out.
5 $8x + 7 + x + 4 = 11x + 5 − x − 4$; $x = 10$
6 Check students' explanations.

Answers Lessons 6.1 – 6.5

6.4 Setting up equations
Exercise 6H
1 90p
2 **a** 1.5 **b** 2
3 **a** 1.5 cm **b** 6.75 cm^2
4 17
5 8
6 **a** $8c - 10 = 56$ **b** £8.25
7 **a** B: 450 cars, C: 450 cars, D: 300 cars
   b 800 **c** 750
8 Length is 5.5 m, width is 2.5 m and area is 13.75 m^2. Carpet costs £123.75.
9 3 years
10 9 years
11 3 cm
12 5
13 **a** $4x + 40 = 180$ **b** $x = 35°$
14 **a** $\frac{x+10}{5} = 9.50$ **b** £37.50
15 **a** 15 **b** −1
   c $2(n + 3)$, $2(n + 3) - 5$
   d $2(n + 3) - 5 = n$, $2n + 6 - 5 = n$, $2n + 1 = n$, $n = -1$
16 No, as $x + x + 2 + x + 4 + x + 6 = 360$ gives $x = 87°$ so the consecutive numbers (87, 89, 91, 93) are not even but odd.
17 $4x + 18 = 3x + 1 + 50$, $x = 33$
   Large bottle 1.5 litres, small bottle 1 litre.

6.5 Trial and improvement
Exercise 6I
1 **a** 4 and 5 **b** 4 and 5 **c** 2 and 3
2 $x = 3.5$
3 $x = 3.7$
4 $x = 2.5$
5 $x = 1.5$
6 **a** $x = 2.4$ **b** $x = 2.8$ **c** $x = 3.2$
7 **a** Area = $x(x + 5) = 100$
   b Width = 7.8 cm, length = 12.8 cm
8 $x = 5.8$
9 Volume = $2x^2(x + 8) = 500$, $x^3 + 8x^2 = 250$, $x = 4 \Rightarrow 192$, $x = 5 \Rightarrow 325$, $x = 4.4 \Rightarrow 240.064$, $x = 4.5 \Rightarrow 253.125$, $x = 4.45 \Rightarrow 246.541\,125$, so dimensions are 4.5 cm, 9 cm and 12.5 cm.

Examination questions
1   32
2   **a i** $y^2 + 4y - 5$
    ii Odd × odd + 4 × odd − odd = odd + even − odd = even
    b $2y(x - 3y)$
3   $x^2 + x - 12$
4   **a** Identity, (Formula), Equation, Expression **b** Add any even number, multiply by any odd number
5   **a** 3.5 **b** 2.2 **c** 1.5
6   **a i** y^9 **ii** y^5 **iii** y^{14} **b i** y^{14} **ii** y^5
7   6
8   $5x(x + 4)$
9   **a** $x = 20$ **b** $y = \frac{1}{3}$ **c** $2ab(3b - 1)$
10 $5x + 16$
11 1.25
12 **a** 5 **b** 6
13 2.4
14 **a** 5 **b** 27
15 **a** $x(x + 7)$ **b** $15x + 40$ **c** $4x + 9$

Chapter 7 Algebra: Expressions and equations 2

Overview

| 7.1 Simultaneous equations | 7.2 Rearranging formulae |

Lesson 7.1 is concerned with the solving and setting up of simultaneous equations. The chapter finishes with transposing formulae in Lesson 7.2.

Context
Solving equations, dealing effectively with algebraic expressions, and rearranging formulae is the basis of much of mathematics and science, and of many other subjects ranging from economics to architecture.

Curriculum references
KS4 Programme of Study references
2.1c Simplify the situation or problem in order to represent it mathematically, using appropriate variables, symbols, diagrams and models.
2.2j Reason inductively, deduce and prove.
2.2o Record methods, solutions and conclusions.
2.4d look for equivalence in relation to both the different approaches to the problem and different problems with similar structures.
3.1e Linear, quadratic and other expressions and equations.

Linear specification
N4.2 Distinguish in meaning between the words 'equation', 'formula' and 'expression'.
N5.1 Manipulate algebraic expressions by collecting like terms, by multiplying a single term over a bracket and by taking out common factors.
N5.4h Set up and solve simple linear equations including simultaneous equations.
N5.6 Change the subject of a formula.

Functional Skills standards
R2 Identify the situation or problems and identify the mathematical methods needed to solve them.
A1 Apply a range of mathematics to find solutions.
I1 Interpret and communicate solutions to multistage practical problems in familiar and unfamiliar contexts and situations.
I2 Draw conclusions and provide mathematical justifications.

PLTS
Independent enquirers explore issues, events or problems from different perspectives. **Creative thinkers** ask questions to extend their thinking. **Reflective learners** assess themselves and others, identifying opportunities and achievements.
Team workers collaborate with others to work towards common goals. **Self-managers** work towards goals, showing initiative, commitment and perseverance. **Effective participators** discuss issues of concern, seeking resolution where needed.

APP
Algebra L6.2 Construct and solve linear equations with integer coefficients, using an appropriate method. **L7.2** Use algebraic and graphical methods to solve simultaneous linear equations in two variables. **L7.4** Use formulae from mathematics and other subjects; substitute numbers into expressions and formulae; derive a formula and, in simple cases, change its subject. **L8.2** Manipulate algebraic formulae, equations and expressions, finding common factors and multiplying two linear expressions. **L8.3** Derive and use more complex formulae and change the subject of a formula. **Using and applying mathematics L8.3** Select and combine known facts and problem-solving strategies to solve problems of increasing complexity.

© HarperCollins*Publishers* 2012

Chapter 7 Algebra: Expressions and equations 2

Route mapping

		Grades	
Exercise	C	B	A
A		1	2
B		all	
C		1–2	3–4
D	1–15	16–24	

Overview test

Questions 1–3 are grade G, questions 4–11 are grade F, questions 12–19 are grade E and question 20 is grade D.

1. $2 + 3 \times 4 =$
2. Put brackets in this calculation to make the answer 49.
   $5 \div 2 \times 3 \div 4 = 49$
3. Put brackets in this calculation to make the answer 3.
   $19 - 2 \times 4 + 4 = 3$
4. $-7 - (-3) =$
5. $4(2 + 3) =$
6. $5(4 - 1) =$
7. Solve for x: $5x = 25$
8. Solve for x: $x + 8 = 10$
9. Solve for x: $x - 6 = 12$
10. Solve for x: $12x = 3$
11. Solve for x: $\dfrac{x}{5} = 2$
12. $x(4 + 2) =$
13. $3x(6 - 2) =$
14. $10 + (2x + 4x) =$
15. $3 \times 2a =$
16. $2p \times 4p =$
17. Solve for x: $x + 6 = 3$
18. Solve for x: $\dfrac{2}{x} = 4$
19. Solve for x: $8 - x = 12$
20. $a^3 \times a =$

Answers to overview test

1. 14
2. $(5 + 2) \times (3 + 4) = 49$
3. $19 - 2 \times (4 + 4) = 3$
4. -4
5. 20
6. 15
7. 5
8. 2
9. 18
10. 0.25
11. 10
12. $6x$
13. $12x$
14. $10 + 6x$
15. $6a$
16. $8p^2$
17. -3
18. 0.5
19. -4
20. a^4

Why this chapter matters

Cross-curricular: This work has links with the sciences, Art and Crafts, Physical Education and languages as well as having a historical aspect.

Introduction: Algebra can seem remote from everyday arithmetic and problems but, in reality, it is part of our everyday world. It allows us to carry out calculations in standard ways and to describe those calculations to others. The introductory page gives a brief summary to the development of algebra.

Discussion points: Do you use mathematics outside of maths lessons? How do you calculate a taxi fare? How are electricity and gas bills calculated? Can you describe the processes in simple and accurate ways? Is mathematics a universal language?

Plenary: Draw out the usefulness of expressing formulae in algebraic terms that can be understood by all mathematicians, wherever they are.

Answers to the quick check test can be found at the end of the chapter.

7.1 Simultaneous equations

Curriculum references

Functional Skills standards
A1 Apply a range of mathematics to find solutions.

PLTS
Independent enquirers, Creative thinkers, Reflective learners, Team workers, Self-managers, Effective participators

APP
Algebra L7.2

Collins references
- Student Book pages 154–160
- Interactive Book: Paper Animation

Learning objectives
- Solve simultaneous linear equations in two variables. [Grade B–A]

Learning outcomes
- Students **must** be able to solve straightforward simultaneous equations. [Grade B]
- Students **should** be able to solve more complicated simultaneous equations. [Grade B]
- Students **could** be able to select the most efficient method for solving particular simultaneous equations. [Grade B]

Key words
- balance the coefficients
- simultaneous equations
- check
- substitute
- coefficient
- variable
- eliminate

Prior knowledge
Students must be able to rearrange and solve basic equations such as $4x - 6 = 2$.

Common mistakes and remediation
Students may become confused with negatives, e.g. when subtracting the two equations, and one of the numbers is already a negative.
Not labelling equations and explaining steps can lead to confusion and errors; encourage students to label equations and give a full explanation of the steps they take, as good practice.

Useful tips
Advise students always to try to use the elimination method in favour of the substitution method, as this is the main type of GCSE question.

Functional maths and problem-solving help
Exercises 7A, 7B and 7C in this lesson do not include functional maths (FM) questions. Exercises 7A and 7C each include a problem-solving (PS) question. The process skills required are in the translation of the given information into pairs of equations.
Note: The exercises deal with the methods of solving simultaneous equations. These start with pairs of equations that do not need to have coefficients balanced, then move on to pairs of equations where only one equation needs to be balanced, and finally to pairs of equations where both pairs need balancing.
Depending on the ability of students, some of the questions could be worked through quickly or missed altogether. **Less able** students will need to work their way through the process slowly, assimilating each step before moving on to the next.

© HarperCollins*Publishers* 2012

Lesson 7.1 Simultaneous equations

Starter
- Make sure students have their books closed.
- Ask, "If $x + y = 10$, what is the value of x?" Assuming someone ventures a number, say 5, ask, "What is the value of y when $x = 5$?" Write down some of the pairs of solutions; make sure $x = 4$ and $y = 6$ are included. Get students to realise that there are many (indeed, infinitely many) values for both x and y.
- Ask the same of $2x + y = 14$. Write down possible values of x and y; make sure $x = 4$ and $y = 6$ are included. Point out that $x = 4$ and $y = 6$ satisfy both of the equations at the same time, that is, *simultaneously*.
- Some students might appreciate being shown a quick sketch of the two graphs, showing that they intersect at the point (4, 6).

Main lesson activity
Elimination method, Substitution method
- Explain that there are six steps to consider when solving simultaneous equations.
  - Step 1: Make the coefficients of one of the variables the same.
  - Step 2: Depending on the signs of the balanced variables, add or subtract the equations to eliminate this variable.
  - Step 3: This will leave a linear equation in the other variable; solve this equation, using the methods in previous sections.
  - Step 4: Having found the value of one variable, substitute it back into one of the original equations to set up a linear equation in the other variable.
  - Step 5: Solve this linear equation, using the methods of the previous sections.
  - Step 6: Substitute both values into *both* of the *original* equations to check that they are correct.
- It is essential to get students into good habits in solving these equations, as there are so many ways to make errors.
- Emphasise the need to label equations and write down the operations performed on each one. This not only helps the student to keep track of their work, but also helps examiners to follow the students' work.
- Work through the introduction and the section on the elimination method, including Examples 1 and 2.
- Next, work through the solution of simultaneous equations by the substitution method, and Example 3, just before Exercise 7A. **Less able** students can leave this out but it is essential for **more able** students, as they will need the skill later when solving pairs of simultaneous equations where one is linear and one is non-linear.
- Students can now work through Exercise 7A, in which question 2 is PS.

Balancing coefficients, Balancing coefficients in both equations
- Continue with the methods of balancing equation, working through Examples 4 and 15, before students do Exercise 7B, in which question 2 is an assessing understanding (AU) question. Go through Example 6, then students can do Exercise 7C, in which question 2 is PS and 3 and 4 are AU questions.

Plenary
- Work through the PS and AU questions in Exercises 7A, 7B and 7C.

Answers to Exercises 7A, 7B and 7C can be found at the end of this chapter.

7.2 Rearranging formulae

Curriculum references

Functional Skills standards
A1 Apply a range of mathematics to find solutions.

PLTS
Independent enquirers, Creative thinkers, Reflective learners, Team workers, Self-managers, Effective participators

APP
Algebra L8.3

Collins references
- Student Book pages 160–162
- Interactive Book: Paper Animation

Learning objectives
- Rearrange formulae, using the same methods as for solving equations. [Grade C–B]

Learning outcomes
- Students **must** be able to change the subject of a simple formula. [Grade C]
- Students **should** be able to use simultaneous equations in practical situations. [Grade C]
- Students **could** be able to change the subject of a formula that contains non-linear terms. [Grade B]

Key words
- expression
- transpose
- rearrange
- variable
- substitute

Prior knowledge
Students need to have the skills to be able to solve basic algebraic equations.

Common mistakes and remediation
The usual rearrangement errors made when solving equations will be common, so stress the importance of making the same change to both sides every time.

Useful tips
Remind students that, in algebra, the letters are just numbers with unknown values – there is no real difference between $2x - 4 = 6$ and $wx - y = z$.

Functional maths and problem-solving help
Questions 13 and 14 of Exercise 7D are functional maths (FM). These questions are structured, with hints to help students access this topic, which they may find difficult. The main process skill required here is to read the question carefully and plan a step-by-step strategy to solve it. Most students will benefit from class discussions on some of the more complex questions.
In Exercise 7D, question 15 is problem solving (PS).

© HarperCollins*Publishers* 2012

Lesson 7.2 Rearranging formulae

Starter
- Ask the following questions:
  - Solve the equation (that is, make *x* the subject): $2x - 4 = 6$
  - Make *x* the subject of: $2x - y = 6$
  - Make *x* the subject of: $2x - y = z$
  - Make *x* the subject of: $wx - y = z$
- Emphasise that all four equations are essentially the same.

Main lesson activity
- Give students the formula for the area of a rectangle: $A = lw$. Ask them to give you the rule for working out the length *l* when you know the area *A* and the width *w*.
- If this is too difficult, use a numerical example, such as: what is the length of a rectangle with a width of 4 cm and an area of 24 cm^2? Show the rule numerically or algebraically:
  $$l = \frac{A}{w} \text{ or } 24 \div 4.$$
- Repeat with the formula $y = x + 2$ and ask how to find *x* if we know *y*. ($y = x - 2$)
- Establish that the process is the same as solving an equation, except that at each stage, rather than getting a numerical value, we get an expression.
- Go through the text and Examples 7, 8 and 9 to give more practice in rearranging both sides of formulae.
- Students can now do Exercise 7D, in which questions 13 and 14 are FM and 15 is PS. Advise **less able** students to complete questions 1–15 first, then move on to 16–24, working with a partner.
- **Less able** students find this topic difficult. Talking through the questions in the exercise and identifying the inverse operation will be useful for all students.

Plenary
- Ask students for the relationship between speed, distance and time.
- They should know that distance = speed × time.
- Show them the D, S, T road sign. Students may have seen this before.

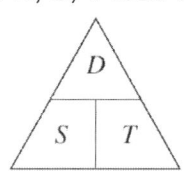

  (triangle diagram with D on top, S and T below)

- Either ask or show how to obtain the formulae for speed and time from this.

- Point out that this format can be used for many other formulae such as $A = lw$, which links back to the main lesson activity.

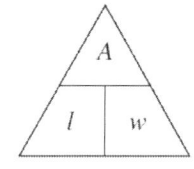

  (triangle diagram with A on top, *l* and *w* below)

Answers
Answers to Exercise 7A can be found at the end of this chapter

© HarperCollins*Publishers* 2012

Functional maths
Walking using Naismith's rule

Curriculum references

Functional Skills standards
R2 Identify the situation or problems and identify the mathematical methods needed to solve them.
I1 Interpret and communicate solutions to multistage practical problems in familiar and unfamiliar contexts and situations.

PLTS
Independent enquirers, Self-managers, Creative thinkers

APP
Using and applying mathematics L8.3

Collins references
- Student Book pages 168–169
- Interactive Book: Paper Animation

Learning objectives
- Explore the effects of varying values.
- Calculate accurately.
- Estimate, approximate and check working.

Key words
- accuracy
- formula
- estimation
- time

Functional maths help
Students will require guidance before they use the formula. Although **more able** students should be able to understand the different uses of the letter *m* in the table, make sure that **less able** students also understand the uses. Students will need to be able to substitute into formulae and work formulae backwards; that is, to calculate values, and then use the same formula to find the values of different variables, given the values of other variables and the final value of the formula.
Students will need a calculator to complete the activities, and may need to be reminded that there are 60 minutes in an hour and that the calculator works in the decimal system.

Lesson plans
Build
Discuss with students the things they need to consider before setting off for a walk in remote areas. This discussion will focus on safety issues, such as making sure they have correct clothing and equipment and that they have enough time to complete the walk before dark.
Activity 1
- Talk through the formula. Establish what each term represents and how long it will take to walk D km in minutes. $\frac{H}{10}$ is the time taken to ascend H metres assuming 1 minute to ascend 10 m.
- Go through Day 1 and fill in the table, using Naismith's rule. Students can then copy and fill in the table. This could be done as a group activity.
- Ask students to compare the rule and actual times for the given walk and write a short report on the differences.
- The last part (Task 3) should only be used if students can make their own enquiries. Students will need to look up the distance and climb details of the pathway up Ben Nevis.
- (The total return distance is 16 km (10 miles) and you will be climbing to a height of 1344 m (4406 ft). A respectable time is four hours up and three hours down. Refer to the 'Answers by Naismith's rule' in the table at the top of the next page.)

© HarperCollins*Publishers* 2012

Functional maths Walking using Naismith's rule

Day	Distance	Height (m)	Time (minutes)	Time (h/m)	Start	Breaks	Finish
1	16	250	217	3 h 37 m	10.00	2 h	3.37 pm
2	18	0	216	3 h 36 m	10.00	1 h 30 m	3.06 pm
3	11	340	166	2 h 46 m	09.30	2 h 30 m	2.46 pm
4	13	100	166	2 h 46 m	10.30	2 h 30 m	3.46 pm
5	14	120	180	3 h 0 m	10.30	2 h 30 m	4.00 pm

Naismith's rule gives quicker times but this assumes a reasonable level of fitness. The table in the Student Book may be more realistic for younger people, especially if they are carrying tents, food and water.

Activity 2 (Extension task)
Ask students to produce a report that compares and contrasts the walking times, for some of Britain's most famous walks. They should compare: Ben Nevis, Snowdon, Helvelyn and the Pennine Way.
Tell students that they are required to do research on the internet to find the distances of these walks, and then investigate the length of time these walks are expected to take, based on weather conditions, starting times and groups of walkers. (Point out that one walk cannot be done in one day, and is therefore classed as an expedition.)
The following questions could be asked.
- Should one walk in poor weather? (Justification for answers should be provided.)
- If you decide to walk in poor weather, what equipment should you take?
- Should you leave a copy of your itinerary with a trusted individual?
- What precautions can you take for a weather change?
- Are the individuals fit? Are they competitive walkers? Or are they walking just for fun?

Ask **more able** students to tell the class about their assumptions, which could include the following.
- If every group member is fit then the walk should be quicker.
- If the group includes young students or older individuals, then fitness may (or may not) be a problem. What could be done about this?

Students should show appreciation of these ideas, such as using mountain rescue in their reports.

Apply
Students will still need the opportunity to discuss the different parts of the formula. **Less able** students may need some support in looking at the units involved in the formula. They should be able to look at the different times and make comparisons, thereby clearly applying their own skills. **More able** students who have acquired skills should be able to look at the last question and begin to solve the problem, and make recommendations to a friend.

Master
In order to demonstrate mastery of the learning objectives, students should be able to:
- Use the formula efficiently.
- Deal with the different units.
- Discuss the issues relating to fitness and estimation of actual times.
- Plan their own walk and suggest an estimated time for the walk, and then look at sensible starting and finishing times, bearing in mind that no one wants to walk in the dark.

Plenary
Look up 'Naismith's Rule' and 'Tranter's correction' on the internet. Write a brief comparison of the values in the table and the values likely to be calculated using Naismith's rule.
Ask: if you were to walk up Ben Nevis, how long do you think it would take and when would you do the walk? Ask students to explain their suggestions.

© HarperCollins*Publishers* 2012

Answers Lessons 7.1 – 7.2

Quick check
1. a $2x + 12$ b $4x - 12$ c $12x - 6$ d $x = 8$ e $x = 24$ f $x = 15$
2. a $5y$ b $4x - 3$ c $-x - 4$
3. a $6x$ b $8y^2$ c $2c^3$
4. a $x = 1$ b $x = 3$ c $x = 9$

7.1 Simultaneous equations
Exercise 7A
1. a $x = 4, y = 1$ b $x = 1, y = 4$
   c $x = 3, y = 1$ d $x = 5, y = -2$
   e $x = 7, y = 1$ f $x = 5, y = \frac{1}{2}$
   g $x = 4\frac{1}{2}, y = 1\frac{1}{2}$ h $x = -2, y = 4$
   i $x = 2\frac{1}{4}, y = -1\frac{1}{2}$ j $x = 2\frac{1}{4}, y = 6\frac{1}{2}$
   k $x = 4, y = 3$ l $x = 5, y = 3$
2. a 3 is the first term. The next term is $2 \times a + b$, which equals 14.
   b $14a + b = 47$ c $a = 3, b = 5$
   d 146, 443

Exercise 7B
1. a $x = 2, y = -3$ b $x = 7, y = 3$
   c $x = 4, y = 1$ d $x = 2, y = 5$
   e $x = 4, y = -3$ f $x = 1, y = 7$
   g $x = 2\frac{1}{2}, y = 1\frac{1}{2}$ h $x = -1, y = 2\frac{1}{2}$
   i $x = 6, y = 3$ j $x = \frac{1}{2}, y = -\frac{3}{4}$
   k $x = -1, y = 5$ l $x = 1\frac{1}{2}, y = \frac{3}{4}$
   m $x = 1\frac{1}{2}, y = \frac{3}{4}$
2. a They are the same equation. Divide the first by 2 and it is the second, so they have an infinite number of solutions.
   b Double the second equation to get $6x + 2y = 14$ and subtract to get $9 = 14$. The left-hand sides are the same if the second is doubled so they cannot have different values.

Exercise 7C
1. a $x = 5, y = 1$ b $x = 3, y = 8$
   c $x = 9, y = 1$ d $x = 7, y = 3$
   e $x = 4, y = 2$ f $x = 6, y = 5$
   g $x = 3, y = -2$ h $x = 2, y = \frac{1}{2}$
   i $x = -2, y = -3$ j $x = -1, y = 2\frac{1}{2}$
   k $x = 2\frac{1}{2}, y = -\frac{1}{2}$
   l $x = -1\frac{1}{2}, y = 4\frac{1}{2}$
   m $x = -\frac{1}{2}, y = -6\frac{1}{2}$ n $x = 3\frac{1}{2}, y = 1\frac{1}{2}$
   o $x = -2\frac{1}{2}, y = -3\frac{1}{2}$
2. (1, 2) is the solution to equations A and C; (1, 3) is the solution to equations A and D; (2, 1) is the solution to B and C; (3, 3) is the solution to B and D.
3. Intersection points are (1, 3), (6, 0) and (2, 4). Area is 2 cm².
4. Intersection points are (0, 3), (6, 0) and (4, 1). Area is 6 cm².

Answers Lessons 7.1 – 7.2

7.2 Rearranging formulae
Exercise 7D

1. $k = \dfrac{T}{3}$
2. $y = X + 1$
3. $p = 3Q$
4. $r = \dfrac{A-9}{4}$
5. $n = \dfrac{W+1}{3}$
6. a $m = p - t$ b $t = p - m$
7. $m = gv$ or $m = g \times v$
8. $m = \sqrt{t}$
9. $r = \dfrac{C}{2\pi}$
10. $b = \dfrac{A}{h}$
11. $l = \dfrac{P - 2w}{2}$
12. $p = \sqrt{m-2}$
13. a $-40 - 32 = -72$, $-72 \div 9 = -8$
    $5 \times -8 = -40$
    b $68 - 32 = 36$, $36 \div 9 = 4$,
    $4 \times 5 = 20$
    c $F = \dfrac{9}{5}C + 32$
14. a $5x = 9y + 75$, $y = \dfrac{5x - 75}{9}$
    b 25p
15. Average speeds: outward journey = 72 km/h, return journey = 63 km/h, taking 2 hours. He was held up for 15 minutes.
16. a $a = \dfrac{v-u}{t}$ b $t = \dfrac{v-u}{a}$
17. $d = \sqrt{\dfrac{4A}{\pi}}$
18. a $n = \dfrac{W-t}{3}$ b $t = W - 3n$
19. a $y = \dfrac{x+w}{5}$ b $w = 5y - x$
20. $p = \sqrt{\dfrac{k}{2}}$
21. a $t = u^2 - v$ b $u = \sqrt{v+t}$
22. a $m = k - u^2$ b $n = \sqrt{k - m}$
23. $r = \sqrt{\dfrac{T}{5}}$
24. a $w = K - 5n^2$ b $n = \sqrt{\dfrac{K-w}{5}}$

Examination questions

1. $x = 8$, $y = -2$
2. a $4a + 3 = 2b + 5$, $4a - 2b = 2$,
   $2a - b = 1$
   b $a = 2.25$, $b = 3.5$
3. a $2y + x + 1 + 11 = 11 + 2x + 2y + y$,
   $x + 1 = 2x + y$, $1 = x + y$
   b $2y + x + 1 + 11 = x + 1 + 2x + y$,
   $2y + 11 = 2x + y$, $2x - y = 11$
   c $x = 4$, $y = -3$
   d Clockwise from 11: (11), 2, −3, 8, 5, −6
4. pen = 11p, ruler = 21p
5. $b = \dfrac{2A - ah}{h}$
6. $x = \dfrac{y - 7}{3}$
7. $x = -1$, $y = -4$
8. a $x = wy + t$ b $x = 4$, $y = 5$
9. $p = \sqrt{n-6}$
10. $x = 4$, $y = 1.5$

Chapter 8 Geometry: Pythagoras and trigonometry

Overview

> 8.1 Pythagoras' theorem
> 8.2 Finding a shorter side
> 8.3 Applying Pythagoras' theorem in real situations
> 8.4 Pythagoras' theorem in three dimensions

This chapter covers Pythagoras' theorem and the trigonometry of right-angled triangles with their associated applications.

Context
Trigonometric ratios and Pythagoras' theorem have many applications in the real world. Navigation and surveying rely on trigonometric results and techniques. Sines and cosines are a natural way to describe waves and they are essential in the fields of electricity and electronics.
This chapter gives students an introduction to the basic ideas of a huge subject.

Curriculum references
KS4 Programme of Study references
1.3b Understanding that mathematics is used as a tool in a wide range of contexts.
1.3c Recognising the rich historical and cultural roots of mathematics.
2.1c Simplify the situation or problem in order to represent it mathematically using appropriate variables, symbols, diagrams and models.
2.2l Calculate accurately, using mental methods or calculating devices as appropriate.
4d Work on problems that arise in other subjects and in contexts beyond the school.

Linear specification
G2.1h Use Pythagoras' theorem and extend to use in 3D.

Functional Skills standards
R2 Identify the situation or problems and identify the mathematical methods needed to solve them.
A1 Apply a range of mathematics to find solutions.
A2 Use appropriate checking procedures and evaluate their effectiveness at each stage.
I2 Draw conclusions and provide mathematical justifications.

PLTS references
Independent enquirers identify questions to answer and problems to resolve; analyse and evaluate information, judging its relevance and value. **Creative thinkers** generate ideas and explore possibilities. **Reflective learners** review progress, acting on the outcomes. **Effective participators** discuss issues of concern, seeking resolution where needed.

APP
Shape, space and measure L7.1 Understand and apply Pythagoras' theorem when solving problems in 2D. **L8.2** Understand and use trigonometrical relationships in right-angled triangles, and use these to solve problems, including those involving bearings. **Using and applying mathematics L7.3** Justify generalisations, arguments or solutions.

Chapter 8 Geometry: Pythagoras and trigonometry

Route mapping

Exercise	Grades C	B	A	A*
A	all			
B	all			
C		all		
D		all		
E			1–7	8–9

Overview test

Questions 1 and 2 are grade F, questions 3 to 6 are grade E.

1. Write down the first ten square numbers.
2. Work these out.
   a $4^2 + 9^2$ b $12^2 - 3^2$ c 9×6^2
3. Round these numbers to one decimal place.
   a 4.793 b 18.0165 c 20.96
4. Round these numbers to three significant figures.
   a 2.684 b 12.678 c 0.1552
5. Use your calculator to work out these square roots. Give your answers to three significant figures.
   a $\sqrt{30}$ b $\sqrt{285}$ c $\sqrt{0.58}$
6. a Using a scale of 1 cm to represent 2 km, draw a scale diagram of the journey of a boat that sails 6 km north and then 8 km west.
   b By measuring your scale drawing, find the distance in a direct line back from the boat's finishing point to its starting point.

Answers to overview test

1. 1, 4, 9, 16, 25, 36, 49, 64, 81, 100
2. a 97 b 135 c 324
3. a 4.8 b 18.0 c 21.0
4. a 2.68 b 12.7 c 0.155
5. a 5.48 b 16.9 c 0.762
6. a Correct drawing, with line north 3 cm and line west 4 cm long
   b 10 km

Why this chapter matters

Cross-curricular: This chapter links to the sciences, Design and Technology and ICT.
Introduction: Students may have met Pythagoras' theorem in KS3. It has been used for thousands of years. Trigonometry has ancient origins. Its name comes from Greek words meaning 'measuring triangles'.
Discussion points: Check students' knowledge of squares and square roots by asking for examples. Check that they can use a calculator to find square roots. Ask if anyone knows about 'Pythagorean triples' and introduce a few, such as 3, 4, 5; 5, 12, 13; 7, 24, 25.
Plenary: Use the Introductory page in the Student Book to lead into this chapter.

Answers to the quick check test can be found at the end of the chapter.

© HarperCollins*Publishers* 2012

8.1 Pythagoras' theorem

Curriculum references

Functional Skills standards
A1 Apply a range of mathematics to find solutions.

PLTS
Independent enquirers, Creative thinkers, Reflective learners, Effective participators

APP
Shape, space and measure L7.1

Collins references
- Student Book pages 172–173
- Interactive Book: Paper Animation

Learning objectives
- Calculate the length of the hypotenuse in a right-angled triangle. [Grade C]

Learning outcomes
- Students **must** be able to calculate a hypotenuse using Pythagoras' theorem. [Grade C]
- Students **should** be able to give answers to an appropriate degree of accuracy. [Grade C]
- Students **could** be able to demonstrate why Pythagoras' theorem is correct. [Grade C]

Key words
- hypotenuse
- Pythagoras' theorem

Prior knowledge
Students should know how to find the square and square root of a number, and it will be helpful if they are familiar with square numbers. Students should also know how to round numbers correctly, and they should have an appreciation of what is a suitable level of accuracy given the context of the problem involved.

Common mistakes and remediation
Perhaps the most common mistake students make is to forget to take the square root to find the final answer. They can generally avoid doing this if they pay attention to the layout of solutions. Students also try to apply the theorem in unsuitable situations, such as when the triangle in question is not known to be a right-angled triangle.
Emphasise that Pythagoras' theorem only applies to right-angled triangles.

Useful tips
Pythagoras' theorem may be familiar to students. If this is the case, this lesson will be revision.

Functional maths and problem-solving help
There are no functional maths (FM) questions in Exercise 8A. The diagram in question 10, which is the problem solving (PS) question, gives a visual demonstration of Pythagoras' theorem. Make sure you emphasise that this is a result about areas as well as a result about numbers.

Starter
- Display a target board of whole numbers and ask students to find the square numbers.
- Ask for any known square numbers that are missing from the board. Try to elicit the first 12 square numbers, with some other commonly known numbers such as 400, 625, 1 000 000.
- Ask for a definition of square numbers. Ask for squares of some simple decimal numbers such as 0.5 and 0.1.

© HarperCollins*Publishers* 2012

Lesson 8.1 Pythagoras' theorem

Main lesson activity
- Ask students if they have heard of Pythagoras' theorem, which concerns a triangle. They might remember it in the form $a^2 + b^2 = c^2$ or they might remember it as a fact about squares. Try to encourage students to give a clear and precise description.
- Now ask how they know it is true. They are unlikely to recall a proof.
- Give them a copy of this dissection. Make sure they understand that it shows squares on the sides of a right-angled triangle.

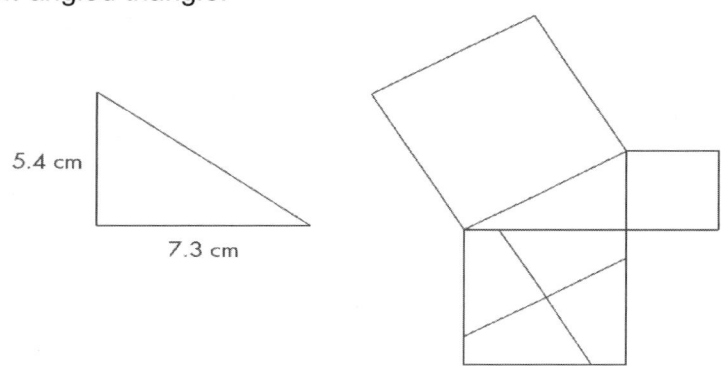

- Students can work in pairs to cut out the five smaller pieces and fit them into the largest square. This demonstrates Pythagoras' theorem. By labelling the sides of a right-angled triangle as a, b and c, deduce that $a^2 + b^2 = c^2$.
- Draw this triangle on the board.
- Discuss how to find the missing side. Ask students to do the required calculation.
- Discuss the appropriate way to round the calculator answer. In this case, 1 decimal place is appropriate, because the lengths are given to that degree of accuracy. It gives 9.1 cm.
- If necessary, work through Example 1 with students.
- Now students can complete Exercise 8A, in which the PS question is 10. (It may be useful to have a plenary during the lesson to look at this question, as it is important for all students to think about it.)

Plenary
- Draw a right-angled triangle with a semi-circle on each side. Ask if Pythagoras' theorem is true for semi-circles as well as squares. (yes)
- If students want to check a numerical example, give the sides as 6 cm, 8 cm and 10 cm. Encourage discussion, asking students to justify their answers. Challenge **more able** students with the question: What about other shapes? (In fact, it is true for any similar shapes.)

Answers
Answers to Exercise 8A can be found at the end of this chapter.

© HarperCollinsPublishers 2012

8.2 Finding a shorter side

Curriculum references

Functional Skills standards
A1 Apply a range of mathematics to find solutions.

PLTS
Independent enquirers, Creative thinkers, Reflective learners, Effective participators

APP
Shape, space and measure L7.1

Collins references
- Student Book pages 174–175
- Interactive Book: Multiple Choice Questions

Learning objectives
- Calculate the length of a shorter side in a right-angled triangle. [Grade C]

Learning outcomes
- Students **must** be able to use Pythagoras' theorem to find the third side of a right-angled triangle. [Grade C]
- Students **should** be able to give answers to an appropriate degree of accuracy. [Grade C]
- Students **could** be able to find sets of whole numbers which satisfy Pythagoras' equation. [Grade C]

Key words
- Pythagoras' theorem

Prior knowledge
As in Lesson 8.1, students should know how to find the square and square root of a number, and be familiar with square numbers. Students should know how to round numbers correctly, and have an appreciation of what is a suitable level of accuracy given the context of the problem involved. For this lesson, students should also be familiar with using Pythagoras' theorem to find the length of the hypotenuse of a right-angled triangle.

Common mistakes and remediation
Students often confuse the two methods shown in Lesson 8.1 and in this lesson. This is usually due to a lack of care when identifying the hypotenuse. Encourage students to label the hypotenuse before starting the question. The common mistakes highlighted in Lesson 8.1 also apply to this lesson.

Useful tips
If students are familiar with Pythagoras' theorem from KS3, let them complete this lesson quickly, allowing you to give more emphasis to Lesson 8.3.

Functional maths and problem-solving help
There are no functional maths (FM) questions in Exercise 8B. Question 4 is a problem-solving (PS) question, and gives students (preferably in pairs) an opportunity to look for Pythagorean triples.

© HarperCollins*Publishers* 2012

Lesson 8.2 Finding a shorter side

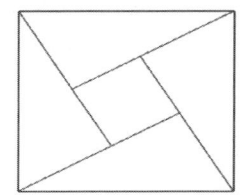

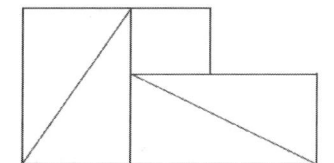

Starter
- Draw these diagrams as accurately as possible on the board.
- Ask: How do these diagrams show that Pythagoras' theorem is true? (The diagram on the right can be divided into two squares.)
- Make sure students can see that these are the squares on the shorter sides of the triangle. You may need to explain this carefully to **less able** students.

Main lesson activity
- Draw a right-angled triangle with hypotenuse 13.6 cm and shortest side 8.4 cm.
- Ask how the third side can be found.
- Make sure students understand that this can be done by subtraction. The answer is 10.7 cm to 1 decimal place.
- Look at Example 2 with students.
- Now students can complete Exercise 8B, in which the PS question is 4. Provide **less able** students with support, as needed.

Plenary
- Draw a triangle with sides 8, 15 and 17 but do not mark a right angle. Show that $8^2 + 15^2 = 17^2$. Ask if this means that the triangle is right-angled.
- This is the inverse of Pythagoras' theorem. Students often do not realise this is different.
  Right-angled triangle $\rightarrow$ sum of squares property
  Is different from
  Sum of square property $\rightarrow$ right-angled triangle
  It is in fact true.
- Briefly look at any answers students have found to question 4.

Answers
Answers to Exercise 8B can be found at the end of this chapter.

© HarperCollins*Publishers* 2012

8.3 Applying Pythagoras' theorem in real situations

Curriculum references

Functional Skills standards
I2 Draw conclusions and provide mathematical justifications.

PLTS
Independent enquirers, Creative thinkers, Reflective learners, Effective participators

APP
Shape, space and measure L7.1

Collins references
- Student Book pages 176–180
- Interactive Book: Multiple Choice Questions

Learning objectives
- Solve problems using Pythagoras' theorem. [Grade C–B]

Learning outcomes
- Students **must** be able to use Pythagoras' theorem to find lengths in practical situations. [Grade C]
- Students **should** be able to justify answers to problems involving Pythagoras' theorem. [Grade C]
- Students **could** be able to use Pythagoras' theorem in isosceles triangles. [Grade B]

Key words
- isosceles triangle
- Pythagoras' theorem

Prior knowledge
Students should know how to use Pythagoras' theorem to find either the hypotenuse or a shorter side. This lesson provides a wide variety of situations in which problems can be solved using Pythagoras' theorem. The extra skills and knowledge required include:
- the use of compass directions
- the use of coordinates
- finding the area of a rectangle
- the properties of isosceles triangles.

Common mistakes and remediation
Mistakes can arise when students do not identify a right-angled triangle or when they do not identify the hypotenuse correctly. Say that drawing a clear diagram at the start will help to avoid these mistakes.

Useful tips
There are two exercises in this lesson. It may be preferable to break your teaching with the first exercise.

Functional maths and problem-solving help
It this lesson students are given opportunities to apply their knowledge in realistic situations. Many questions in Exercises 8C and 8D are functional maths (FM) or problem solving (PS). There is also further practice in these areas and opportunities for discussion during the lesson.

Lesson 8.3 Applying Pythagoras' theorem in real situations

Starter
- Tell students that when builders want to construct a wall that is perpendicular to an existing wall, they use the '3, 4, 5 rule'. Ask students for suggestions as to how they think this rule works.
- Explain that builders generally use a rod, such as a metre ruler, as a measuring unit. Marking off 3, 4 and 5 lengths in a triangle with the 3 or 4 along the existing wall will give the required right angle. (Remember that a 3, 4, 5 triangle is right-angled.)
- You could demonstrate this in the classroom using string instead of rods.
- Explain that scientists believe that the Egyptians used this method when building the pyramids (to get the base edges perpendicular) – and it is still used today.

Main lesson activity
- Use Example 3 as a starting point. Work through it with the class. From the initial statement of the problem, draw a diagram. With students' input, identify the required side and use the appropriate form of Pythagoras' theorem to find it.
- Now students can complete Exercise 8C. The last question is suitable for **more able** students. The FM questions are 1, 5, 7 and 10, the PS questions are 3, 12 and 13, and the assessing understanding (AU) questions are 4, 8, 9 and 11.
- Now draw an equilateral triangle with sides 12 cm on the board. Ask students how they can work out the area. Discuss suggestions and remind them that the height must be found.

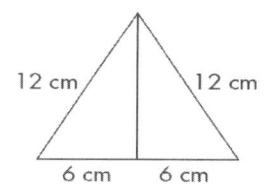

- Work through the calculation. (Height = $\sqrt{108}$ = 10.4 cm, area = 6 × height = 62.4 cm^2)
- Explain that this method can be used in any isosceles triangle. Example 4 gives a further example if required.
- Now students can complete Exercise 8D, in which the FM question is 8 and the PS questions are 3, 4, 6 and 7.
- **Less able** students can work with a partner on questions they may find difficult. Provide support as needed.

Plenary
- Discuss students' answers to question 7 in Exercise 8D. The equilateral triangle has the largest area. This will seem likely if students try different examples.

Answers
Answers to Exercises 8C and 8D can be found at the end of this chapter.

8.4 Pythagoras' theorem in three dimensions

Curriculum references

Functional Skills standards
R2 Identify the situation or problems and identify the mathematical methods needed to solve them.

PLTS
Independent enquirers, Creative thinkers, Reflective learners, Effective participators

APP
Shape, space and measure L8.2

Collins references
- Student Book pages 181–183

Learning objectives
- Use Pythagoras' theorem in problems involving three dimensions [Grade A–A*]

Learning outcomes
- Students **must** be able to use Pythagoras' theorem to find lengths in three dimensions. [Grade A]
- Students **should** be able to solve practical problems involving Pythagoras' theorem is three dimensions. [Grade A]
- Students **could** be able to solve more complex problems using Pythagoras' theorem in three dimensions. [Grade A*]

Key words
- 3D
- Pythagoras' theorem

Prior knowledge
Students need to know how to use Pythagoras' theorem to find any missing side of a right-angled triangle.

Common mistakes and remediation
Mistakes can arise when students do not identify a right-angled triangle from the 3D figure or when they do not identify the hypotenuse correctly. Say that drawing a clear diagram of each triangle needed will help to avoid these mistakes.

Useful tips
As more than one calculation is often involved when solving 3D problems, remind students not to round answers to intermediate steps but only at the final stage.

Functional maths and problem-solving help
In this lesson students apply their knowledge in three-dimensional situations. Questions 5 and 6 in Exercise 8E are functional maths (FM) and problem solving (PS), respectively. There are also opportunities for discussion during the lesson.

© HarperCollins*Publishers* 2012

Lesson 8.4 Pythagoras' theorem in three dimensions

Starter
- Draw this diagram on the board.

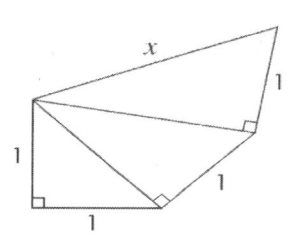

- Ask students to find *x*. They could work in pairs.
- After a short time ask for answers and methods. It can be done by using Pythagoras' theorem three times to find $x = 2$.
- In fact the intermediate lengths are $\sqrt{2}$ and $\sqrt{3}$. If students round their answers at each stage they will not get exactly 2. This is an opportunity to stress the importance of not rounding intermediate answers.
- Ask students to generalise the result. (If 1 is replaced by *a*, then *x* will equal $2a$.)

Main lesson activity
- Ask students to imagine a wire going from one corner of the room to the opposite corner. How can you find the length?
- Take suggestions. Students need to understand that they can find the length by using Pythagoras' theorem twice: once to find the diagonal of the floor and then using that as one side of another right-angled triangle.
- Work through Example 5, which provides another example and shows how to set out a solution.
- Now students can complete Exercise 8E, which is more difficult than the preceding exercise. **Less able** students may need extra support.
- In this exercise, the FM question is 5, the PS question is 6 and the assessing understanding (AU) questions are 2–4, 7 and 8.

Plenary
- Have an unopened drinks can available. Ask students how long a drinking straw must be so as not to disappear inside the can. Measure any lengths students suggest against the unopened can.
- The triangle to use here is that formed by the height and the diameter of the can. This will give a minimum length.

Answers
Answers to Exercise 8E can be found at the end of this chapter.

© HarperCollins*Publishers* 2012

Problem solving
Map work using Pythagoras

Curriculum references

Functional Skills standards
A1 Apply a range of mathematics to find solutions.
I2 Draw conclusions and provide mathematical justifications.

PLTS
Reflective learners, Creative thinkers

APP
Shape, space and measure L7.1; Using and applying mathematics L7.3

Collins references
- Student Book pages 188–189

Learning objectives
- Recognise Pythagoras' theorem as the tool for solving a problem, in the context of map work.

Key words
- estimate
- Pythagoras' theorem

Functional maths help
Students often make errors by misinterpreting scales and therefore using wrong lengths for their estimations. Encourage students to look for and identify triangles, and note where the right-angled triangle is in each situation.

Lesson plans
Build
The main learning point of this activity is for students to discover that they can estimate distances in ways other than using the scale on a map – in this case, using Pythagoras' theorem with small right-angled triangles.
You may need to help students see exactly where the right-angle triangles are and then to interpret the scale to find the lengths of the sides in these triangles.
You may also need to work through some examples with students, on coordinate axes and using Pythagoras' theorem, to reinforce the ideas before they attempt to use the map.

Activity 1
Refer students to 'Getting started' in the Student Book.
- They need to write down the square of each number. (25, 0.01, 0.16, 0.0009) Then they should think of a number that has a square between 0.1 and 0.001. (examples are 0.04, 0.05, ……, 0.09, 0.1, 0.2, 0.3.)
- Next, they need to write down the square root of each number. (4, 9, 0.1, 0.158) Then ask them to think of a number that has a square root between 50 and 60. (an example is 7.5 squared = 56.25)
- Finally, ask students to draw (plot) the points and to measure the distance from point A to point B. (Students' measurements will vary.)

Activity 2
Refer students to question 1 of 'Your task' in the Student Book.
Ask them to write five questions that are similar to the examples provided in the 'Example' box. Then ask them to swap exercise books with the person next to them, and answer each other's questions, making sure they show their working for each question clearly. Finally, ask students to swap their work again and mark each other's answers, making sure they provide constructive feedback.
More able students should be able to complete this task without your help, but **less able** students may need you to provide them with questions, as follows.
Here are some examples of questions students could provide.

© HarperCollins*Publishers* 2012

Problem solving   Map work using Pythagoras

Use the above technique (as shown in the Student Book 'Example' text) to estimate the distances between:
- Edale and Bamford (about 9.5 km)
- Edale and Bradwell (about 6.5 km)
- Bradwell and Stanage Edge (just over 8 km)
- Dove Holes and Tideswell (just over 8 km)
- Tideswell and Bamford (10 km)

Activity 3
Once students have completed Activity 2, ask them to refer to question 2 of 'Your task'.
Ask them to plan a walk with a circular route that is between 20 km and 35 km long. They need to take into account the average person's walking speed per hour, and estimate the time it would take to complete the route.
An example is as follows.
I am planning a walk that will take me from Hathersage to Stanage Edge, then to Hope, then to Edale, then to Tideswell, then to Eyam and finally back to Hathersage.
Estimate how far this walk will be and how long it will take me, if my walking speed is estimated at 4.5 miles per hour. (between 46 and 47 km, six-and-a-half hours)

Apply
Students should study the map, and read the all the text in the Student Book carefully, including the 'Example' and 'Additional information' text. Support students in identifying the right-angled triangles they need to use in their calculations.
Work through a couple more examples with students before they write and answer their own questions. You could also work through the examples provided in Activity 2, once students have completed it. This will help **less able** students, who may have done these calculations, to see where they were correct or where they made mistakes.
Make sure students know the fact that 5 miles is equal to approximately 8 kilometres, and you may need to provide support with changing 4.5 m/h to km/h.

Master
In order to demonstrate mastery of the learning objectives, students should be able to:
- Identify right-angled triangles and use them to estimate the distance between two points.
- Mark another student's work and provide constructive feedback on how to improve the work.
- Plan a route; estimate the length of the route, and the time it will take to walk the route.

Plenary
Talk about walking in the countryside. Have a class discussion about any walks students may have done, why they did these walks, where the walks were, how long they walked for and what distance they covered. How did they measure the distances on these occasions?
Discuss the health-giving aspects of walking. Suggest that walking, in the countryside or not, is beneficial to everyone as exercise.

© HarperCollins*Publishers* 2012

Answers Lessons 8.1 – 8.4

Quick check
1. 5.3
2. 246.5
3. 0.6
4. 2.8
5. 16.1
6. 0.7

8.1 Pythagoras' theorem
Exercise 8A
1. 10.3 cm
2. 5.9 cm
3. 8.5 cm
4. 20.6 cm
5. 18.6 cm
6. 17.5 cm
7. 5 cm
8. 13 cm
9. 10 cm
10. The square in the first diagram and the two squares in the second have the same area.

8.2 Finding a shorter side
Exercise 8B
1. a 15 cm b 14.7 cm
   c 6.3 cm d 18.3 cm
2. a 20.8 m b 15.5 cm
   c 15.5 m d 12.4 cm
3. a 5 m b 6 m
   c 3 m d 50 cm
4. There are infinitely many possibilities, e.g. any multiple of 3, 4, 5 such as 6, 8, 10; 9, 12, 15; 12, 16, 20; multiples of 5, 12, 13 and of 8, 15, 17.
5. 42.6 cm

8.3 Applying Pythagoras' theorem in real situations
Exercise 8C
1. No. The foot of the ladder is about 6.6 m from the wall.
2. 2.06 m
3. 11.3 m
4. About 17 minutes, assuming it travels at the same speed all the way.
5. 127 m − 99.6 m = 27.4 m
6. 4.58 m
7. a 3.87 m b 1.74 m
8. 3.16 m
9. 13 units

10. a 4.85 m
    b 4.83 m (There is only a small difference.)
11. Yes, because $24^2 + 7^2 = 25^2$
12. 6 cm
13. Greater than 20 cm (no width) and less than 28.3 cm (a square)

Exercise 8D
1. a 32.2 cm^2
   b 2.83 cm^2
   c 50.0 cm^2
2. 22.2 cm^2
3. 15.6 cm^2
4. a

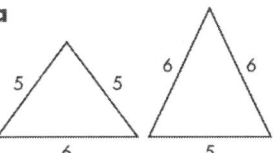

   b The areas are 12 cm^2 and 13.6 cm^2 respectively, so the triangle with 6 cm, 6 cm, 5 cm sides has the greater area.
5. a

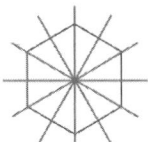

   b 166.3 cm^2
6. 259.8 cm^2
7. a No, areas vary from 24.5 cm^2 to 27.7 cm^2.
   b No, the equilateral triangle gives the largest area.
   c The closer the isosceles triangle gets to an equilateral triangle the larger its area becomes.
8. 19.8 m^2 or 20 m^2
9. 48 cm^2
10. a 10 cm b 26 cm c 9.6 cm

© HarperCollinsPublishers 2012

8.4 Pythagoras' theorem in three dimensions
Exercise 8E
1. **a i** 14.4 cm **ii** 13 cm **iii** 9.4 cm
   b 15.2 cm
2. No, 6.6 m is longest length.
3. **a** 20.6 cm **b** 15.0 cm
4. 21.3 cm
5. **a** 8.49 m **b** 9 m
6. 17.3 cm
7. 20.6 cm
8. **a** 11.3 cm **b** 7 cm
   c 8.06 cm
9. **a** 50.0 cm **b** 54.8 cm
   c 48.3 cm **d** 27.0 cm

Examination questions
1. 110 cm^2
2. 10.8 cm
3. 12 cm^2
4. 24.1 cm
5. 13.6 cm (3 sf)
6. Height^2 = 6^2 − 2^2 = 32. Height = $\sqrt{32}$
   Area = 0.5 × 4 × $\sqrt{32}$ = 2$\sqrt{32}$ = 2$\sqrt{16 \times 2}$ = 2 × 4 × $\sqrt{2}$ = 8$\sqrt{2}$
7. **a** AX and BY are both at right angles to AB so they are parallel. Hence ABYX has one pair of parallel sides.
   b Form a right-angled triangle with XY as the hypotenuse.
   This has sides XY = 8 cm and a short side of 2 cm. 8^2 − 2^2 = 60.
   $\sqrt{60}$ = 7.7459 = 7.75 to 3 sf.

Chapter 9 Geometry: Angles

Overview

> 9.1 Special triangles and quadrilaterals
>
> 9.2 Angles in polygons

In this chapter, students will work out the sizes of angles in special triangles and quadrilaterals and they will work out the sizes of interior and exterior angles in polygons.

Context
Understanding angles is essential, as they are used when constructing anything from a cork for a bottle to a building. Students should be prepared to offer logical reasons for their choice of values of missing angles. Dissuade **less able** students from guessing these values.

Curriculum references
KS4 Programme of Study references
1.1b Communicating mathematics effectively.
2.2o Record methods, solutions and conclusions.
2.3c Appreciate the strength of empirical evidence and distinguish between evidence and proof.
3.2a Properties of 2D shapes.
4a Develop confidence in an increasing range of methods and techniques.

Linear specification
G1.2 Understand and use the angle properties of parallel and intersecting lines, triangles and quadrilaterals.
G1.3 Calculate and use the sums of the interior and exterior angles of polygons.
G1.4 Recall the properties and definitions of special types of quadrilateral, including square, rectangle, parallelogram, trapezium, kite and rhombus.
G2.3h Simple geometrical proofs.

Functional Skills standards
R2 Identify the situation or problems and identify the mathematical methods needed to solve them.
A1 Apply a range of mathematics to find solutions.
I2 Draw conclusions and provide mathematical justifications.

PLTS references
Independent enquirers identify questions to answer and problems to resolve; explore issues, events or problems from different perspectives; analyse and evaluate information, judging its relevance and value. **Creative thinkers** connect their own and others' ideas and experiences in inventive ways. **Effective participants** discuss issues of concern, seeking resolution where needed. **Team workers** collaborate with others to work towards common goals. **Self-managers** work towards goals, showing initiative, commitment and perseverance.

APP
Shape, space and measure L6.2 Solve geometrical problems using properties of angles, of parallel and intersecting lines, and of triangles and other polygons.

Chapter 9 Geometry: Angles

Route mapping

Exercise	Grades			
	E	D	C	B
A	1–2	3	4–11	
B			1–7	8–18

Overview test
Questions 1a and 1b are grade E, questions 1c, 1d and 1e are grade D, questions 2a and 2b are grade E and questions 2c and 2d are grade D.

1 Find the value of x in each of these diagrams.

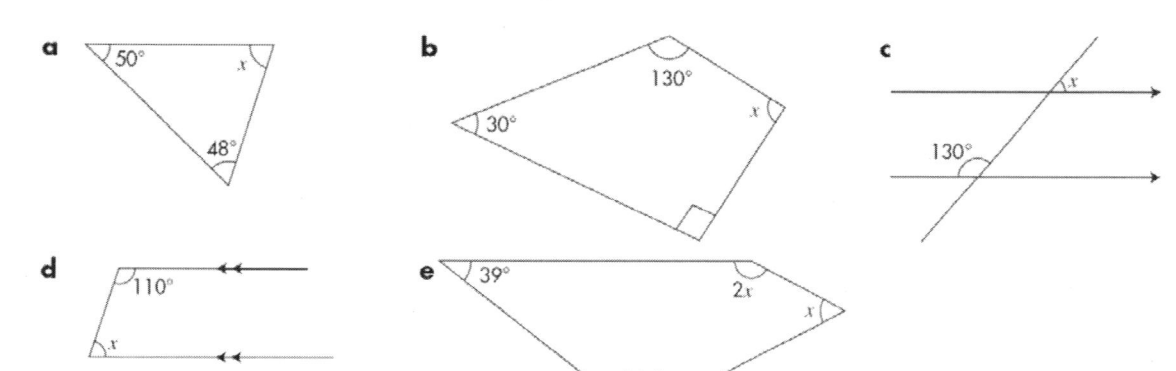

a [diagram: triangle with angles 50°, 48°, x]

b [diagram: quadrilateral with angles 130°, 30°, x, right angle]

c [diagram: two parallel lines with transversal, angles 130° and x]

d [diagram: parallel lines with angles 110° and x]

e [diagram: triangle with angles 39°, 114°, $2x$, x]

2 Solve each of these equations to find the value of x.
  a $5x + 60 = 180$
  b $3x + 20 + 5x - 60 = 360$
  c $7x - 80 = 4x - 20$
  d $3(2x - 10) = 180$

Answers to overview test
1 a 82° b 110° c 50° d 70° e 69°
2 a 24 b 50 c 20 d 35

Why this chapter matters
Cross-curricular: This topic has links with the sciences, Design and Technology, History and Arts and Crafts.
Introduction: Discuss ancient civilisations and how they used angles in constructions. Introduce the idea, put forward by most historians, that the ancient Babylonians thought of the 'circle' of the year as consisting of 360 days.
Discussion points: How did the ancient approximation of a circle affect the unit angle we still use today? What is the difference between a 'degree' and an 'angle'? How do we use angles in circle theorems?
Plenary: Make sure students recall the basic facts about triangles and other polygons.

Answers to the quick check test can be found at the end of the chapter.

© HarperCollins*Publishers* 2012

9.1 Special triangles and quadrilaterals

Curriculum references

Functional Skills standards
A1 Apply a range of mathematics to find solutions.

PLTS
Independent enquirers, Creative thinkers, Effective participators

APP
Shape, space and measure L6.2

Collins references
- Student Book pages 193–196
- Interactive Book: 10 Quick Questions

Learning objectives
- Work out the sizes of angles in triangles and quadrilaterals.
  [Grade E–D]

Learning outcomes
- Students **must** be able to calculate angles in isosceles triangles. [Grade E]
- Students **should** be able to calculate angles in special quadrilaterals. [Grade D]
- Students **could** be able to solve algebraic problems in a geometrical context. [Grade D]

Key words
- equilateral triangle
- isosceles triangle
- kite
- parallelogram
- rhombus
- trapezium

Prior knowledge
Students should know that the sum of the angles in any triangle is 180°, that in an isosceles triangle two of the sides are equal, and that the sum of the angles in a quadrilateral is 360°. Students should also know that when a transversal cuts *parallel lines* the alternate angles are equal, *corresponding angles* are equal and allied (or interior) angles are supplementary (have a sum of 180°).

Common mistakes and remediation
Some students may be confused between alternate and corresponding angles. Constant reference to them during the lesson would be beneficial.
Note: Remind students that in examinations it is not acceptable to refer to alternate angles as Z angles and corresponding angles as F angles.

Useful tips
Remind students that diagrams in mathematical textbooks and in examination papers are not drawn accurately, unless it is stated that they are. Therefore, they should never try to measure angles in diagrams.

Functional maths and problem-solving help
When solving geometrical problems, students should always make sure that the sides and angles in their own diagrams look realistic. There are no functional maths (FM) questions in Exercise 9A. However, question 9 places the problem in the context of making a kite. Students simply need to apply what they know about the mathematical kite shape. For problem-solving (PS) questions 2, 7 and 10, students need to identify the information they are given and apply it.
Note: This will be a revision lesson for most students. Direct them to the questions that suit their ability.

Lesson 9.1 Special triangles and quadrilaterals

Starter
- Draw on the board a pair of parallel lines cut by a transversal.
- Ask a student to mark a pair of alternate angles on the diagram. Repeat for a pair of corresponding angles and a pair of allied angles. Note that knowledge of allied angles is not a requirement for GCSE, but students will find it useful.
- Encourage **less able** students to copy the following into their notes.

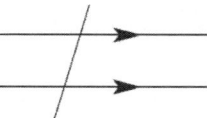

Angles in parallel lines

a and b are equal	a and b are equal	$a + b = 180°$
a and b are alternate angles	a and b are corresponding angles	a and b are allied angles

Main lesson activity
- Draw an equilateral triangle and an isosceles triangle on the board. Go through the geometrical properties of both shapes, ensuring students are familiar with them. Go through Example 1.
- Draw a parallelogram on the board. Go through the geometrical properties of the parallelogram and encourage students to record all the facts for future reference. Go through Example 2.
- Remind **less able** students about the mathematical conventions for parallel lines, equal sides and naming angles. Remind **more able** students that the diagonals of the parallelogram create alternate angles and vertically opposite angles.
- Draw a rhombus on the board. Go through its geometrical properties and encourage students to record all the facts for future reference. Rhombuses are often called *diamonds*. This is usually accepted in GCSE questions.
- Draw a kite on the board. Go through its geometrical properties and encourage students to record all the facts for future reference.
- Draw a trapezium on the board. Go through its geometrical properties and encourage students to record all the facts for future reference.
- Tell **more able** students that if the sloping sides are equal, the shape is an *isosceles trapezium*.
- Briefly revise the symmetry of the four quadrilaterals, since this is often a problem area for **less able** students.
- Now students can complete Exercise 9A, in which questions 2, 7 and 10 are PS and 4 and 11 are assessing understanding (AU) questions. If necessary, allow **less able** students work on the questions in pairs.

Shape	Number of lines of symmetry	Order of rotational symmetry
Parallelogram	0	2
Rhombus	2	2
Kite	1	0
Trapezium	0	0

Plenary
- Draw on the board an isosceles triangle, a trapezium, a parallelogram, a kite and a rhombus (without marking identical angles and sides).
- Ask students (in turn) to mark pairs of identical angles or sides on the diagrams until all possibilities have been covered.

Answers
Answers to Exercise 9A can be found at the end of this chapter.

© HarperCollins*Publishers* 2012

9.2 Angles in polygons

Curriculum references

Functional Skills standards
A1 Apply a range of mathematics to find solutions.

PLTS
Independent enquirers, Creative thinkers, Effective participators

APP
Shape, space and measure L6.2

Collins references
- Student Book pages 197–201
- Interactive Book: Paper Animation

Learning objectives
- Work out the sizes of interior angles and exterior angles in a polygon. [Grade C]

Learning outcomes
- Students **must** be able to calculate the sum of the angles of a polygon. [Grade C]
- Students **should** be able to calculate the angles of a regular polygon. [Grade C]
- Students **could** be able to solve problems about the angles of polygons. [Grade C]

Key words
- decagon
- exterior angle
- heptagon
- hexagon
- interior angle
- nonagon
- octagon
- pentagon
- polygon
- regular polygon

Prior knowledge
Students should know that the sum of the interior angles of a triangle is 180° and the sum of the interior angles in a quadrilateral is 360°.

Common mistakes and remediation
Students often mistakenly think that the sum of an exterior angle and an interior angle is 360°. Throughout the lesson, reinforce that it is 180°.

Useful tips
It is helpful to have available a collection of regular polygons or templates, which can be purchased commercially. Stress that, for *any polygon*, an exterior angle and its interior angle add up to 180°.

Functional maths and problem-solving help
There are no functional maths (FM) questions in Exercise 9B. Advise students, when they are doing problem solving (PS) questions 11 and 15, to check the results for interior and exterior angles in the tables in the text preceding the exercise.
Note: The work done in this lesson requires at least two sessions.

Starter
- Ask: What is a three-sided shape called? a four-sided shape? a five-sided shape? ... a 10-sided shape? (Students are expected to know the names of these polygons for GCSE examinations.)
- **More able** students might ask about polygons with more than 10 sides (e.g. 11 sides: *hendecagon*, 12 sides: *dodecagon*, 20 sides: *icosagon*). They could do their own research into this.

Lesson 9.2 Angles in polygons

Main lesson activity
- Draw a quadrilateral and a pentagon on the board. Show students how to split them into triangles to find the sum of the interior angles. Ask students to do the same for a hexagon and a heptagon. They should use convex polygons, but not regular ones. Encourage students to draw their own polygons and split them into triangles. Ensure that students draw all the diagonals from one vertex of the polygon.
- To help **less able** students, provide them with a worksheet of ready-drawn polygons
- Ask students to work out the sum of the interior angles for octagons, nonagons and decagons without drawing the shapes. Ask them to record their results in a table.
- **Less able** students should notice the formula in words: sum of interior angles = 180 × (number of sides − 2)
- **More able** students should write down the algebraic formula:
  $S = 180(n - 2)°$, where S is the sum of the interior angles and n is the number of sides.
- Encourage **more able** students to check that the result still holds for concave polygons. These are polygons in which at least one internal angle is greater than 180°, e.g. a concave hexagon is shown here. It can still be split into four triangles.
- Define a *regular polygon* as one in which all the sides are equal and all the angles are equal. Ask students to find the size of each interior angle for the regular polygons and record their results in a table.
- Define an exterior angle of a polygon. Make sure students know that the sum of an interior angle and its exterior angle is 180°. Ask them to find the size of each exterior angle for the regular polygons in their tables.
- **Less able** students should notice that the formula in words for finding each exterior angle and interior angle in a regular polygon is:

  exterior angle = $\dfrac{360°}{\text{number of sides}}$

  interior angle = $180° - \dfrac{\text{exterior angle}}{\text{number of sides}}$

- **More able** students should write down the algebraic formulae for any regular polygon:

  $E = \dfrac{360°}{n}$ $\qquad\qquad I = 180° - E = 180° - \dfrac{360°}{n}$

  where E is the exterior angle, I is the interior angle and n is the number of sides.
- Point out that when students need to find the size of an interior angle of a regular polygon, it is easier to find the size of the exterior angle first. Go through Example 3 with students.
- Now students can do Exercise 9B, in which questions 11 and 15 are PS and 8 and 13 are assessing understanding (AU) questions. Advise **less able** students to do the earlier questions in pairs until they are confident enough to complete the exercise on their own.

Plenary
- Display a table for regular polygons with headings 'Number of sides', 'Exterior angle' and 'Interior angle'.
- Ask individuals to complete rows of the table for shapes with three to 10 sides, and with n sides.
- Ask: What happens to the interior angle of a regular polygon as the number of sides increases? Why?

Answers
Answers to Exercise 9B can be found at the end of this chapter.

© HarperCollinsPublishers 2012

Functional maths
Tessellations

Curriculum references

Functional Skills standards
R2 Identify the situation or problems and identify the mathematical methods needed to solve them
R3 Select a range of mathematics to find solutions.
A1 Apply a range of mathematics to find solutions.
I2 Draw conclusions and provide mathematical justifications.

PLTS
Creative thinkers, Reflective learners

APP
Using and applying mathematics L5.1; Shape, space and measure L8.1

Collins references
- Student Book pages 206–207

Learning objectives
- Understand tessellations.
- Understand and use transformations.

Key words
- tessellation
- rotation
- reflection
- translation
- glide reflection
- congruency

Functional maths help
The task will develop students' knowledge of art and design and their understanding of the concept of tessellations. Tessellations are around us in everyday life. Brick patterns, floor tile patterns, mosaics, etc. Some of these are used for decorative effect. As a multi-cultural reference there are many tessellation patterns used in Islamic Art. The Alhambra in Granada, Spain has the finest examples of tessellation art in the world. There are many websites and Youtube™ videos available to show students some of the spectacular decoration of the Moorish Palace originally built in the 14th century.

Students can find various tessellations by Escher on the internet. They may also be interested in his 'impossible' pictures, such as the staircase and the waterfall, but stress that these do not involve tessellations.

Lesson plans
Build
The task outlined is easy to follow, but students may need help initially in using the drawing tools provided in the software. **Less able** students may need some help in planning and achieving the rotation of sections of the sides. This should lead to some pleasing designs.

Activity 1
Refer students to 'Your task' in the Student Book. Explain that while some shapes such as squares, rectangles and triangles will tessellate easily, other shapes can be much more difficult. In the first activity (steps 1–6 of 'Your task'), students create a non-regular shape that will tessellate. Although this activity can be done on paper, it would be better to use a drawing program such as Microsoft Word™ as the facilities aid the replication and reflection of drawing components.

Activity 2
In Activity 1 students produced a non-regular shape that will tessellate. In this activity they decorate their shapes and position them to produce an Escher-type tessellation. Point out that although the shapes will tessellate no matter what decoration is applied, it may be better to limit the range of decorative elements used, to reduce the complexity of the task.

© HarperCollins*Publishers* 2012

Functional maths Tessellations

Activity 3 (extension)
Less able students may try the first part of this activity. **More able** students may attempt to complete it.
Remind students about the three common transformations with which they are familiar: reflection, rotation and translation.
Ask students how many different wallpaper patterns they think there are. They may be surprised by this question, but explain that wallpaper with a pattern has to be matched up as it is put up on the wall, strip by strip. Therefore, there must be a regular pattern within the design so this can happen, even when the design on an individual strip may appear to be random.
Students will probably think there must still be millions of possibilities. In fact there are only 17. The mathematics behind this is very complex and beyond the scope of GCSE but the principle can be demonstrated by a simple example, using a shape such as a rhombus.
Again, this activity is probably best done via a drawing program.

Start with a rhombus. Make several copies and, by rotating, reflecting and translating, produce a variety of patterns.

Repeat with other quadrilaterals such as parallelograms and kites.
Let students experiment, to see how many variations they can produce.

Most able students could investigate how to produce patterns that would work as wallpaper and match across, from strip to strip.

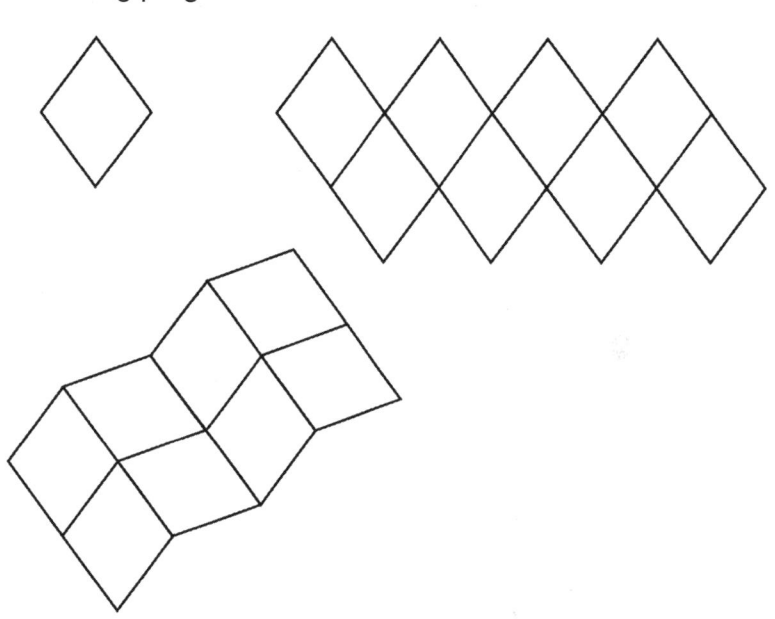

Apply
Discuss the idea of tessellations, and the process of producing a 'complicated' shape based on a simple shape that is known to tessellate. Ask students to describe the process of taking small parts away from one side and attaching them to another, and to explain why this should produce a new shape that will tessellate with itself. **Less able** students should keep their shapes simple. Challenge **more able** students to produce more complicated designs.

Master
Students should be able to:
- Understand tessellations and be able to create shapes that will tessellate, starting from a simple shape such as a rhombus or kite.
- Understand and use transformations to produce tessellating shapes and patterns.

Plenary
Look up the 17 wallpaper groups on the internet. Explain that students will need another transformation, the glide reflection, to produce all 17 possibilities. There are many examples in art and nature of these. Show students some of these and ask if they can see the transformations that are involved.

© HarperCollins*Publishers* 2012

Answers Lessons 9.1 – 9.2

Quick check
1 $a = 50°$
2 $b = 140°$
3 $c = d = 65°$

9.1 Special triangles and quadrilaterals
Exercise 9A
1 $a = b = 70°$, $c = 50°$, $d = 80°$,
$e = 55°$, $f = 70°$, $g = h = 57.5°$
2

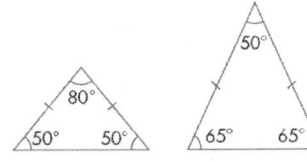

3 **a** $a = 110°$, $b = 55°$
  b $c = e = 105°$, $d = 75°$
  c $f = 135°$, $g = 25°$
  d $e = f = 94°$
  e $j = l = 105°$, $k = 75°$
  f $m = o = 49°$, $n = 131°$
4 40°, 40°, 100°
5 $a = b = 65°$, $c = d = 115°$,
$e = f = 65°$, $g = 80°$, $h = 60°$,
$i = 60°$, $j = 60°$, $k = 20°$
6 **a** $x = 25°$, $y = 15°$
  b $x = 7°$, $y = 31°$
  c $x = 60°$, $y = 30°$
7 **a** $x = 50°$: 60°, 70°, 120°, 110° – possibly trapezium
  b $x = 60°$: 50°, 130°, 50°, 130° – parallelogram or isosceles trapezium
  c $x = 30°$: 20°, 60°, 140°, 140° – possibly kite
  d $x = 20°$: 90°, 90°, 90°, 90° – square or rectangle
8 52°
9 Both 129°
10 $y = 360° − 4x$
11 **a** 65°
  b Trapezium, angle A + angle D = 180° and angle B + angle C = 180°

9.2 Angles in polygons
Exercise 9B
1 **a** 1440° **b** 2340°
  c 17 640° **d** 7740°
2 **a** 150° **b** 162°
  c 140° **d** 174°
3 **a** 9 **b** 15
  c 102 **d** 50
4 **a** 15 **b** 36
  c 24 **d** 72
5 **a** 12 **b** 9
  c 20 **d** 40
6 **a** 130° **b** 95° **c** 130°
7 **a** 50° **b** 40° **c** 59°
8 Hexagon
9 100°
10 141°
11 **a** Octagon **b** 89°
12 **a i** 71° **ii** 109° **iii** Equal
  b If S = sum of the two opposite interior angles, then $S + I = 180°$ (angles in a triangle), and we know $E + I = 180°$ (angles on a straight line), so $S + I = E + I$, therefore $S = E$
13 $a = 144°$
14 Three angles are 135° and two angles are 67.5°.
15 88°; $\dfrac{1440° − 5 \times 200°}{5}$
16 **a** 36° **b** 10
17 8
18 45°

Examination questions

1. **a** Because the quadrilateral can be divided into two triangles (by joining opposite vertices), and the angle sum of each triangle is 180°.

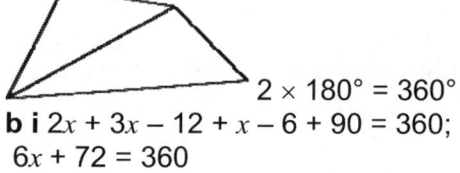

   $2 \times 180° = 360°$

   b i $2x + 3x - 12 + x - 6 + 90 = 360$; $6x + 72 = 360$

   ii $x = 48°$, largest angle is 132°

2. **a** 360° ÷ 6 or angle in an equilateral triangle
   b 720° **c** 20°

3. 72°

4. 28°

5. **a** c **b** d **c** g

6. **a** Statement A is true.
   b 36°

7. **a** 25° **b** 50° alternate to $2x$

10 Geometry: Constructions

Overview

| 10.1 Constructing triangles | 10.3 Defining a locus |
| 10.2 Bisectors | 10.4 Loci problems |

In this chapter, students will develop their skills of construction. The first of the four lessons covers using compasses, a protractor and a straight edge to construct triangles. This lesson will help them to construct bisectors of lines and angles, and angles of 60° and 90°. Students will also draw, describe, and solve real-life problems with loci.

Context

It is essential to understand angles, as they are used in almost all constructions. In order to create constructions, drawing skills are needed. The material in this chapter has applications in accurately drawing structures that involve triangles, such as bridges and roofs. It also applies to any situation that requires determining the location of a facility such as television, radio and telephone masts, radar stations, water sprinkler systems on golf courses and movement sensors on security lighting.

Curriculum references
KS4 Programme of Study references
1.1c Selecting appropriate mathematical tools and methods, including ICT.
2.2k Make accurate mathematical diagrams, graphs and constructions on paper and on screen.
3.1h Graphs of simple loci.
3.2a Properties and mensuration of 2D and 3D shapes.
4a Develop confidence in an increasing range of methods and techniques.

Linear specification
G3.9 Draw triangles and other 2D shapes using a ruler and a protractor.
G3.10 Use straight edge and a pair of compasses to do constructions.
G3.11 Construct loci.

Functional Skills standards
R2 Identify the situation or problems and identify the mathematical methods needed to solve them.
R3 Select a range of mathematics to find solutions.
A1 Apply a range of mathematics to find solutions.
I2 Draw conclusions and provide mathematical justifications.

PLTS
Independent enquirers identify questions to answer and problems to resolve; explore issues, events or problems from different perspectives; analyse and evaluate information, judging its relevance and value. **Team workers** collaborate with others to work towards common goals. **Creative thinkers** connect their own and others' ideas and experiences in inventive ways. **Effective participators** discuss issues of concern, seeking resolution where needed. **Reflective learners** review progress, acting on the outcomes; evaluate experiences and learning to inform future progress. **Self-managers** work towards goals, showing initiative, commitment and perseverance.

APP
Shape, space and measure L6.2 Solve geometrical problems using properties of angles, of parallel and intersecting lines, and of triangles and other polygons; **L6.8** Use straight edge and compasses to do standard constructions; **L7.4** Find the locus of a point that moves according to a given rule, both by reasoning and using ICT. **Using and applying mathematics L7.3** Justify generalisations, arguments or solutions; **L8.5** Examine generalisations or solutions reached in an activity, commenting constructively on the reasoning and logic or the process employed, or the results obtained.

© HarperCollins*Publishers* 2012

Chapter 10 Geometry: Constructions

Route mapping

Exercise	Grades D	Grades C
A	all	
B		all
C		all
D		all

Overview test
Questions 1 and 2 a–c are grade E, and questions 2 d–g are grade D.
1 Using a pencil, ruler, protractor and a pair of compasses, draw:
   a a triangle with two sides of 5 cm and the included angle of 60°
   b a triangle with all three sides of length 6 cm.
2 This is a scale drawing of a rectangle.

10 cm

0.5 cm

Copy and complete these sentences.
   a If the scale is 1 cm : 10 cm, the rectangle is ... long and ... deep.
   b If the scale is 1 cm : 15 m, the rectangle is ... long.
   c If the scale is 1 cm : 12 km, the rectangle is ... deep.
   d If the scale is 1 : 20, the rectangle is ... long.
   e If the scale is 1 : 4000, the rectangle is ... deep.
   f If the scale is 1 : 100 000, the rectangle is ... deep.
   g If the scale is 1 : 2 000 000, the rectangle is ... long.

Answers to overview test
1 a An equilateral triangle of side length 5 cm b An equilateral triangle of side length 6 cm
2 a 100 cm (1 m), 5 cm b 150 m c 6 km d 200 cm (2 m) e 2000 cm (20 m)
   f 50 000 cm (500 m or 0.5 km) g 200 km (20 000 000 cm)
Note: Question 2 is revision work for Lesson 8.4 (Loci problems).

Why this chapter matters
Cross-curricular: This topic links with Science and Design and Technology, and may also be relevant to Arts and Crafts.
Introduction: In almost any field, the need for accuracy in design is clear. Whether it is the intricate detail required in nanotechnology, building self-assembly furniture, designing a garden or even decorating a cake, if plans are not drawn up precisely, the finished product will not be as good as it could be.
Discussion points: Why is it important to use a sharp pencil? What happens if plans are photographically enlarged? What happens to the thickness of the lines? Why is it better to measure a distance with a pair of compasses along a ruler, rather than just laying the ruler along the line? Try measuring a line with a ruler. Move your head slightly, from side to side and backward and forwards. Is the measurement always the same?
Plenary: Look at plans and blueprints, if any are available, and also the diagrams in the Student Book. Discuss the fine detail on them.

Answers to the quick check test can be found at the end of the chapter.

10.1 Constructing triangles

Curriculum references

Functional Skills standards
A1 Apply a range of mathematics to find solutions.

PLTS
Independent enquirers, Creative thinkers, Effective participators

APP
Shape, space and measure L6.8

Collins references
- Student Book pages 210–214

Learning objectives
- Construct triangles, using compasses, a protractor and a straight edge. [Grade D]

Learning outcomes
- Students **must** be able to construct a triangle when given two sides and the included angle. [Grade D]
- Students **should** be able to construct a triangle when given three sides. [Grade D]
- Students **could** be able to give reasons for whether it is possible to draw a triangle or not. [Grade D]

Key words
- angle
- compasses
- construct
- side

Prior knowledge
Students should be able to draw and measure straight lines, and use a protractor to draw and measure angles accurately. They should also be able to open a pair of compasses and set them to a given radius and draw arcs from a fixed point.

Common mistakes and remediation
Most errors in this type of construction are due to poor accuracy in drawing triangles. Remind students of this.

Useful tips
Tell the class that using poor equipment and not taking care with their drawings can make them lose marks in what are usually fairly easy GCSE questions. Students need to make sure their compasses are in good condition, with the joint tight but smooth. They should keep their pencils sharp, and rulers and protractors clean; not scratched or worn. Tell students that it is better to choose a ruler with a small border before the zero, since trying to align the point of the compasses to the very edge of a slippery piece of plastic can cause problems.

Functional maths and problem-solving help
Although not designated as such, 7 is a functional maths (FM) question is based on an old example that is still in practice today. It may be worthwhile finding out if the grounds staff at your school or college are familiar with this technique. Otherwise, a practical activity based on this technique may be appropriate.

Most of this section is revision work. Direct **more able** and **less able** students to questions in Exercise10A that are suited to their ability. **Less able** students may need assistance in using the scale suggested to make the construction practical. They may find it helpful to work on squared paper for problem-solving (PS) question 8.

Lesson 10.1 Constructing triangles

Starter
- Students will almost certainly have done this activity before, but it is a useful exercise in using compasses.
- Ask students to draw a circle of any radius, to fit the page. Say: keeping the compasses set to the same radius, place the point anywhere on the circumference and draw an arc to cut the circle in two places. Repeat, using as centres the points where the arcs cut the circle. Draw six arcs.
- Say that if they have drawn the arcs accurately, they will produce a flower pattern with sharp pointed petals.

Main lesson activity
- Have plenty of plain paper available, as students will use a lot in these lessons. Make sure they all have a pair of compasses fitted with a sharp pencil, a ruler, a protractor and another sharp pencil for drawing.
- Go through Example 1, the construction of a triangle with three sides known. Then ask students to construct a few more triangles, giving them the three sides each time. Emphasise to the **less able** students that they should always start by drawing the longest side as the base. Ask students to measure the largest angle (opposite the longest side), for them to compare their accuracy and to practise using a protractor.
- When students are confident about constructing triangles, given all three sides, go through Example 2 in which they are given two sides and the included angle. Again, draw the longest side first, then use the protractor to measure the angle and draw in a faint line. Remind **less able** students where to place the protractor to draw the given angle. Finally, they should use compasses to mark off the length of the second given line, to find the third corner of the triangle. Ask the class to measure the length of the new line as a means to check accuracy. Set them to draw more triangles from two sides and the included angle.
- Now go through Example 3, constructing a triangle in which two angles and a side are known. Again, ask them to measure the lengths of the two other sides as a means to check accuracy. Remind students that, since they know the angle sum of a triangle (180°), they can always work out the third angle. Set them a few more examples of drawing triangles, given two angles and one side.
- Discuss with **more able** students why they would not necessarily be able to draw the triangle if they just knew two sides and one non-included angle. Give an example, such as AC = 8 cm, BC = 5 cm, ∠ BAC = 35°, and let them draw it, to see the ambiguity for themselves.
- Now students can complete Exercise 10A, in which question 8 is PS and 9 is an assessing understanding (AU) question. **Less able** students may work in pairs to complete these questions.

Plenary
- Ask students to draw a line 5 cm long. To do this, they should draw a straight line, then open their compasses so that the point is on zero on a ruler, and the pencil point is on 5 cm. They should mark a point near the end of the line they have drawn, and mark the 5 cm-length with an arc.
- Now ask them to draw arcs above the line, centred on each end of the 5 cm-line they have marked. Then they should join each of the points on the line to the point where the arcs cross. Ask what type of triangle they have drawn. Let them measure all the angles to check. (An equilateral triangle, 60°)

Answers
Answers to Exercise 10A can be found at the end of this chapter.

© HarperCollins*Publishers* 2012

10.2 Bisectors

Curriculum references

Functional Skills standards
A1 Apply a range of mathematics to find solutions.

PLTS
Independent enquirers, Creative thinkers, Effective participators

APP
Shape, space and measure L6.8

Collins references
- Student Book pages 215–219
- Interactive Book: Common Misconceptions

Learning objectives
- Construct the bisectors of lines and angles. [Grade C]
- Construct angles of 60° and 90°. [Grade C]

Learning outcomes
- Students **must** be able to bisect a line. [Grade C]
- Students **should** be able to bisect an angle. [Grade C]
- Students **could** be able to accurately carry out constructions. [Grade C]

Key words
- angle bisector
- bisect
- line bisector
- perpendicular bisector

Prior knowledge
Students should be able to use compasses to draw arcs and circles accurately.
Students should also be able to bisect a line.

Common mistakes and remediation
Students may just measure lengths rather than construct them. (Many examiners would say that 'Arcs mean marks'.) Less able students often erase the construction lines. Remind them of these things.

Useful tips
Tell students that to create thin and accurate construction lines, they should always use very sharp pencils.

Functional maths and problem-solving help
Functional maths (FM) question 10 of Exercise 10B hinges on drawing a perpendicular bisector between Norwich and Bristol on a map of the UK, and then looking for a perpendicular line from Birmingham to that bisector. You will need some prepared A4 or smaller maps of the UK to enable students to attempt this question. If the map illustrates areas rather than points for cities, they should pinpoint the nominal centre of each town as a point representing it. Problem-solving (PS) question 11 is similar to the earlier question 5, but approaches the idea from the opposite starting point.

Lesson 10.2 Bisectors

Starter
- Discuss how to bisect a line (cut it into two equal parts), then demonstrate the wrong way to bisect a line. Draw a straight line, measure half-way along and make a mark. Use a protractor, centred on this mark, to make an angle of 90°.
- Stress that in an examination, if students use this method to bisect a line, i.e. by measuring and adding a perpendicular bisector (the line at right angles), they would lose all marks for the question.
- Now demonstrate on the board the correct method to bisect a straight line, using a pair of compasses and a straight edge.

Main lesson activity
- Make sure each student has a pair of compasses fitted with a sharp pencil, a straight edge and another sharp pencil for drawing. Also make sure you have plenty of plain paper. These constructions are detailed in the Student Book.
- Go through how to construct a line bisector, reminding students that the arcs drawn are construction lines and so should be faint, but still visible. Ask students to draw a few straight lines and then bisect them all.
- Then talk about an angle bisector; go through how to construct this. They will need to draw quite a few angles of their own to practise this skill and they could use a protractor to check their accuracy.
- Now go through constructing an angle of 60°. **More able** students could try to construct an angle of 30° at this point by constructing an angle of 60° and then bisecting it.
- Now show all students how to construct a perpendicular line from a point on a line and practise this with a few perpendiculars from various lines on their plain paper. **Less able** students may omit drawing a perpendicular from a point to the line at this stage, but **more able** students should attempt it.
- Students can now work their way through Exercise 10B. They will need simple maps of the UK, showing the main cities, for several of the questions, including FM question 10. Question 11 is a PS question and 12 is an assessing understanding (AU) question.

Plenary
- Provide a board pen, attached to a piece of string at least 50 cm long. Draw an angle of 120° and ask a volunteer to bisect it, using the board pen and the string.
- Ask another volunteer to bisect one of the resulting 60° angles; a third to bisect one of the 30° angles and a fourth to use a board protractor to check the size of the 15° angle produced.
- Draw a horizontal line (40–50 cm long) on the board. Ask another volunteer to bisect the line, again using the board pen and string. Ask a final volunteer to bisect one of the two halves.

Answers
Answers to Exercise 10B can be found at the end of this chapter.

© HarperCollins*Publishers* 2012

10.3 Defining a locus

Curriculum references

Functional Skills standards
R2 Identify the situation or problems and identify the mathematical methods needed to solve them.

PLTS
Independent enquirers, Creative thinkers, Effective participators

APP
Shape, space and measure L7.4

Collins references
- Student Book pages 219–221
- Interactive Book: Matching Pairs

Learning objectives
- Draw a locus for a given rule. [Grade C]

Learning outcomes
- Students **must** be able to find the locus of points satisfying one condition. [Grade C]
- Students **should** be able to find a region satisfying particular conditions. [Grade C]
- Students **could** be able to describe or identify loci. [Grade C]

Key words
- equidistant
- loci
- locus

Prior knowledge
Students need to be able to construct the perpendicular bisector of a line, although it will frequently be shown as two points, without the line joining them.

Common mistakes and remediation
Students may not recognise the difference between the distance 'from a point' and the distance 'from a line'. Another problem is lack of accuracy. Remind them to watch these things.

Useful tips
Stress that the distance 'from a point' is fixed, since the point is fixed. The distance 'from a line' depends on the point from which it is measured. In this context it should always be the shortest distance, so involves a perpendicular line.

Functional maths and problem-solving help
This lesson is about skills, ensuring that students are able to construct loci accurately, and leads into the functional mathematics, which is more apparent in the next lesson. Question 3 of Exercise 10C starts to move the topic into functional maths (FM) with the idea of a practical situation. Students should tackle problem-solving (PS) question 7 in stages, eliminating the unwanted areas until they are left with the solution.

Starter
- Pick a point in the classroom, or go outside if space is limited. Ask a student to move in such a way as to stay two metres from the point. Discuss the shape of the path traced out.
- Now choose two points. Ask a student to move so that he or she remains the same distance from the two points. Discuss the shape of the path traced out. Now choose three points, A, B and C and repeat, then discuss.

© HarperCollins*Publishers* 2012

Lesson 10.3 Defining a locus

Main lesson activity
- Provide plain paper and make sure all students have rulers. Ask them to mark a dot (labelled C) on the paper, about 10 cm down from the top and in from the left-hand side. Now ask students to mark four more dots on the paper, all 4 cm away from point C, and label each one P. Check that students have the right idea. (**Less able** students may be confused, and may put the points 4 cm away from each other.) Now ask them to mark another four points, each 4 cm away from point C, still labelling them all P; then ask them to mark four more points.
- Unless a student has already remarked on this, ask what is so special about this set of points P they are marking. (They are making a circle around point C.) Confirm this fact and explain that this circle is the path or locus (plural loci) of point P. The locus of point P, 4 cm from point C, is a circle, centre C, of radius 4 cm. Ask students to mark another point, D, on their paper and find the locus of point P that is always 3 cm away from point D. Discuss what they have drawn and tell them they should all have a circle, radius 3 cm, centred on point D. Refer to Example 4 to reinforce this idea.
- Ask students to mark two points, A and B, on their paper and find the locus of point P that is always the same distance (equidistant) from A and B. Lead them to see that this will be a line perpendicular to the imaginary line joining the two points, the perpendicular bisector. Refer to Example 5 to reinforce this idea.
- Ask students to think about a straight wall or fence between two fixed boundaries. Ask about the locus of a point that must remain one metre from the wall or fence and encourage them to realise that it will be a line parallel to the wall and one metre from it. Discuss this carefully with **less able** students, who may not realise they need to consider the perpendicular distance from the wall to be one metre. Do not rush the point, as it is vital that they realise why the distance must be perpendicular. Refer to Example 6 to reinforce this idea.
- Now ask students to draw a straight line 4 cm long somewhere in the space on their paper. Ask them to mark four points 3 cm away from this line, labelling each one P. Watch carefully what **less able** students do, and make sure they measure the perpendicular distance from the line to be 3 cm. Provide extra support if needed.
- Now ask students to mark another four points P, then four more. To lead to useful discussion, discuss students' shapes and where their points are. Some points will only be marked on one side of the line, but stress that they can be on either side. At each end of the line, the points form a semicircle. Discuss this in detail, as there are many issues, even for **more able** students. The locus could be described as a racecourse shape. Ask students to draw another line 3 cm long and to draw the locus of a point P that is always 2 cm away from this line. Again, they should draw a racecourse shape. Example 7 reinforces this idea. Also work through Example 8.
- Now students can complete Exercise 10C, in which question 3 is FM, 7 is PS and 8 is an assessing understanding (AU) question. This work is not too difficult, once students have grasped the basic ideas.

Plenary
- Ask students to imagine a bird tethered by a leash one metre long to a point on a perch at least one metre high. What is the locus of the possible points that the bird can reach in flight? (a sphere, radius one metre)
- Now ask them to imagine the bird tethered by a leash one metre long to a perch three metres long. This time, the tether is attached to a ring that can slide along the full length of the perch. What is the locus of the possible points that the bird can reach in flight now? (a cylinder, radius one metre and length three metres, with a hemisphere, radius one metre, on each end)

Answers
Answers to Exercise 10C can be found at the end of this chapter.

© HarperCollins*Publishers* 2012

10.4 Loci problems

Curriculum references

Functional Skills standards
R2 Identify the situation or problems and identify the mathematical methods needed to solve them.

PLTS
Independent enquirers, Creative thinkers, Effective participators

APP
Shape, space and measure L7.4

Collins references
- Student Book pages 222–226
- Interactive Book: Common Misconceptions

Learning objectives
- Solve practical problems using loci. [Grade C]

Learning outcomes
- Students **must** be able to find simple loci in practical situations. [Grade C]
- Students **should** be able to find more complex loci. [Grade C]
- Students **could** be able to solve realistic problems by constructing a locus. [Grade C]

Key words
- loci
- scale

Prior knowledge
Students should be able to use a pair of compasses to draw a circle and know that the distance from the centre to any point on the circumference is the same, equal to the radius.

Common mistakes and remediation
Students may have problems with scale, and so need to take care to ensure that they use the correct scale. They should also take care to interpret 'closer to' and 'further away' correctly.

Useful tips
Remind students that when they are considering a locus of points a fixed distance from a line, there may be points on both sides. **More able** students could draw loci that involve 'closer to' or 'further away' as dotted lines, to indicate that points on the line are not included.

Functional maths and problem-solving help
The first seven questions of Exercise 10D lead the way to the remaining eight functional maths (FM) questions in the exercise. They are problems set in real-life contexts, simplified but relevant to the types of problem that arise in various situations. Students need to be aware that, for these types of problems, they are applying the loci techniques they learned in the previous lesson.
Problem-solving (PS) question 17 is based on a man navigating a course to fly from the south to the north. Students need to interpret the question as finding a locus for a point that remains equidistant from the two cities, namely Bristol and London.
Less able students may have difficulty with questions 10 onwards. They need to read the problems carefully, interpreting the relevant loci that are needed and then drawing them on the maps.

Lesson 10.4 Loci problems

Starter
- Provide plain paper and ask students to put three small crosses on their paper, to represent local towns, e.g. Barnsley, Rotherham and Doncaster. Make sure they choose places that are not all in a line. Accuracy at this stage is not really important. Ask students to label the points, then to find all the points that are the same distance between, say Barnsley, and Rotherham. This will be the perpendicular bisector of the line drawn between Barnsley and Rotherham. Then ask them to find all the points that are the same distance from Rotherham and Doncaster.
- Discuss the lines they have drawn and talk about what is special about the point of intersection of both lines. Draw out that this point is equidistant from all three towns.

Main lesson activity
- Ensure that every student has a pair of compasses fitted with a sharp pencil, as well as another sharp pencil and a ruler.
- Talk through Example 9 to demonstrate a practical locus problem. Make sure students understand what the problem is in the first place, before going to the solution. **Less able** students should do this work themselves, on their own plain paper, to comprehend fully what it is they are doing. They can trace the points of the towns onto their plain paper.
- Talk through Example 10, which is a different type of locus problem. Again **less able** students will do better if they draw this solution for themselves.
- Go through Example 11 only with **more able** students, as the idea of joining three loci together is not necessary for GCSE but is useful to them for a fuller appreciation of some practical difficulties.
- Now students can complete Exercise 10D. Ensure that you have plenty of squared paper available for questions 1–7, as students need to copy the diagrams and positions given by the squares. You will also need plenty of UK maps, duplicated from the Student Book, for questions 10–15 and 17. In this exercise, all questions up to 15 are FM, 17 is a PS question and 18 is an assessing understanding (AU) question. If students are working well in the first 16 questions, they can move on to 17 and 18, for the practice in these questions. However, **less able** students may need help.

Plenary
- To provoke discussion and comment, ask: What is the locus of a point that is one metre away from the walls, ceiling and floor of this classroom?
- There are two regions to consider, one inside the room, which will give a locus of a smaller cuboid shape, and the other outside the room (over and under), which will give an interesting shape of a cuboid with rounded corners. Discuss the shape of these corners, which will be quarter spheres.

Answers
Answers to Exercise 10D can be found at the end of this chapter.

© HarperCollins*Publishers* 2012

Problem solving
Planning a football pitch

Curriculum references

Functional Skills standards
R3 Select a range of mathematics to find solutions.
A1 Apply a range of mathematics to find solutions.

PLTS
Reflective learners, Self-managers, Team workers

APP
Using and applying mathematics L7.3

Collins references
- Student Book pages 232–233

Learning objectives
- Construct perpendicular lines.
- Construct defined loci.
- Solve problems using construction skills.

Key words
- circle
- dimensions
- perpendicular
- right angle
- semicircle

Functional maths help
Most students will be familiar with the design of a football pitch. However, to guide students and ensure that they draw a comprehensive diagram, show a photo of a football pitch and ask them to identify all the elements in it. To reinforce the context, mention that pitches are regularly marked out for matches. Arenas such as Wembley have the turf re-laid regularly; the pitches must be marked afresh each time.

Discuss and compare sizes of the local club pitch, the school pitch and Premier League pitches. Most students will then be able to use their construction skills independently to complete the task.

Lesson plans
Build
Discuss the various sports pitches with which students may be familiar. Sketch two or three of these and then ask for comments on the shapes and dimensions. Ask why sports pitches are usually drawn to fixed layouts. Discuss who decides on the layouts, and how they are drawn on the playing field. Ask students about the bodies that govern football and introduce the idea of a FIFA standard football pitch.

Activity 1
Refer students to the 'Getting started' questions. As a class, discuss the shapes commonly seen on sports pitches. At this stage, do not limit the discussion to football pitches. Accept answers that are also relevant to other sports, such as tennis and athletics. Students should identify semicircles, triangles, rectangles and arcs.

Ask students to work out how each shape on these playing areas would be constructed and identify the equipment that would be required on a real field. Ensure students are confident in selecting the right equipment for drawing these shapes on a scale drawing. If necessary, make suggestions, such as a pair of compasses for a semicircle.

Students should be comfortable with using a ruler and a pair of compasses to construct the basic shapes required. However, for **less able** students, take time to remind students how to use a pair of compasses effectively, holding them without altering the radius. Demonstrate this to students. Discuss any angles or curved areas that appear on sports pitches. Students might identify arcs on sports pitches such as discus arenas, as well as on the more common football pitches.

© HarperCollins*Publishers* 2012

Problem solving Planning a football pitch

Activity 2
If possible, take students outside to measure any available sports pitches. Alternatively, ask students to come to the lesson, having measured at least one element of a local or school playing field or courts. Then ask students to draw the pitches they have seen (or discussed). The drawings need not be accurate at this stage.

Activity 3
Discuss with students how sports pitches are laid out. Ask how the grounds staff can construct right angles and how they would draw circular curves. If necessary, remind students about knotted ropes, to form right-angled triangles. Draw out the idea of using a white-line marker tied to a rope, of the required length, which is attached to a stake or peg in the ground to draw circles. Discuss any other ideas that students mention.

Activity 4
Ask students to read the FIFA specifications in the fact box carefully. You might then ask students to do further research into the make-up of a football pitch and FIFA's specifications.
Now refer students to question 1 of 'Your task', in which they should use the information they have gathered to construct an accurate scale drawing of a FIFA-standard football pitch.

You may need to give **less able** students a scale to work to, e.g. 2 mm represents 1 m. **More able** students may choose their own scale. Allow students to use A4 or A3 paper. **Less able** students may benefit from using squared paper.

Students who are building their functional skills will need help with breaking down the problem into manageable chunks. You may need to lead them through the criteria, one at a time. Use a series of prompts, for example:
- What size pitch have you chosen?
- Using the scale discussed, draw the touchlines and centre line.
- Add in the centre circle. At what size should this be drawn?
- Move on to the corner arcs, penalty area, goal area (off the pitch) and penalty spot.

The scale drawings should be as shown below.

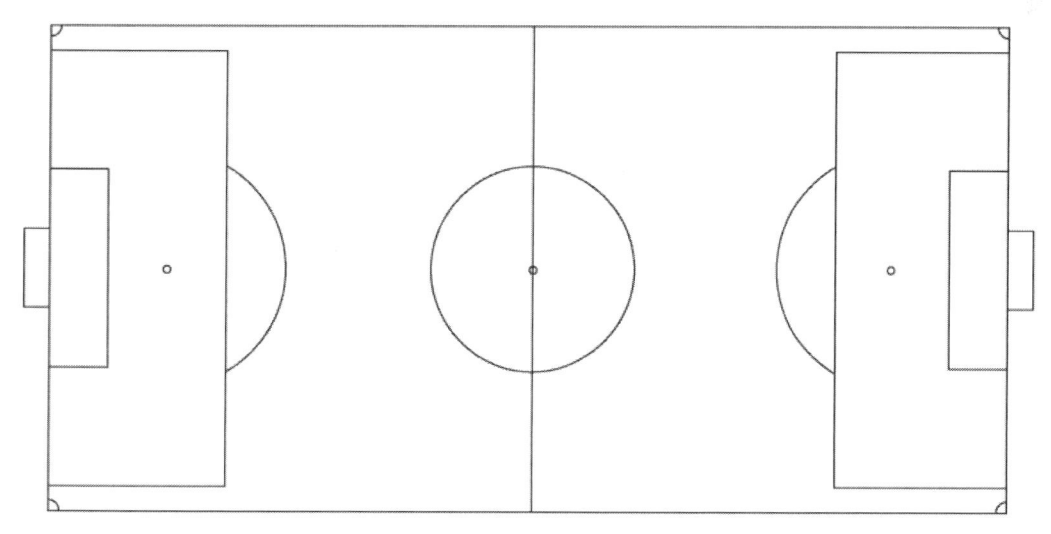

Check students' measurements.

© HarperCollins*Publishers* 2012

Problem solving Planning a football pitch

Activity 5
Refer students to question 2 of 'Your task'. More able students should be able to decide where to place sprinklers to ensure that the pitch is fully watered. A sprinkler can spray a circular area with a radius of 9 m. There are many places to position the sprinklers, but they should be 18 m apart to achieve maximum coverage. Note that there is no overlap between the areas covered by the sprinklers, so parts of the pitch will not be watered. Students should develop their own strategy and justifications. They must present their solutions as scale drawings. There are many possible answers. This diagram (also on the accompanying CD-ROM) shows one possible solution.

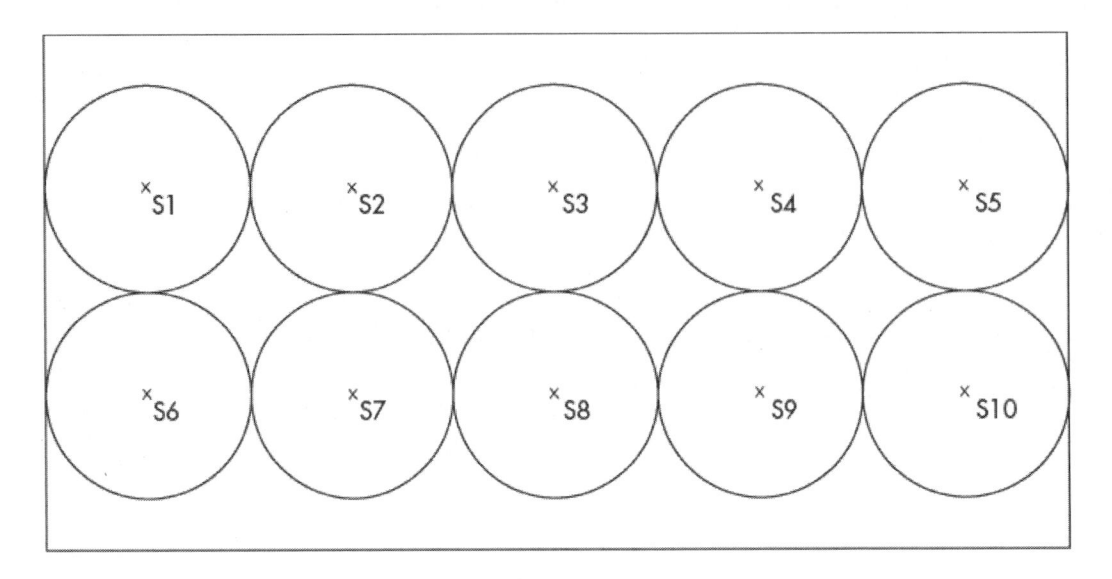

Apply
Have a class discussion about the shapes that one can see on sports fields. Then ask students to draw sketches of these shapes as they appear on fields of different sports. Take time to ensure that all students know all the elements of a FIFA-standard football pitch, before asking them to work through the specifications, in small groups. This gives students the opportunity to provide peer support, helping each other to decide which measurements and playing areas should be drawn next.

Master
In order to demonstrate mastery of the learning objectives, students should be able to:
- Demonstrate confidence in using geometric construction equipment.
- Construct a pitch that fits all the required regulations.
- Position the minimum number of sprinklers to cover the maximum amount of pitch.

Plenary
Look at the pitch used in American baseball. Ask students to think about how they would use geometric construction equipment to draw an accurate diagram of the pitch. Or, ask students to investigate a sport of their choice and show how they would draw an accurate diagram using as many construction skills as possible.

Answers Lessons 10.1 – 10.4

Quick check
1   **a** 6 cm **b** 7.5 cm **c** 11 cm
2   **a** 30° **b** 135°

10.1 Constructing triangles
Exercise 10A
1   **a** BC = 2.9 cm, ∠B = 53°, ∠C = 92°
    b EF = 7.4 cm, ED = 6.8 cm, ∠E = 50°
    c ∠G = 105°, ∠H = 29°, ∠I = 46°
    d ∠J = 48°, ∠L = 32°, JK = 4.3 cm
    e ∠N = 55°, ON = OM = 7 cm
    f ∠P = 51°, ∠R = 39°, QP = 5.7 cm
2   **a** Students can check one another's triangles.
    b ∠ABC = 44°, ∠BCA = 79°, ∠CAB = 57°
3   **a** 5.9 cm **b** 18.8 cm²
4   BC = 2.6 cm, 7.8 cm
5   **a** 4.5 cm **b** 11.25 cm²
6   **a** 4.3 cm **b** 34.5 cm²
7   **a** Right-angled triangle constructed with sides 3, 4, 5 and 4.5, 6, 7.5, and scale marked 1 cm : 1 m.
    b Right-angled triangle constructed with 12 equally spaced dots.
8   An equilateral triangle of side 4 cm.
9   Even with all three angles, you need to know at least one length.

10.2 Bisectors
Exercise 10B
1–9 Practical work; check students' constructions.
10 Leicester
11 The centre of the circle
12 Start with a base line AB; then construct a perpendicular to the line from point A. At point B, construct an angle of 60°. Ensure that the line for this 60° angle crosses the perpendicular line; where they meet will be the final point C.
13–15 Practical work; check students' constructions.

10.3 Defining a locus
Exercise 10C
1   Circle with radius:
    a 2 cm **b** 4 cm **c** 5 cm
2   **a**   **b**

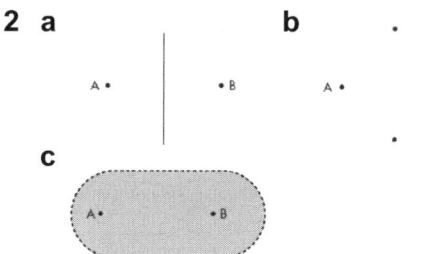

    c

3   **a** Circle with radius 4 m   **b**

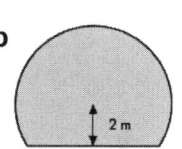

4   **a**   **b**   **c**
    d   **e**   **f**

5

6

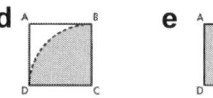

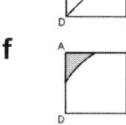

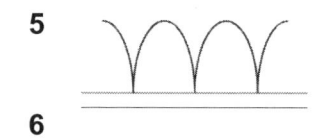

7   Construction of the bisector of angle BAC and the perpendicular bisector of line AC. (see below right.)

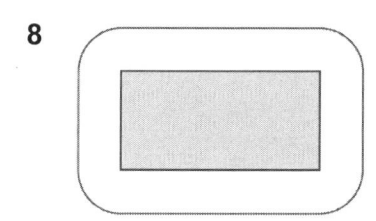

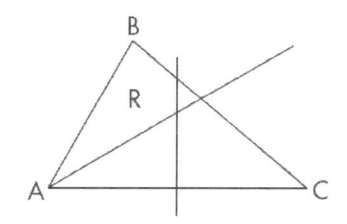

8

Answers Lessons 10.1 – 10.4

10.4 Loci problems
Exercise 10D

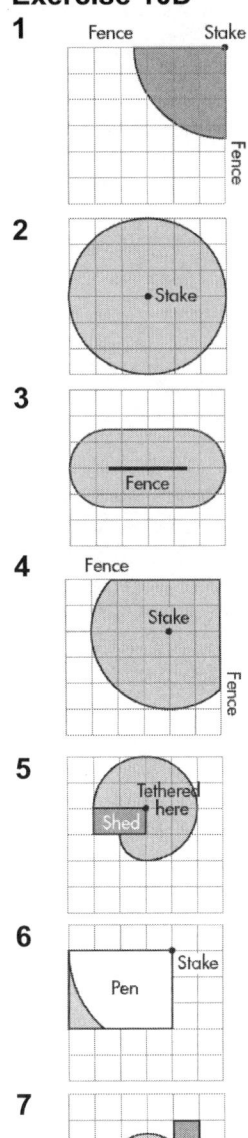

1. [Sketch: quarter circle in corner bounded by Fence and Stake]

2. [Sketch: full circle around Stake]

3. [Sketch: stadium-shape around Fence]

4. [Sketch: shape around Stake bounded by Fence]

5. [Sketch: circle around Tethered here point, next to Shed]

6. [Sketch: quarter-circle Pen near Stake]

7. [Sketch: circle overlapping square around Stake]

8. **a** Sketch should show a circle of radius 6 cm around London and one of radius 4 cm around Glasgow.
   b No
   c Yes

9. **a** Yes
   b Sketch should show a circle of radius 4 cm around Leeds and one of radius 4 cm around Exeter. The area where they overlap should be shaded.

10. **a** This is the perpendicular bisector of the line from York to Birmingham. It should pass just below Manchester and just through the top of Norwich.
    b Sketch should show a circle of radius 7 cm around Glasgow and one of radius 5 cm around London.
    c The transmitter can be built anywhere on line constructed in part a that is within the area shown in part b.

11. Sketch should show two circles around Birmingham, one of radius 3 cm and one of radius 5 cm. The area of good reception is the area between the two circles.

12. Sketch should show a circle of radius 6 cm around Glasgow, two circles around York, one of radius 4 cm and one of radius 6 cm and a circle around London of radius 8 cm. The small area in the Irish Sea that is between the two circles around York and inside both the circle around Glasgow and the circle around London is where the boat can be.

13. Sketch should show two circles around Newcastle upon Tyne, one of radius 4 cm and one of radius 6 cm, and two circles around Bristol, one of radius 3 cm and one of radius 5 cm. The area that is between both pairs of circles is the area that should be shaded.

14. Sketch should show the perpendicular bisector of the line running from Newcastle upon Tyne to Manchester and that of the line running from Sheffield to Norwich. Where the lines cross is where the oil rig is located.

15. Sketch should show the perpendicular bisector of the line running from Glasgow to Norwich and that of the line running from Norwich to Exeter. Where the lines cross is where Fred's house is.

16. Sketch should show the bisectors of the angles made by the piers and the sea wall at points A and B. These are the paths of each boat.

17. Leeds

18. On a map, draw a straight line from Newcastle to Bristol, construct the line bisector, then the search will be anywhere on the sea along that line.

© HarperCollins*Publishers* 2012

Answers Lessons 10.1 – 10.4

Examination questions
1 Check students' drawings, top angle should be 110°.

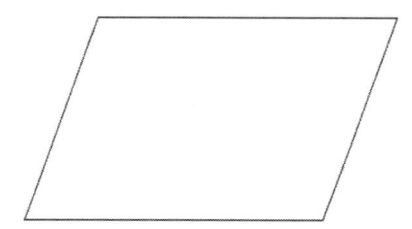

2 Check students' drawings, top angles should be 90° and 130°.

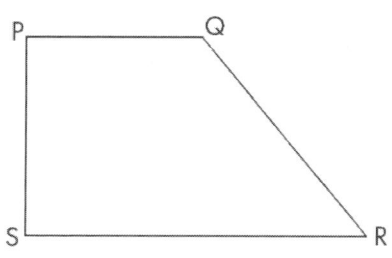

3 a

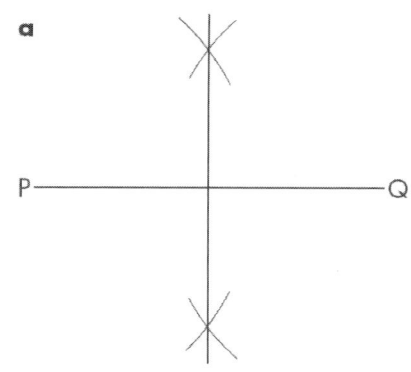

  b An equal distance from P and Q.

4

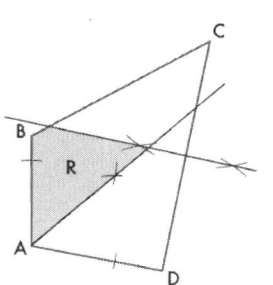

5

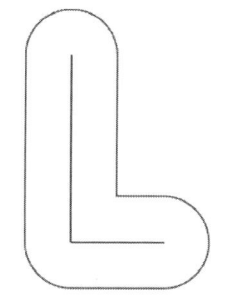

6 a

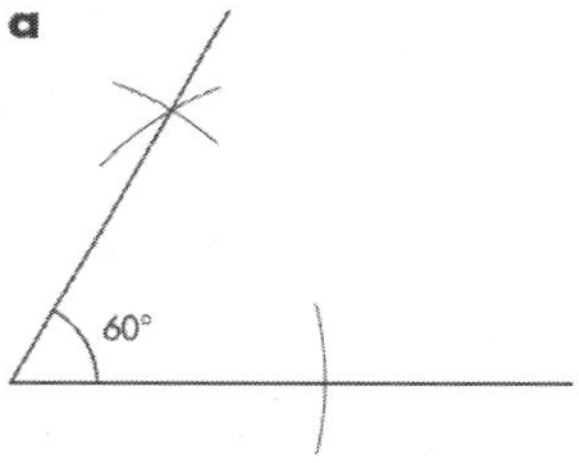

  b

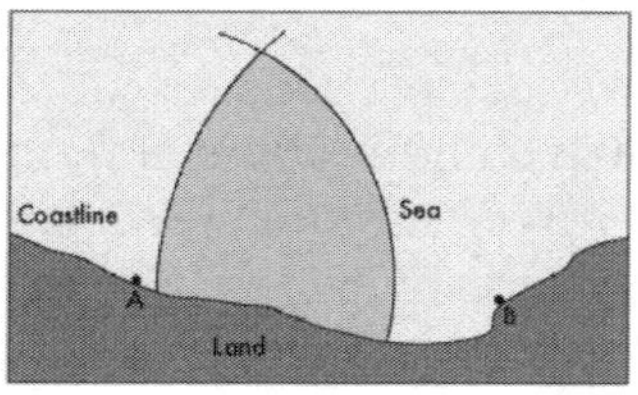

7 a

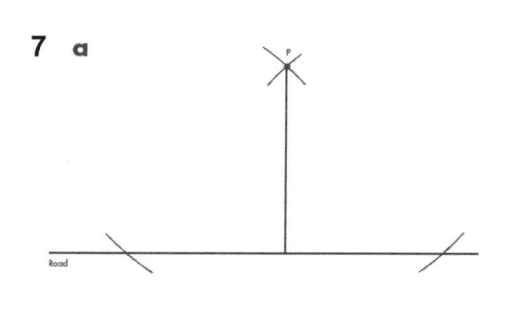

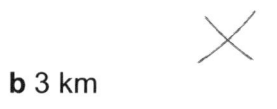

  b 3 km

© HarperCollinsPublishers 2012

Chapter 11 Geometry: Transformation geometry

Overview

11.1 Congruent triangles	11.4 Rotations
11.2 Translations	11.5 Enlargements
11.3 Reflections	11.6 Combined transformations

The lessons in this chapter cover two main sections in transformation geometry. The first lesson covers the work required at this level on congruent triangles with the corresponding rules for congruency. The remaining lessons cover the work on transformations required at this level, namely translations, reflections, rotations, enlargements and combined transformations.

Context
This chapter is geometrical in nature, so students are likely to recognise few 'practical' applications other than reflection in relation to using a mirror. In fact, transformations are important and occur everywhere around us – from facial symmetry in biology through to the idea of isometrics in chemistry. Islamic art is particularly notable for its highly geometric designs and dependence on mathematics and geometric transformations.

Curriculum references
KS4 Programme of Study references
1.2b Using existing mathematical knowledge to create solutions to unfamiliar problems.
2.2k Make accurate mathematical diagrams, graphs and constructions on paper and on screen.
3.2d Properties and combinations of transformations.
4a Develop confidence in an increasing range of methods and techniques.

Linear specification
G1.7h Describe and transform 2D shapes using single or combined rotations, reflections, translations, or enlargements by a scale factor, including positive fractional or negative ones. Distinguish properties that are preserved under particular transformations.
G1.8 Understand congruence and similarity.

Functional Skills standards
R1 Understand routine and non-routine problems in familiar and unfamiliar contexts and situations.
R3 Select a range of mathematics to find solutions.
A1 Apply a range of mathematics to find solutions.
A2 Use appropriate checking procedures and evaluate their effectiveness at each stage.
I2 Draw conclusions and provide mathematical justifications.

PLTS
Independent enquirers identify questions to answer and problems to resolve; explore issues, events or problems from different perspectives; analyse and evaluate information, judging its relevance and value. **Self-managers** work towards goals, showing initiative, commitment and perseverance. **Creative thinkers** connect their own and others' ideas and experiences in inventive ways. **Effective participators** discuss issues of concern, seeking resolution where needed. **Team workers** collaborate with others to work towards common goals.

© HarperCollins*Publishers* 2012

Chapter 11 Geometry: Transformation geometry

APP
Shape, space and measure L6.6 Enlarge 2D shapes, given a positive whole-number scale factor; **L6.7** Know that translations, rotations and reflections preserve length and angle and map objects onto congruent images; **L7.3** Enlarge 2d shapes, given a centre of enlargement and fractional scale factor, on paper and using ICT; recognise the similarity of the resulting shapes. **L8.1** Understand and use congruence and mathematical similarity. **Using and applying mathematics L8.3** Select and combine known facts and problem-solving strategies to solve problems of increasing complexity.

Route mapping

Exercise	D	C	B	A
A			1–4	5–8
B		all		
C	1–7	8–12		
D	1–3	4–12		
E	1–4	5–6	7–9	
F	1–2	3–8	9	

Overview test
Questions 1 and 2 are grade F and questions 3 and 4 are grade E.
1 Draw the lines of symmetry on the following shapes.
   a   b   c   d

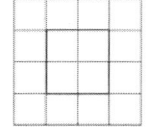

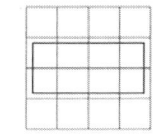

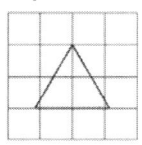

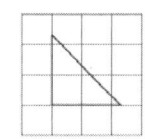

2 Write down the order of rotational symmetry of the shapes in question 1.
3 a State the coordinates of the point of intersection of the lines $x = 3$ and $y = -1$.
  b State the coordinates of the point of intersection of the lines $x = 3$ and $y = x$.
  c State the coordinates of the point of intersection of the lines $x = 3$ and $y = -x$.

Answers to overview test
1 a   b   c   d

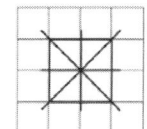

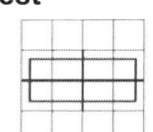

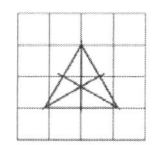

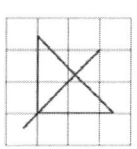

2 a 4 b 2 c 3 d 1   3 a (3, −1) b (3, 3) c (3, −3)

Why this chapter matters
Cross-curricular: This topic links to the sciences, Design and Technology and also Arts and Crafts.
Introduction: Students will recognise few practical applications of transformations other than perhaps reflection in relation to using a mirror. However, encourage them to realise that transformations are an important part of our lives and occur all around us. Discuss and, with students, carry out the instructions in the introductory text of the Student Book.
Discussion points: How can you show that two triangles are congruent? How do you translate a 2D shape? Or reflect a 2D shape in a mirror line? How do you rotate a 2D shape about a point? Or enlarge a 2D shape by a scale factor?
Plenary: Suggest that students do extra research about the Möbius strip, mentioned in the introductory text, and also try to find as many information about transformations and examples as they can.

Answers to the quick check test can be found in at the end of the chapter.

© HarperCollins*Publishers* 2012

11.1 Congruent triangles

Curriculum references

Functional Skills standards
I2 Draw conclusions and provide mathematical justifications.

PLTS
Independent enquirers, Creative thinkers, Effective participators

APP
Shape, space and measure L8.1

Collins references
- Student Book pages 236–238
- Interactive Book: Matching Pairs

Learning objectives
- Show that two triangles are congruent. [Grade B–A]

Learning outcomes
- Students **must** be able to identify congruent triangles. [Grade B]
- Students **should** be able to state the reason why two triangles are congruent. [Grade B]
- Students **could** be able to make reasoned arguments involving congruent triangles. [Grade A]

Key words
- congruent

Prior knowledge
Students should be able to recognise congruent shapes and understand the meaning of congruency.

Common mistakes and remediation
Students often think that to prove congruency it is enough to show that the three sets of angles are the same in both triangles. This is untrue, as it merely shows that the two triangles are similar. So it is important to reinforce the four conditions that establish congruency, regularly.

Useful tips
Students could use tracing paper to check if shapes are congruent. Remind them that if they have three corresponding measurements (including one side) that are the same in two triangles, the triangles are congruent.

Functional maths and problem-solving help
There are no functional maths (FM) questions in Exercise 11A, since it is difficult at this stage to set the theory into a relevant and sensible context. Ensure that students read problem-solving (PS) question 7 carefully, noting all the information they are given.

Starter
- Remind students of the definition of congruent shapes, i.e. identical shapes. Draw these two pairs of triangles on the board and ask if the shapes are congruent. Most students should agree.

- Draw another pair of triangles, one a reflection of the other. Students may be unsure as to whether these triangles are congruent.

© HarperCollins*Publishers* 2012

Lesson 11.1 Congruent triangles

- Use this simple exercise to emphasise the importance of clearly understanding what it means if shapes are congruent or identical. Stress that it does not matter that one triangle is reflected; the sides and angles remain identical and therefore these triangles are congruent.

Main lesson activity
- The following conditions necessary for congruent triangles leads to formal proofs. **Less able** students could omit it. Remind students that if two triangles are congruent they are identical, with three corresponding angles equal and three corresponding sides equal. In fact, three identical measurements on both triangles will show that they are congruent. There are four conditions to show that two triangles are congruent.
- Work through **Condition 1** in the Student Book. Draw the triangles on the board labelling them ABC and XYZ and ask students to list the pairs of equal angles. Show **more able** students the convention for showing congruent triangles for Condition 1. △ ABC ≡ △ XYZ (SSS)
- Now work through **Condition 2**. Again draw the triangles on the board labelling them ABC and XYZ. Ask for pairs of equal sides and angles. Show **more able** students the convention for showing congruent triangles for Condition 2. △ ABC ≡ △ XYZ (SAS)
- Look at the two possible diagrams for **Condition 3**, drawing them on the board. Label the triangles ABC and XYZ and ask for the third angle and pairs of equal sides. Show **more able** students the convention for showing congruent triangles for Condition 3. △ ABC ≡ △ XYZ (ASA) or △ ABC ≡ △ XYZ (AAS)
- Finally work through **Condition 4**, again displaying the triangles and labelling them ABC and XYZ. Ask the students to give the pairs of equal angles and sides. Show **more able** students the convention for showing congruent triangles for Condition 4. △ ABC ≡ △ XYZ (RHS)
- Look at Example 1 with the students. Now show **more able** students how to prove and set out a geometrical problem, using congruent triangles. In the diagram, AB is parallel to CD and AE = ED.
  Prove that triangles ABE and CDE are congruent.
  AE = ED (given)
  ∠ BAE = ∠ CDE (alternate angles)
  ∠ AEB = ∠ CED (opposite angles)
  So △ ABE ≡ △ CDE (ASA)
- Now students can complete Exercise 11A, in which question 7 is PS and 8 is an assessing understanding (AU) question. **Less able** students can complete the questions in pairs. Provide support as needed.

Plenary
- Ask students to consider whether these triangles are congruent.
- The answer is not obvious, as none of the congruency rules initially seem to fit.
  However, the remaining angles to be found must both be acute so that there will be no ambiguity when calculating the non-included angle, and hence none regarding the included angle. So, according to the SAS rule these triangles are congruent.

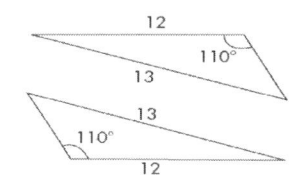

Answers
Answers to Exercise 11A can be found at the end of this chapter.

11.2 Translations

Curriculum references

Functional Skills standards
A2 Use appropriate checking procedures and evaluate their effectiveness at each stage.

PLTS
Independent enquirers, Creative thinkers, Effective participators

APP
Shape, space and measure L6.7

Collins references
- Student Book pages 239–241
- Interactive Book: Paper Animation

Learning objectives
- Translate a 2D shape. [Grade C]

Learning outcomes
- Students **must** be able to use a vector to describe a translation. [Grade C]
- Students **should** be able to draw the result of a translation. [Grade C]
- Students **could** be able to use vectors in practical situations. [Grade C]

Key words
- transformation
- translation
- vector

Prior knowledge
Students need very little prior knowledge to begin basic work on transformations. Recent experience of work involving Cartesian coordinates might be useful, but is not essential.

Common mistakes and remediation
Less able students may be confused by the terms 'transformation' and 'translation'. Ensure that they understand that a translation is a specific type of transformation, as are a rotation and a reflection.

Useful tips
To build confidence, encourage students to use the terms, 'transformation' and 'translation', in the correct context. Remind them that when a triangle has simply been translated, it has not been reflected or rotated.

Functional maths and problem-solving help
Functional maths (FM) questions 7 and 8 of Exercise 11B are straightforward and structured. Ensure that students read problem-solving (PS) questions 4 and 5 carefully. If they do this, then these questions should also be straightforward.

Starter
- Begin by discussing the general idea of the transformation of an object, and what it means. Many students will have come across different types of transformations before, so ask them to name specific types.
- Emphasise that students will study each of the four basic transformations – translation, reflection, rotation and enlargement – in turn. They need to be able to recognise, describe and perform these transformations.

Main lesson activity
- Explain that a translation is a transformation that moves a shape from one position to another.
- Triangle A has been translated to triangle B.

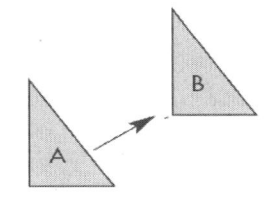

Point out that under the translation, the triangle keeps its orientation: it has not been reflected or rotated. Also, triangle A is congruent to triangle B. Say, "Triangle A has been mapped onto triangle B by a translation." Explain that this just means one triangle has been changed by being slid into another position.

© HarperCollinsPublishers 2012

Lesson 11.2 Translations

- Discuss how to describe different translations on a coordinate grid.

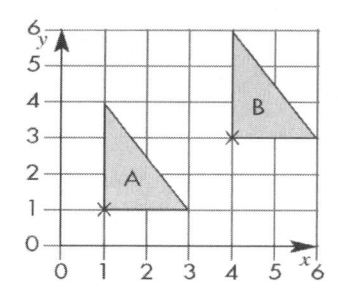

Put a cross on a vertex on both triangles. Explain that to move from its position in triangle A to triangle B the cross moves 3 squares to the right and 2 squares up. Show students that it is the same movement for the other two vertices.

- This movement or translation is represented by the vector $\binom{3}{2}$.
  A vector has two distinct components:

  $\binom{x}{y}$ → number of squares moved horizontally
  → number of squares moved vertically

  If x is positive, it describes a movement to the right.
  If x is negative, it describes a movement to the left.
  If y is positive, it describes a movement up.
  If y is negative, it describes a movement down.
- Take students through Example 2, explaining how this works.
  Reinforce with **less able** students that a vector is not a coordinate pair.
- Now students can complete Exercise 11B. Suggest that **less able** students complete questions 1–3 in pairs before moving on to complete the exercise. Provide support as needed.

Plenary
- Draw on the board any two congruent shapes, for example:

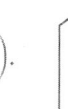

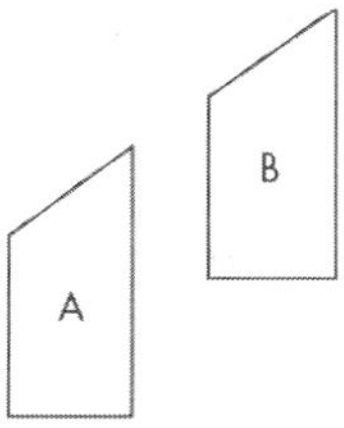

- Tell students that shape A is mapped onto shape B by the vector $\binom{x}{y}$.

- Ask students to give the vector that would map shape B onto shape A.
  Answer: $\binom{-x}{-y}$.

Answers
Answers to Exercise 11B can be found at the end of this chapter.

© HarperCollins*Publishers* 2012

11.3 Reflections

Curriculum references

Functional Skills standards
A2 Use appropriate checking procedures and evaluate their effectiveness at each stage.

PLTS
Independent enquirers, Creative thinkers, Effective participators

APP
Shape, space and measure L6.7

Collins references
- Student Book pages 242–245
- Interactive Book: Matching Pairs

Learning objectives
- Reflect a 2D shape in a mirror line. [Grade D–C]

Learning outcomes
- Students **must** be able to reflect a shape on a coordinate grid. [Grade D]
- Students **should** be able to draw the result of a reflection on a coordinate grid. [Grade D]
- Students **could** be able to find and describe the results of repeated reflections. [Grade C]

Key words
- image
- object
- mirror line
- reflection

Prior knowledge
Students should have completed Lesson 11.2 on translations and be familiar with equations of lines of the type $y = x$, $y = -x$, $y = a$ and $x = b$.

Common mistakes and remediation
Students often draw the image and the object at different distances from the mirror line. When reflecting shapes in diagonal lines, they may also draw them like this.

Demonstrate that if they use a horizontal line it is much easier to see that the reflection should be like this.

Useful tips
When reflecting shapes in diagonal lines, it is easier if students turn the page so that the mirror line is horizontal or vertical.

Functional maths and problem-solving help
There are no functional maths (FM) questions in Exercise 11C. Question 1 shows how designs can be built up using reflection. In this question, encourage **more able** students to extend the lines being used as mirror lines to diagonals outside the shape.
Ensure that students read problem-solving (PS) questions 5, 11 and 12 carefully, which should then be quite straightforward.

Starter
- Draw a set of axes, each labelled from −5 to 5. Draw some horizontal and vertical lines and ask students for their equations.
  Make sure they understand that lines of the form $x = b$ are vertical and those of the form $y = a$ are horizontal. Finish with $y = x$ and $y = -x$.

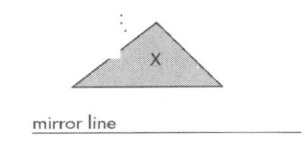

© HarperCollins*Publishers* 2012

Lesson 11.3 Reflections

Main lesson activity
- Most students will have covered this topic before, but less able students may find this revision section useful.
- Explain that a reflection is a transformation that changes a shape into a mirror image of itself.
  The original shape is the object and the reflected shape is the image.
  Triangle X has been reflected in the mirror line to give triangle Y.
  Point out that, under the reflection, each vertex on the object is the same distance from the mirror line as each corresponding vertex on the image. Point out that triangle X is congruent to triangle Y. Say: Triangle X has been mapped onto triangle Y by a reflection in the mirror line. Look at Example 3 in the Student Book.
- Discuss how to describe different reflections on a coordinate grid. Establish that the best way to do this is to give the equation of the mirror line. Demonstrate, with an example.
- Triangle B is the reflection of triangle A in the line $x = -1$.
  Triangle C is the reflection of triangle A in the line $y = 1$.

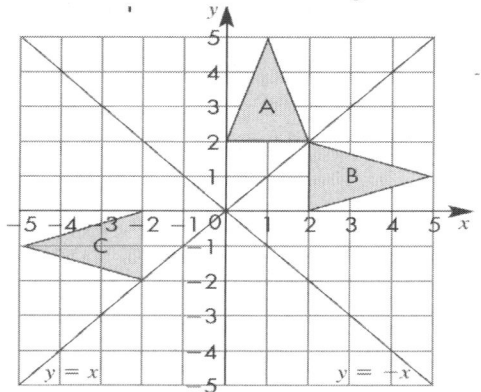

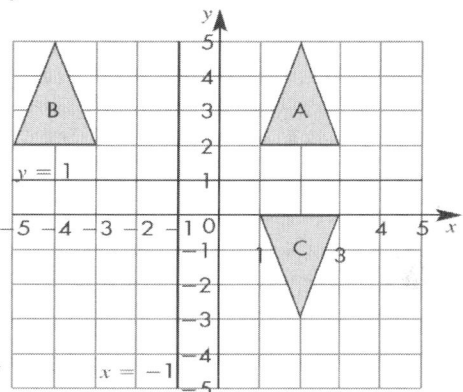

- **Less able** students find it more difficult to reflect shapes in a diagonal mirror line. Demonstrate with another example.
  Triangle B is the reflection of triangle A in the line $y = x$.
  Triangle C is the reflection of triangle A in the line $y = -x$.
- Explain how to count across the squares from a vertex to the mirror line and then count across the same number of squares to find the position of the reflection of that vertex.
- Now students can complete Exercise 11C, in which 5, 11 and 12 are PS questions and 3, 4 and 6 are assessing understanding (AU) questions. Suggest that **less able** students complete questions 1–4 in pairs before moving on to complete the exercise.

Plenary
- Ask students to reflect some points in the line $y = x$ and write down the coordinates of the object points with their image points. Ask them what they notice.
- Ask **more able** students, who may have covered this in the exercise, to explain to the class that if the point (a, b) is reflected in the line $y = x$, then the image point is (b, a).

Answers
Answers to Exercise 11C can be found at the end of this chapter.

© HarperCollinsPublishers 2012

11.4 Rotations

Curriculum references

Functional Skills standards
A2 Use appropriate checking procedures and evaluate their effectiveness at each stage.

PLTS
Independent enquirers, Creative thinkers, Effective participators

APP
Shape, space and measure L6.7

Collins references
- Student Book pages 245–247
- Interactive Book: Multiple Choice Questions

Learning objectives
- Rotate a 2D shape about a point. [Grade D–C]

Learning outcomes
- Students **must** be able to rotate a shape on squared paper. [Grade D]
- Students **should** be able to rotate a shape on a coordinate grid. [Grade C]
- Students **could** be able to describe a rotation that will take one shape onto another. [Grade C]

Key words
- angle of rotation
- anticlockwise
- centre of rotation
- clockwise
- rotation

Prior knowledge
Ideally, students should have covered Lessons 11.2 and 11.3 on translations and reflections.

Common mistakes and remediation
When describing a rotation, students often forget that they need to provide three pieces of information: the centre of rotation, the angle of rotation and the direction of the rotation. Stress this point during the lesson.

Useful tips
Encourage students to use tracing paper for any problems involving a rotation and ask them to note that this is allowed in examinations.

Functional maths and problem-solving help
There are no functional maths (FM) questions in Exercise 11D. Question 1 extends the similar question in the last lesson by using rotation in addition to reflection to create a design. Encourage students to use tracing paper if they find the question difficult.
Ensure that students read problem-solving (PS) questions 3, 9 and 10 carefully, noting all the information they are given. The questions are quite straightforward.

Starter
- Ask students to draw a T-shape, like this.
- Ask them to draw another T-shape to add to the first, to make a shape with rotational symmetry of order 2.
  Here are some examples of what students may come up with.
- Ask students what the difference is between the diagrams. Establish that they are rotations, and the concept of the centre of rotation.
  Mark the centres on the diagrams.

© HarperCollins*Publishers* 2012

Lesson 11.4 Rotations

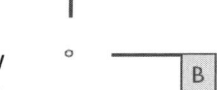

Main lesson activity
- Most students will have covered this topic before, but **less able** students may find this revision section useful.
- Explain that a rotation is a transformation that changes a shape by turning it about a fixed point. Again, the original shape is the object and the reflected shape is the image. 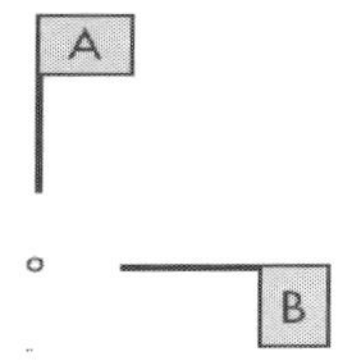
- Flag A has been rotated about the point O to give flag B. Show students how to trace flag A and then turn it about O to give flag B. The fixed point O is the centre of rotation. Point out that flag A is congruent to flag B.
- Stress that to describe a rotation it is important to give the angle and the direction of the turn. The direction is specified by either a clockwise turn or an anticlockwise turn. Here flag A has been rotated 90° clockwise.
- Explain that describing a rotation requires three pieces of information: the centre of rotation, the angle of rotation and the direction of the rotation. Say: Flag A has been mapped onto flag B by a rotation of 90° clockwise about the point O. Note that in examinations, the angle of rotation is usually 90° or 180°.
- Look at Example 4 in the Student Book for another example.
- Discuss how to describe different rotations on a coordinate grid. Establish that the best way to do this is to give the centre of rotation as a coordinate point. Demonstrate with an example, using tracing paper. 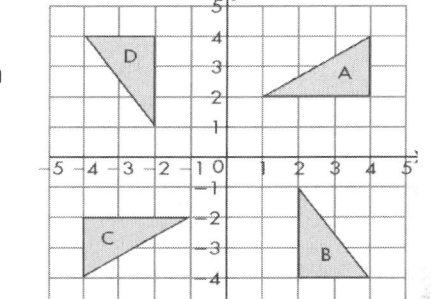
- Triangle B is a rotation of triangle A by 90° clockwise about (0, 0). Note that it is acceptable to describe (0, 0) as the origin.
- Triangle C is a rotation of triangle A by 180° about (0, 0). Note that there is no need for a direction for a 180° rotation.
- Triangle D is a rotation of triangle A by 90° anticlockwise about (0, 0). Note that 270° clockwise is also acceptable.
- **Less able** students find it more difficult to rotate shapes about points other than the origin.

  Demonstrate with another example.
  Triangle B is a rotation of triangle A by 90° clockwise about (0, −1).
  Triangle C is a rotation of triangle A by 180° about (1, 1). 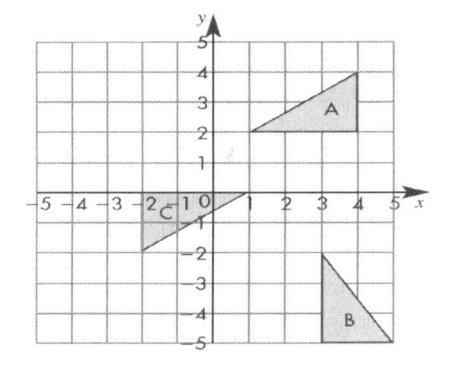
- Now students can complete Exercise 11D. Suggest that **less able** students complete the questions in pairs.

Plenary
- Ask students to rotate some points through 180° about the origin and write down the coordinates of the object points with their image points. Ask them what they notice.
- Ask **more able** students, who may have covered this idea in the exercise, to explain to the class that if the point (a, b) is rotated through 180° about the origin, then the image point is $(-a, -b)$.

Answers
Answers to Exercise 11D can be found at the end of this chapter.

© HarperCollins*Publishers* 2012

169

11.5 Enlargements

Curriculum references

Functional Skills standards
A2 Use appropriate checking procedures and evaluate their effectiveness at each stage.

PLTS
Independent enquirers, Creative thinkers, Effective participators

APP
Shape, space and measure L6.6

Collins references
- Student Book pages 248–253
- Interactive Book: 10 Quick Questions

Learning objectives
- Enlarge a 2D shape by a scale factor. [Grade D–B]

Learning outcomes
- Students **must** be able to enlarge a shape with a positive integer scale factor. [Grade D]
- Students **should** be able to enlarge a shape with a fractional scale factor. [Grade C]
- Students **could** be able to enlarge a shape with a fractional scale factor. [Grade B]

Key words
- centre of enlargement
- enlargement
- scale factor

Prior knowledge
Ideally, students should have covered Lessons 11.2–11.4 on translations, reflections and rotations.

Common mistakes and remediation
Often, students do not use the centre of enlargement that is provided. Encourage them to mark the given centre clearly each time.

Useful tips
Suggest that when they have drawn an enlargement, they should add the rays to check that they pass through the centre of enlargement. Refer students to the following information as per the plenary.

Transformation	Information needed
Translation	column vector
Reflection	equation of mirror line
Rotation	angle, direction, centre of rotation
Enlargement	scale factor, centre of enlargement

Functional maths and problem-solving help
There are no functional maths (FM) questions in Exercise 11E. Question 1 extends the similar question in the last two lessons by using enlargement in addition to reflection and rotation to create a design. Encourage students to use tracing paper for the reflection and the rotation if they are finding the question difficult. Ensure that students read problem-solving (PS) question 9 carefully, which should then be quite straightforward.

Starter
- Ask students what they think it means to enlarge a shape by a scale factor of 2. Most are likely to have the (correct) idea that the sides of the shapes will double in length.
- Then ask what they think it means to enlarge a shape by a scale factor of $\frac{1}{2}$. Now there may be some uncertain answers. This could provide a useful lead-in to an initial summary of the types of enlargement.

© HarperCollins*Publishers* 2012

Lesson 11.5 Enlargements

Main lesson activity
- Explain that an enlargement is a transformation that changes the size of a shape. Again the original shape is the object and the enlarged shape is the image. See Example 5.
- Draw on the board two triangles, as shown. Explain that the lengths of the sides of triangle Y are twice the lengths of the corresponding sides of triangle X. Say: Triangle Y has been enlarged by a scale factor of 2. Emphasise that triangle X is not congruent to triangle Y. 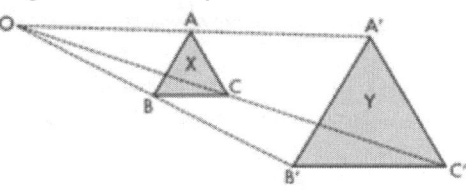
- Explain that the dotted lines passing through the corresponding vertices of the two triangles are rays and they intersect at the centre of enlargement. OA' = 2 × OA, OB' = 2 × OB and OC' = 2 × OC.
- Point out that describing an enlargement requires two pieces of information: the centre of enlargement and a scale factor. Say: Triangle X has been mapped onto triangle Y by an enlargement of a scale factor of 2.
- Go through Example 6 to show students how to use the ray method to enlarge a shape.
- Discuss how to describe enlargements on a coordinate grid. Establish that the best way to do this is to give a coordinate point as the centre of enlargement. Demonstrate with an example.

  Triangle Y is an enlargement of triangle X by a scale factor 3 about (0, 0).
  Show students how to count squares from (0, 0) to each vertex of X and then repeat this three times from (0, 0) to each vertex of Y. 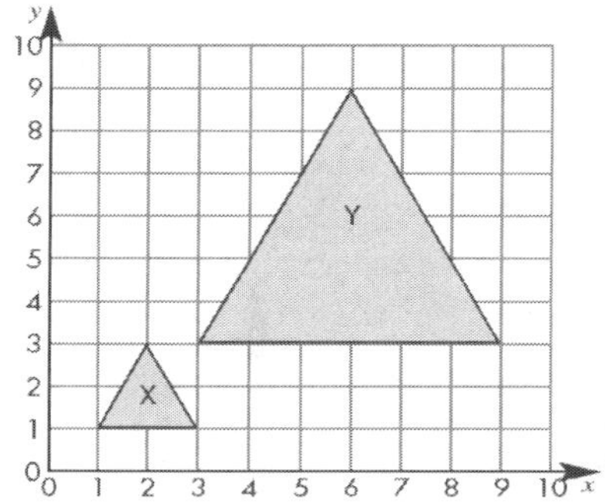
- Point out that as the centre of enlargement is the origin, simply multiplying the coordinates of the vertices of X by 3 gives the vertices of Y.
- Let **more able** students go through Example 7, to see how to enlarge a shape about any coordinate point.
- Explain that enlargements can also make shapes smaller. Use the example on the right to explain that triangle X is an enlargement of triangle Y by scale factor 1/3 about (0, 0).

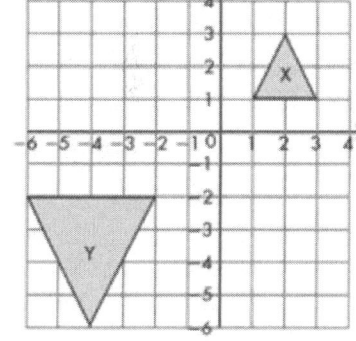

- Show **more able** students how to enlarge shapes by a negative scale factor.
  Triangle Y is an enlargement of triangle X by a scale factor −2 about (0, 0).
  Demonstrate how to count squares from (0, 0) to each vertex of X and then repeat this twice from (0, 0) but in the opposite direction to each vertex of Y.
  Point out that a negative scale factor inverts the original shape on the opposite side of the centre of enlargement. As the centre of enlargement is the origin, multiplying the coordinates of the vertices of X by −2 gives the vertices of Y.
- Let **more able** students go through Example 8, individually or in pairs, to see how to enlarge a shape by a negative scale factor about any coordinate point.
- Discuss Example 9. Now students can do Exercise 11E. **Less able** students can work in pairs.

Plenary
- Summarise, as shown under useful tips, the information needed to describe the four transformations.

Answers
Answers to Exercise 11E can be found at the end of this chapter.

© HarperCollins*Publishers* 2012

171

11.6 Combined transformations

Curriculum references

Functional Skills standards
R3 Select a range of mathematics to find solutions.

PLTS
Independent enquirers, Creative thinkers, Effective participators

APP
Not in assessment criteria.

Collins references
- Student Book pages 253–255
- Interactive Book: Worked Exam Questions

Learning objectives
- Combine transformations. [Grade D–B]

Learning outcomes
- Students **must** be able to describe a sequence of transformations that will map one shape onto another. [Grade C]
- Students **should** be able to draw the results of a sequence of transformations. [Grade C]
- Students **could** be able to solve problems involving sequences of transformations. [Grade B]

Key words
- enlargement
- transformation
- translation
- rotation
- reflection

Prior knowledge
Students should have covered Lessons 11.2–11.5 on translations, reflections, rotations and enlargements.

Common mistakes and remediation
Students often miss out some of the information required to explain rotations and enlargements fully. Throughout the lesson, reinforce the information in the table on the right.

Transformation	Information needed
Translation	column vector
Reflection	equation of mirror line
Rotation	angle, direction, centre of rotation
Enlargement	scale factor, centre of enlargement

Useful tips
Display a summary of all the transformations covered so far. Always have tracing paper available.

Functional maths and problem-solving help
There are no functional maths (FM) questions in Exercise 11F, since it is difficult at this stage to set the theory into a relevant and sensible context. Ensure that students read problem-solving (PS) questions 7 and 8 carefully, and then these questions should be quite straightforward.

Starter
- Display a coordinate grid, with x-axis and y-axis labelled appropriately. Add two squares: A with vertices at (−2, 0), (−2, 2), (0, 0) and (0, 2) and B with vertices (0, 0), (0, 2), (2, 2), (2, 0). Ask students if B is a transformation of A and, if so if they can describe the transformation.
- Ask, "Is it possible to describe it as a translation? Similarly, can it be described as a reflection, a rotation or an enlargement?" Expected answers include:
    - translation by the vector $\binom{2}{0}$
    - reflection in the line $x = 0$ or the y-axis
    - rotation of 90° clockwise about the origin, or 90° anticlockwise about (0, 2)
    - no enlargement, as B is not enlarged.

© HarperCollins*Publishers* 2012

Lesson 11.6 Combined transformations

Main lesson activity
- Explain that this lesson covers all the transformations they have met and combinations of them.
- Consider enlargements first, as they are the only transformations that change the size of the object and, hence, they are easily identified. Having ruled out enlargement, a translation is the next easiest to spot.
- Sometimes a shape may be transformed more than once, by combined transformations. In GCSE examinations only two combined transformations will be expected in a question. Work through two examples.
- **Example 1:** Display this diagram. Give **less able** students a copy or let them draw it themselves on squared paper.
  Ask students to describe fully the *single* transformation that maps triangle A onto triangle B: a rotation of 90° clockwise about the origin or (0, 0). Now ask students to describe fully the *single* transformation that maps triangle B onto triangle C: a reflection in the line $x = 0$ or the y-axis.
  Explain that the *combined* transformation that maps triangle A onto triangle C is: a rotation of 90° clockwise about the origin, followed by a reflection in the line $x = 0$. Now ask students to give the single transformation that maps triangle C onto triangle A. Make sure they give a full description of the reflection: a reflection in the line $y = -x$.

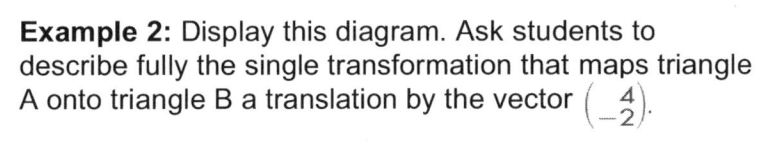

- **Example 2:** Display this diagram. Ask students to describe fully the single transformation that maps triangle A onto triangle B a translation by the vector $\begin{pmatrix} 4 \\ -2 \end{pmatrix}$.

  Now ask students to describe fully the single transformation that maps triangle B onto triangle C: an enlargement by scale factor 2 about the origin or (0, 0). Explain that the combined transformation that maps triangle A onto triangle C is: a translation by the vector $\begin{pmatrix} 4 \\ -2 \end{pmatrix}$ followed by an enlargement by scale factor 2 about the origin. Now ask students to give the single transformation that maps triangle C onto triangle A.
  Make sure they give a full description of the enlargement: an enlargement by scale factor ½ about (−8, 4).

- Students can complete Exercise 11F. **Less able** students can work in pairs.

Plenary
- Ask students to work in pairs to design a poster to summarise all the transformations they have covered.
- If they need help, show this thought-process diagram and ask students to complete it.

Answers
Answers to Exercise 11F can be found at the end of this chapter.

© HarperCollins*Publishers* 2012

173

Problem solving
Developing photographs

Curriculum references

Functional Skills standards
R1 Understand routine and non-routine problems in familiar and unfamiliar contexts and situations.
A1 Apply a range of mathematics to find solutions.

PLTS
Independent enquirers, Self-managers, Team workers

APP
Shape, space and measure L7.3; Using and applying mathematics L8.3.

Collins references
- Student Book pages 262–263

Learning objectives
- Use different scale factors (positive whole number, fractional and negative) to enlarge objects, understanding the effect that the enlargement is having on the image.

Key words
- enlargement
- image
- object
- scale factor

Functional maths help
Activity 3 will work best if students work in pairs. If you take sheets of A3, A4, A5, A6 (and so on, smaller and smaller in size) paper and line them up on the bottom left hand corner so that they all overlap you can clearly see the rays of enlargement, as shown in the diagram on the right.

You can use this method to demonstrate enlargements of photographs. If students do not have access to digital cameras, either provide some similar stimuli or ask them to design their own simple logo or shape to use. Say that they should make sure the photographs are not too complicated. Other areas that students may be interested in the fashion world or nature.

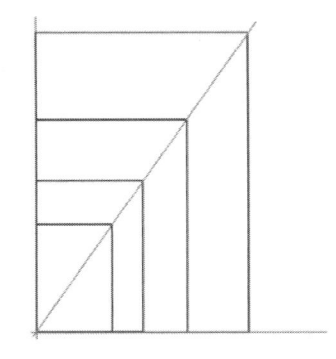

Lesson plans
Build
Start by telling students that enlargement is used in many aspects of life. How many can they think of? They should discuss these with the person sitting next to them, or you could discuss it as a class.
You may need to ask students to draw simple shapes and enlarge them using positive scale factors. Or you could provide enlargements and ask them to look at the ratio of the lines compared to the scale factor. Once students understand this, they should be ready for introduction to the centre of enlargement, as shown in the diagram in 'Functional maths help' (above).
Students can investigate what happens to the image as you move the centre of enlargement to various positions; on the shape, outside the shape, inside the shape, and so on.

Activity 1
Ask students to answer the questions, under the heading 'Getting started' in the Student Book.

Activity 2
Now ask students to look at the images in the continuation of 'Getting started', and to explain which photograph is the odd one out, and why. Then they should look at the images further down, and explain how they can tell that these images are all based on the same original photograph.

Problem solving Developing photographs

Activity 3
Now refer students to 'Your task'. Explain that students must take photographs and make a display or poster to explain enlargement. They also need to answer at least two of the questions that have been provided. Refer students who do not have access to a digital camera to the 'Note' about using an alternative method to create their display or poster.

Apply
Students will probably have an understanding of enlargements with whole number positive scale factors and understand that the lengths double with a scale factor of 2, will triple with a scale factor of 3, and so on. However, **less able** students may struggle with centres of enlargement.
Use this knowledge to support students in constructing enlargements, using a centre. From here, **more able** students should be able to make the link to a scale factor of $\frac{1}{2}$ and other unit fractions.

Support students with their understanding of the links with area and enlargement by looking at growing shapes, where they should try and fit the object inside the image, as shown here.

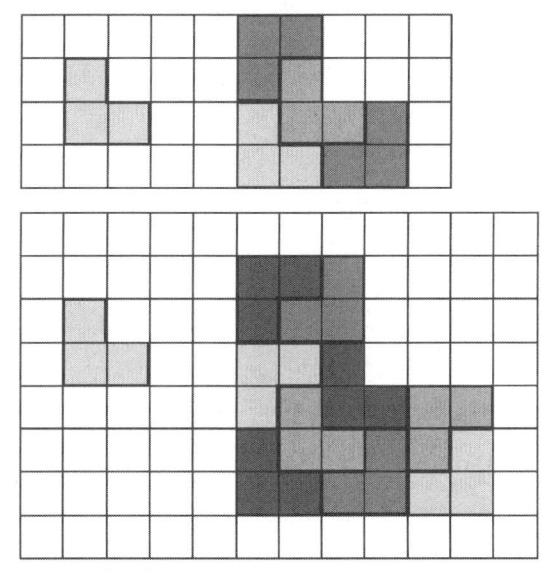

This clearly shows:
Area of image = area of object × scale factor^2

Master
In order to demonstrate mastery of the learning objectives, students should be able to:
- Understand that, when the centre of enlargement is moved, the image is translated in the plane but all images will be congruent.
- Enlarge a shape using a positive scale factor and centre of enlargement (and realise that the two shapes are mathematically similar).
- Know that a scale factor between 0 and 1 will produce a smaller image and that a scale factor greater than one will produce a bigger image.
- Understand how the scale factor links to the changes in lengths of the image, i.e. a scale factor of 2 will double all lengths.
- Understand how the scale factor links to the changes in area of the image and deduce the following formula:
  Area of image = area of object × scale factor^2

Plenary
Ask students to consider what would happen with a negative scale factor.
Ask students to share their displays or posters with the class and explain them.

© HarperCollins*Publishers* 2012

Answers Lessons 11.1 – 11.6

Quick check
Trace shape **a** and check whether it fits exactly on top of the others.

You should find that shape **b** is not congruent to the others.

11.1 Congruent triangles
Exercise 11A
1. a SAS b SSS c ASA
   d RHS e SSS f ASA
2. a SSS; A to R, B to P, C to Q
   b SAS; A to R, B to Q, C to P
3. a 60° b 80° c 40° d 5 cm
4. a 110° b 55° c 85° d 110° e 4 cm
5. SSS or RHS
6. SSS or SAS or RHS
7. For example, use △ADE and △CDG. AD = CD (sides of large square), DE = DG (sides of small square), ∠ADE = ∠CDG (angles sum to 90° with ∠ADG), so △ADE ≡ △CDG (SAS), so AE = CG
8. AB and PQ are the corresponding sides to the 42° angle, but they are not equal in length.

11.2 Translations
Exercise 11B
1. a i $\begin{pmatrix} 1 \\ 3 \end{pmatrix}$ ii $\begin{pmatrix} 4 \\ 2 \end{pmatrix}$ iii $\begin{pmatrix} 2 \\ -1 \end{pmatrix}$
   iv $\begin{pmatrix} 5 \\ 1 \end{pmatrix}$ v $\begin{pmatrix} -1 \\ 6 \end{pmatrix}$ vi $\begin{pmatrix} 4 \\ 6 \end{pmatrix}$
   b i $\begin{pmatrix} -1 \\ -3 \end{pmatrix}$ ii $\begin{pmatrix} 3 \\ -1 \end{pmatrix}$ iii $\begin{pmatrix} 1 \\ -4 \end{pmatrix}$
   iv $\begin{pmatrix} 4 \\ -2 \end{pmatrix}$ v $\begin{pmatrix} -2 \\ 3 \end{pmatrix}$ vi $\begin{pmatrix} 3 \\ 3 \end{pmatrix}$
   c i $\begin{pmatrix} -4 \\ -2 \end{pmatrix}$ ii $\begin{pmatrix} -3 \\ 1 \end{pmatrix}$ iii $\begin{pmatrix} -2 \\ -3 \end{pmatrix}$
   iv $\begin{pmatrix} 1 \\ -1 \end{pmatrix}$ v $\begin{pmatrix} -5 \\ 4 \end{pmatrix}$ vi $\begin{pmatrix} 0 \\ 4 \end{pmatrix}$
   d i $\begin{pmatrix} 3 \\ 2 \end{pmatrix}$ ii $\begin{pmatrix} -4 \\ 2 \end{pmatrix}$ iii $\begin{pmatrix} 5 \\ -4 \end{pmatrix}$
   iv $\begin{pmatrix} -2 \\ -7 \end{pmatrix}$ v $\begin{pmatrix} 5 \\ 0 \end{pmatrix}$ vi $\begin{pmatrix} 1 \\ -5 \end{pmatrix}$

2. [graph showing triangles P, Q, C, A, B, R, S on coordinate axes]

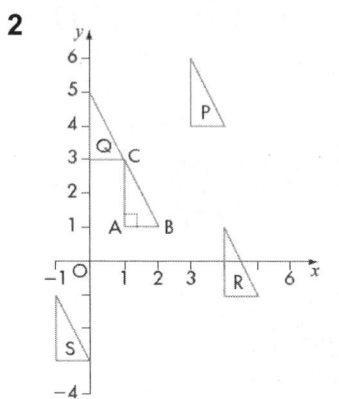

3. a $\begin{pmatrix} -3 \\ -1 \end{pmatrix}$ b $\begin{pmatrix} 4 \\ -4 \end{pmatrix}$ c $\begin{pmatrix} -5 \\ -2 \end{pmatrix}$ d $\begin{pmatrix} 4 \\ 7 \end{pmatrix}$
   e $\begin{pmatrix} -1 \\ 5 \end{pmatrix}$ f $\begin{pmatrix} 1 \\ 6 \end{pmatrix}$ g $\begin{pmatrix} -4 \\ 4 \end{pmatrix}$ h $\begin{pmatrix} -4 \\ -7 \end{pmatrix}$

4. 10 × 10 = 100 (including $\begin{pmatrix} 0 \\ 0 \end{pmatrix}$)

5. Check students' designs for a *Snakes and ladders* board.

6. $\begin{pmatrix} -x \\ -y \end{pmatrix}$

7. $\begin{pmatrix} -300 \\ -500 \end{pmatrix}$

8. $\begin{pmatrix} -1 \\ 4 \end{pmatrix}$

11.3 Reflections
Exercise 11C

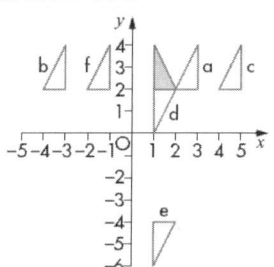

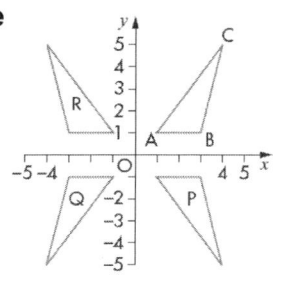

1 [graph showing reflected triangles labeled a, b, c, d, e, f]

2 a–e [graph showing triangles A, B, C, P, Q, R]

f Reflection in the y-axis

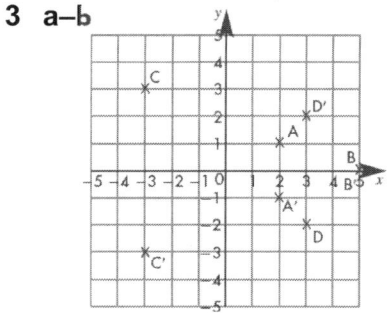

3 a–b [graph showing points A, B, C, D and A', B', C', D']

c y-value changes sign **d** $(a, -b)$

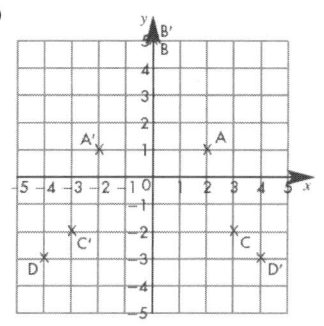

4 a–b [graph showing points A, B, C, D and A', B', C', D']

c x-value changes sign **d** $(-a, b)$

5 Possible answer: Take the centre square as ABCD then reflect this square each time in the line, AB, then BC, then CD and finally AD.

6 $x = -1$

7 Possible answer: [diagram]

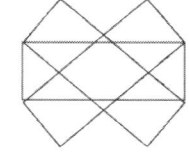

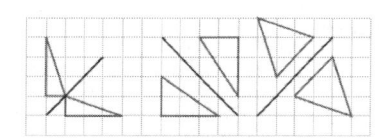

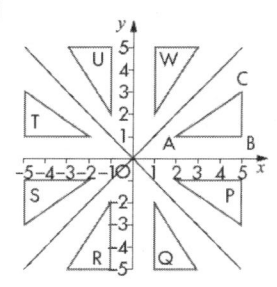

8 [diagram showing reflected triangles]

9 a–i [graph showing triangles A, B, C, P, Q, R, S, T, U, W]

j A reflection in $y = x$

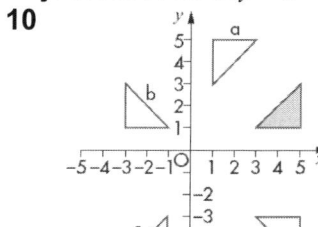

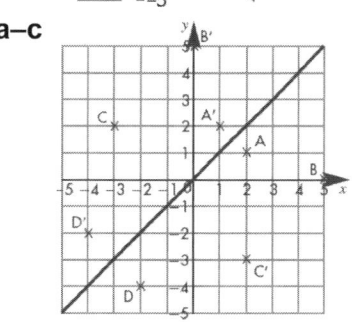

10 [graph showing triangles a, b, c, d]

11 a–c [graph showing points A, B, C, D and A', B', C', D' with line $y = x$]

d Coordinates are reversed: x becomes y and y becomes x

e (b, a)

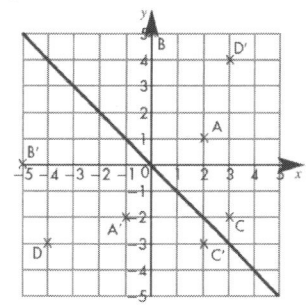

12 a–c [graph showing points A, B, C, D and A', B', C', D' with line $y = -x$]

d Coordinates are reversed and change sign, x becomes $-y$ and y becomes $-x$.

e $(-b, -a)$

Answers Lessons 11.1 – 11.6

11.4 Rotations
Exercise 11D

1 a [diagrams]

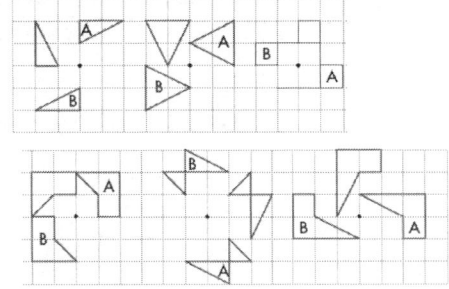

   b i Rotation 90° anticlockwise
   ii Rotation 180°

2 [diagram]

3 Possible answer: If ABCD is the centre square, rotate about A 90° anticlockwise, rotate about new B 180°, now rotate about new C 180°, and finally rotate about new D 180°.

4 [diagram]

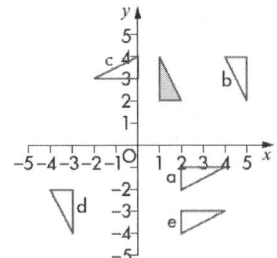

5 a 90° anticlockwise **b** 270° anticlockwise
   c 300° clockwise **d** 260° clockwise

6 a–c i [diagram]

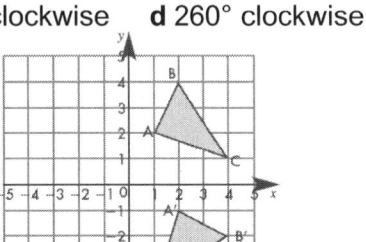

   ii A'(2, −1), B'(4, −2), C'(1, −4)
   iii Original coordinates $(x, y) \Rightarrow (y, -x)$.
   iv Yes

7 i [diagram]

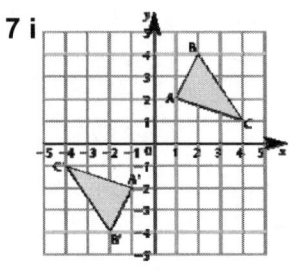

   ii A'(−1, −2), B'(−2, −4), C'(−4, −1)
   iii Original coordinates $(x, y) \Rightarrow (-x, -y)$
   iv Yes

8 i [diagram]

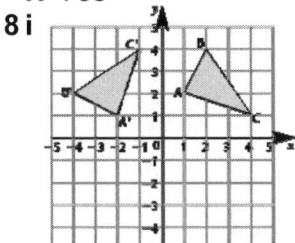

   ii A'(−2, 1), B'(−4, 2), C'(−1, 4)
   iii Original coordinates $(x, y) \Rightarrow (-y, x)$
   iv Yes

9 Show by drawing a shape or use the fact that (a, b) becomes $(a, -b)$ after reflection in the x-axis, and $(a, -b)$ becomes $(-a, -b)$ after reflection in the y-axis, which is equivalent to a single rotation of 180°.

10 Show by drawing a shape or use the fact that (a, b) becomes (b, a) after reflection in the line $y = x$, and (b, a) becomes $(-a, -b)$ after reflection in the line $y = -x$, which is equivalent to a single rotation of 180°.

11 a [diagram]

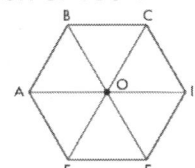

   b i Rotation 60° clockwise about O
   ii Rotation 120° clockwise about O
   iii Rotation 180° about O
   iv Rotation 240° clockwise about O
   c i Rotation 60° clockwise about O
   ii Rotation 180° about O

12 Rotation 90° anticlockwise about (3, −2)

Answers Lessons 11.1 – 11.6

11.5 Enlargements
Exercise 11E

1

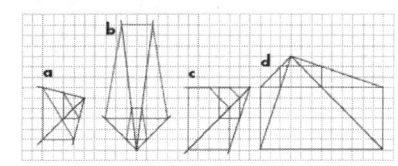

2 a 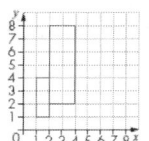 **b**

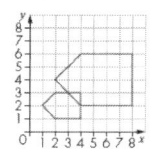

c

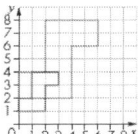

3 a 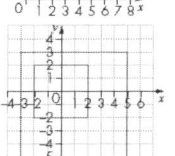 **b**

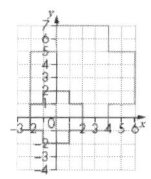

4

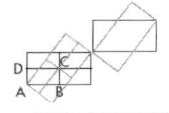

5

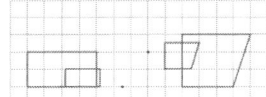

6 a

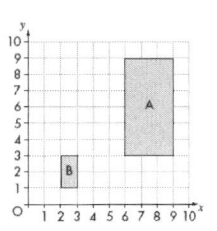

  b 3 : 1
  c 3 : 1
  d 9 : 1

7

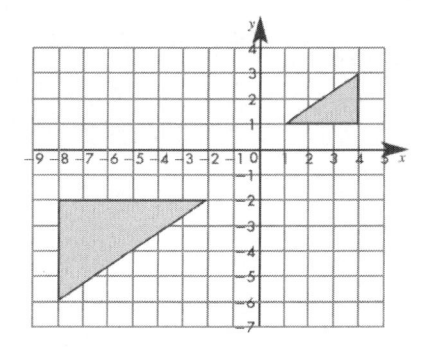

8 a–c

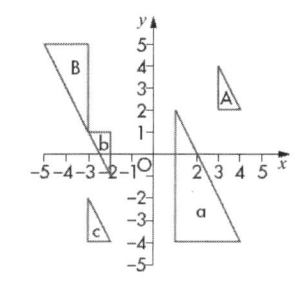

d Scale factor $-\frac{1}{2}$, centre (1, 3)

e Scale factor −2, centre (1, 3)
f Scale factor −1, centre (−2.5, −1.5)
g Scale factor −1, centre (−2.5, −1.5)
h Same centres, and the scale factors are reciprocals of each other.

9 Enlargement, scale factor −2, about (1, 3)

11.6 Combined transformations
Exercise 11F
1. (−4, −3)
2. a (−5, 2) b Reflection in y-axis
3. A: translation $\begin{pmatrix} 1 \\ -2 \end{pmatrix}$

   B: reflection in y-axis
   C: rotation 90° clockwise about (0, 0)
   D: reflection in x = 3
   E: reflection in y = 4
   F: enlargement by scale factor 2, centre (0, 1)
4. a T_1 to T_2: rotation 90° clockwise about (0, 0)
   b T_1 to T_6: rotation 90° anticlockwise about (0, 0)
   c T_2 to T_3: translation $\begin{pmatrix} 2 \\ 2 \end{pmatrix}$
   d T_6 to T_2: rotation 180° about (0, 0)
   e T_6 to T_5: reflection in y-axis
   f T_5 to T_4: translation $\begin{pmatrix} 4 \\ 0 \end{pmatrix}$
5. a–d

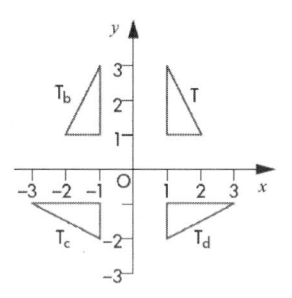

   T_d to T: rotation 90° anticlockwise about (0, 0)
6. (3, 1)
7. Reflection in x-axis, translation $\begin{pmatrix} 0 \\ -5 \end{pmatrix}$, rotation 90° clockwise about (0, 0)
8. Translation $\begin{pmatrix} 0 \\ -8 \end{pmatrix}$, reflection in x-axis, rotation 90° clockwise about (0, 0)
9. a

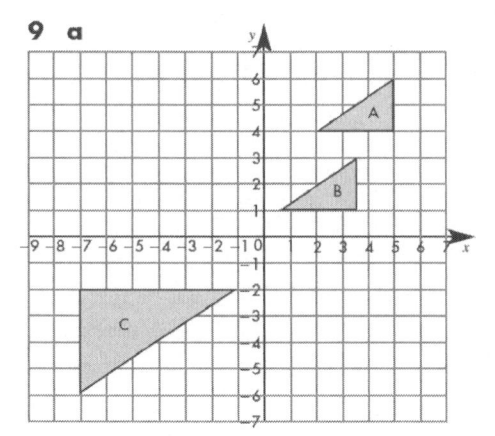

   b Enlargement of scale factor $-\frac{1}{2}$ about (1, 2)

Answers Lessons 11.1 – 11.6

Examination questions

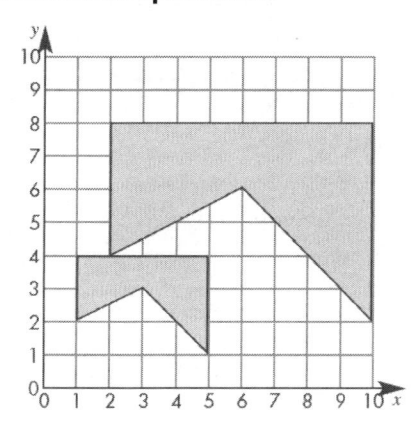

1 [graph shown]

2 a Rotation 90° clockwise about (0, 0)

b $\begin{pmatrix} -5 \\ -4 \end{pmatrix}$

3 a $x = 5$

b $\begin{pmatrix} 5 \\ 0 \end{pmatrix}$

c 180° about (5, 2.5)

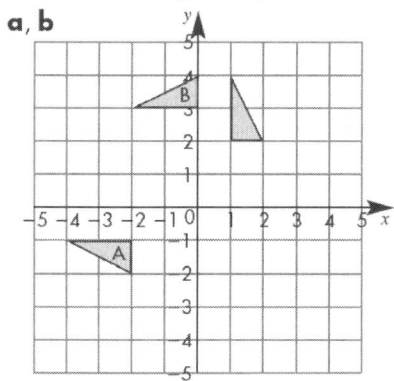

4 a, b [graph shown]

5 a Enlargement scale factor $\frac{1}{3}$ about (−4, 5)

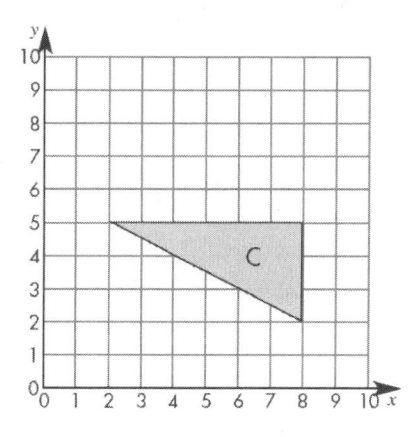

b [graph shown]

6 a Reflection in the line $y = x$

b Rotation of 90° anticlockwise about (1, 1)

c Enlargement scale factor −2 about (0, 0)

7 For example, AB = BC (isosceles triangle), AM = MC (M is midpoint of AC), BM is common, so △ABM ≡ △CBM (SSS)

8 For example, ∠MYZ = ∠NZY (given), YZ is common, MY = NZ (symmetry of isosceles triangle), so △YMZ ≡ △ZNY (SAS)

9 a BE = DF (given), AD = BC (opposite sides of a parallelogram), ∠ADC = ∠EBC (opposite angles in a parallelogram), hence congruent due to SAS. There are other ways to prove this.

b ∠BEF = ∠AFD so EC and AF are parallel, as EB and DF are parallel.

Chapter 12 Statistics: Data handling

Overview

12.1	Averages	12.6	Surveys
12.2	Frequency tables	12.7	Questionnaires
12.3	Grouped data	12.8	The data-handling cycle
12.4	Frequency diagrams	12.9	Other uses of statistics
12.5	Histograms with bars of unequal width	12.10	Sampling

This chapter covers the basic statistical topics required at Higher level. Measures of location of discrete and continuous data are included (apart from median estimate of continuous data), along with displaying data in frequency polygons and histograms. The second half of this chapter covers surveys and related topics.

Context
Averages are used a lot in everyday life, such as in schools for test marks and attendance figures, in shops for daily or weekly takings, and even in sport, for batting averages. The lessons on surveys and sampling are important, as students may be asked to design and conduct a survey or questionnaire in subject areas other than mathematics and, when writing up the task, will have to explain the type of sampling they have chosen.

Curriculum references
KS4 Programme of Study references
1.1a Applying suitable mathematics accurately within the classroom and beyond.
1.2c Posing questions and developing convincing arguments.
2.3d Look at data and find patterns and exceptions.
3.3a The handling data cycle.
3.3b Presentation and analysis of large sets of grouped and ungrouped data, including histograms.
3.3c Measures of central tendency and spread.

Linear specification
S1 Understand and use the statistical problem-solving process.
S2.1 Types of data: qualitative, discrete, continuous. Use of grouped and ungrouped data.
S2.3 Design an experiment or survey.
S2.4 Design data-collection sheets distinguishing between different types of data.
S3.2h Produce charts and diagrams for various data types including line graphs, frequency polygons and histograms with unequal class intervals.
S3.3h Calculate median, mean, range, mode, modal class, quartiles and inter-quartile range.
S4.1 Interpret a wide range of graphs and diagrams and draw conclusions.

Functional Skills standards
R2 Identify the situation or problems and identify the mathematical methods needed to solve them.
R3 Select a range of mathematics to find solutions.
A1 Apply a range of mathematics to find solutions.
I1 Interpret and communicate solutions to multistage practical problems in familiar and unfamiliar contexts and situations.
I2 Draw conclusions and provide mathematical justifications.

PLTS
Independent enquirers identify questions to answer and problems to resolve; analyse and evaluate information, judging its relevance and value; support conclusions, using reasoned arguments and evidence.
Creative thinkers question their own and others' assumptions; adapt ideas as circumstances change.
Reflective learners review progress, acting on the outcomes; invite feedback and deal positively with praise, setbacks and criticism; communicate their learning in relevant ways for different audiences. **Team workers** collaborate with others to work towards common goals; reach agreements, managing discussions to achieve results. **Self-managers** work towards goals, showing initiative, commitment and perseverance; organise time and resources, prioritising actions; respond positively to change, seeking advice and support when needed.

APP
Handling data L5.4 Understand and use the mean of discrete data and compare two simple distributions, using the range and one of mode, median and mean. **L6.1** Design a survey or experiment to capture the necessary data from one or more sources; design, trial and, if necessary, refine data collection sheets; construct tables for large discrete and continuous sets of raw data, choosing suitable class intervals; design

© HarperCollins*Publishers* 2012

Chapter 12 Statistics: Data handling

and use two-way tables. **L7.3** Estimate the mean, median and range of a set of grouped data and determine the modal class, selecting the statistic most appropriate to the line of enquiry. **L7.4** Compare two or more distributions and make inferences, using the shape of the distributions and measures of average and range. **L8.2** Compare two or more distributions and make inferences, using the shape of the distributions and measures of average and spread including median and quartiles. **Using and applying mathematics L7.3** Justify generalisations, arguments or solutions.

Route mapping

Exercise	Grades				
	D	C	B	A	A*
A	1–8	9–17	18	19–20	
B	1–5	6–9	10–12	13	
C		all			
D	all				
E	1–2	3–8			
F				1–5	6–11
G	1–4	5–8			
H	1–2	3–9			
I		all			
J	1–4	5		6	
K	1	2–9			

Overview test

Question 1 is grade G, question 2 is grade F, questions 3 to 6 are grade D and question 7 is grade C.

1. Find the median of each set of numbers.
   a 6, 7, 9, 9, 10 b 6, 7, 9, 10, 10, 20 c 5, 7, 1, 0, 2, 8, 3
2. Find the range of each set of numbers.
   a 6, 0, 4, 4, 5, 8, 3, 2, 2, 6 b 110, 150, 100, 120, 500
3. Find the mean of each set of numbers.
   a 6, 0, 4, 4, 5, 8, 3, 2, 2, 6 b 110, 150, 100, 120, 170
4. Find the mode of each set of numbers.
   a 2, 5, 6, 6, 7, 7, 8, 8, 8
   b 6, 4, 4, 3, 6, 7, 9, 9, 1, 9, 2, 3, 5, 3, 8, 3
5. The ages of the members of an extended family are: 2, 5, 16, 34, 36, 59, 65, 96.
   a Calculate the mean age.
   b Calculate the mean age in 10 years (assuming all are still living).
6. A class was given a mathematics test and the scores obtained are listed here.
   Boys: 5, 45, 48, 60, 60, 64, 64, 67, 69, 79, 87, 88
   Girls: 40, 48, 54, 59, 60, 62, 64, 65, 70, 76, 84, 96
   a Calculate the mean score for
      i the boys ii the girls.
   b Calculate the mean score for the whole class.
7. The mean age of four children is 7.5. Three of the children are aged 3, 4 and 8. Calculate the age of the fourth child.

Answers to overview test
1 a 9 b 9.5 c 3
2 a 8 b 400
3 a 4 b 130
4 a 8 b 3
5 a 39.125 b 49.125
6 a i 61.3 ii 64.8 b 63.1
7 15

Why this chapter matters

Cross-curricular: This links with Science, Geography, History and any subject that includes statistical surveys.
Introduction: Statistical information is used in many areas, from big business to small corner shops. Different data can provide information for government, for planning roads or for restocking the shelves in a supermarket.
Discussion points: How does a clothes shop know which sizes to stock? When planning a party, how do you know how much food to provide? How do advertisers identify their target audience? Why are opinion polls useful? How can people's opinions be surveyed? What would be a useful sample size?
Plenary: Carry out a class survey about an appropriate topic, such as favourite personalities. Use a show of hands to indicate preferences. Then choose a more sensitive subject, such as body weight, and ask whether this would also be a suitable topic for classifying by a show of hands. What would be a more appropriate method?

Answers to the quick check test can be found at the end of the chapter.

© HarperCollins*Publishers* 2012

12.1 Averages

Curriculum references

Functional Skills standards
I2 Draw conclusions and provide mathematical justifications.

PLTS
Independent enquirers, Creative thinkers, Reflective learners, Team workers, Self-managers, Effective participators

APP
Handling data L5.4

Collins references
- Student Book pages 266–269
- Interactive Book: Matching Pairs

Learning objectives
- Use averages. [Grade D–C]
- Solve more complex problems using averages. [Grade D–C]
- Identify the advantages and disadvantages of each type of average and know which one to use in different situations. [Grade D–C]

Learning outcomes
- Students **must** be able to calculate the mean, the median and the mode. [Grade D]
- Students **should** be able to choose the most appropriate average to use. [Grade D]
- Students **could** be able to explain why different averages might be used by different people. [Grade C]

Key words
- mean
- median
- measure of location
- mode

Prior knowledge
Students should know how to calculate the mean, mode, median and range from small sets of discrete data.

Common mistakes and remediation
Students often forget to put data in order before finding the median. They also make careless errors by confusing the three averages. Regularly remind them of the definitions. Students also make errors when finding the median of an even set of numbers. Remind them that they should find the middle two, add, and then halve.

Useful tips
Remind students that outliers often appear in data and these clearly affect the mean, which can have a notable effect. The median and mode are not affected by outliers, which is why they are more useful in some contexts. The mode is the most commonly occurring value. Students might find it helpful to relate the word to 'model', to remind themselves of the modal value. Median is the middle value, in the same way as medium comes between small and large.

Functional maths and problem-solving help
Much of the work in Exercise 12A is functional maths (FM) as it is based on context but often remains open-ended.
Students need to understand when the different averages are used in society and why one is chosen over the other (e.g. for different purposes or simply for the nature of the data).
The problem-solving (PS) question, 15, is structured to lead students through it.

Lesson 12.1 Averages

Starter
- Encourage an initial discussion on averages. Ask students if they know what 'average' means. Say that it is a single number used to represent all the data.
- Ask students why there is a need for three types of averages. Then ask them to describe each type.

Main lesson activity
- Go through the Student Book text to ensure that students remember the meanings of 'mode', 'median' and 'mean'.
- On the board, write a simple list such as 2, 2, 2, 3, 3, 4, 4, 4. Ask students for the mode. They may say there is no mode, or there are two modes 2 and 4. Discuss this aspect of mode, saying that a list can have more than one mode, as shown in the example. There is no mode only if every number has the same frequency.
- Now write on the board a simple list such as 2, 5, 6, 8. Ask students for the median. Discuss the problem of finding the median when there is an even number of numbers. Remind students that they need to find the number between the two middle values, which, in this case, will be between 5 and 6. This is 5.5.
- Ask for the median if the two middle numbers are, e.g. 5 and 9. (7) Remind students of the easy way to find this value: add the numbers and halve the result. Provide more practice questions. Tell **more able** students that the median of a set of data with an even number of values is the mean of the two middle values.
- Discuss the advantages and disadvantages of each type of average, referring to the table in the Student Book. Stress that there are different reasons for using each average and discuss what they are. Remind students that the mean, which is most commonly used, is the only one that takes every value into consideration. Explain that this is the reason why people often, inaccurately, call the mean the average. Mention to **more able** students that the mean is very important in the study of higher statistics, which they might come across at A level.
- Go through Example 1, carefully explaining why the median is the best average to use in this scenario. Make sure all students understand this.
- Now students can do Exercise 12A, in which questions 1, 6, 8 and 16 are FM, question 15 is PS and 4, 5, 7 and 15 are assessing understanding (AU) questions. **Less able** students should complete the questions they can, and then work with a partner to complete the rest.

Plenary
- Ask students to write down five numbers with a range of 4, a mean of 4 and a median of 5. (1, 4, 5, 5, 5) Other answers are possible, e.g. 2, 2, 5, 5, 6.

Answers
Answers to Exercise 12A can be found at the end of this chapter.

© HarperCollins*Publishers* 2012

12.2 Frequency tables

Curriculum references

Functional Skills standards
I2 Draw conclusions and provide mathematical justifications.

PLTS
Independent enquirers, Creative thinkers, Reflective learners, Team workers, Self-managers, Effective participators

APP
Handling data L5.4

Collins references
- Student Book pages 270–274
- Interactive Book: Multiple Choice Questions

Learning objectives
- Calculate the mode and median from a frequency table. [Grade D–B]
- Calculate the mean from a frequency table. [Grade D–B]

Learning outcomes
- Students **must** be able to find a mode, a median or a mean from a frequency table. [Grade D]
- Students **should** be able to use averages to compare two distributions. [Grade D]
- Students **could** be able to solve problems involving frequency tables. [Grade C]

Key words
- frequency table

Prior knowledge
Students should be able to find the mode, mean and median of discrete data.

Common mistakes and remediation
At higher level, students often make the error of using $\bar{x} = \frac{\Sigma fx}{\Sigma x}$ rather than the correct $\bar{x} = \frac{\Sigma fx}{\Sigma f}$. If columns are clearly labelled and the correct formula has been written, this should not occur. When using a calculator, students often make two mistakes, as follows.
- They forget that if they have made a mistake and start again, they should be certain that they have cleared the calculator memory before re-entering any data.
- Calculators can differ, so each student must be sure which way around the calculator accepts the data value and frequency points.

Encourage students to practise using their calculators for statistical calculations.

Useful tips
Encourage **more able** students to see how they can use their calculators to check results, but ignore this for **less able** students. However, in the examinations, all students are expected to show all their working.

Functional maths and problem-solving help
All questions are set in context so consider the whole exercise as functional maths (FM). Encourage students to look at the tables and to write them in columns in their exercise books to assist them in their working. **Less able** students often forget to add the frequencies, and add the first column instead. They may find the mean of the first column rather than using all three columns. Provide plenty of practice and, to reinforce understanding, frequently ask what the final column represents. The problem-solving (PS) question, 8, may be set in a familiar context for some students. Students need to work back from the information provided.

© HarperCollinsPublishers 2012

Lesson 12.2 Frequency tables

Starter
- Ask students for the definition of each type of average, from the previous lesson.
- Select a simple set of numbers in which some are repeated, e.g. 2, 2, 3, 3, 5. Ask students to calculate the mean. Do this with similar sets, leaving each set and answer visible. Then put each set into an appropriate tabular form, introducing students to the formality and notation of labelling the columns as *x*, *f* and *fx*.

Number of children	1	2	3	4
Frequency				

Main lesson activity
- Remind students that when they have a lot of data they should tabulate it rather than work with large lists of numbers. Demonstrate this by asking, "What is the mean number of children in your family home?" Begin a table on the board, making it longer if necessary.
- Collect the data from students, then complete the table. Stress that this is a much better way to collect and show data than making a list. Discuss how to find the mean from this set of data. Explain that it is often better to head the columns 'Number of children' and 'Frequency', then create more columns to help calculate the mean.

Number of children	Frequency	Number of children × Frequency
1	6	
2	14	
3	7	
4	3	

- Add a third column to the table, as shown, but with data from the class.
- Explain why you multiply the first two columns together – to find the total number of children in all the families by finding the total in each row and adding the results. You may need to write each list in full for **less able** students, so that they can see all 30 numbers. Say that this is simply a shorter version of the long list. (In the first row, 1 × 6 is a short way of writing 1+ 1 + 1+ 1 + 1 + 1, and so on.)
- Complete the table and add the numbers in the last column to get a total (67 in the table above), then divide by the total frequency (30) to give the mean. (2.233…)
- Explain that you usually need to round the mean and that the degree of accuracy should be to one more significant figure (sf) than in the data being used. So here, this will be to 2 sf, giving 2.2 children. Introduce **more able** students to sigma notation so that they are familiar with it before A level.
- Remind the class that the mean is not just a number, it should always have a unit (here the unit is children).
- Ask the class how they would find the median from the table. Students should recognise that they need to find the 15th and 16th numbers (once in order) working down the table. In the example, these will both be 2, so the median is 2. Point out that it is easy to find the mode simply by looking for the largest frequency in the table. In this example this is once again 2.
- If **less able** students find the use of frequency tables difficult, use Example 2 to consolidate this process.
- Now students can do Exercise 12B, in which 2d, 6f, 7c, 8, 9 and 11 are FM questions. Question 8 is PS and 3b, 5f and 9 are assessing understanding (AU) questions. **Less able** students may need to work in pairs.

Plenary
- If students have access to scientific calculators, demonstrate how to find a mean, using the statistics mode. Refer to 'Using your calculator' in the Student Book. Demonstrate this approach by checking the answers of some questions in the exercise.

Answers
Answers to Exercise 12B can be found at the end of this chapter.

© HarperCollins*Publishers* 2012

12.3 Grouped data

Curriculum references

Functional Skills standards
I2 Draw conclusions and provide mathematical justifications.

PLTS
Independent enquirers, Creative thinkers, Reflective learners, Team workers, Self-managers, Effective participators

APP
Handling data L7.3

Collins references
- Student Book pages 275–278
- Interactive Book: Multiple Choice Questions

Learning objectives
- Identify the modal group. [Grade C]
- Calculate and estimate the mean from a grouped table. [Grade C]

Learning outcomes
- Students **must** be able to find the modal class and estimate the mean from grouped data. [Grade C]
- Students **should** be able to use averages to compare two sets of grouped data. [Grade C]
- Students **could** be able to draw conclusions from a set of grouped data, with justifications. [Grade C]

Key words
- continuous data
- group
- discrete data
- modal group
- estimated mean

Prior knowledge
Students should know how to calculate the mean from a grouped frequency table containing discrete data, as covered in Lesson 12.2. They should be familiar with the use of inequality symbols $>, \geq, <, \leq$.

Common mistakes and remediation
Students sometimes decide to interpret the phrase 'find an estimate for the mean', as meaning 'guess the mean'. Plenty of practice should help to minimise this error.

Useful tips
In this topic, encourage **more able** students to see how they can use a calculator to check their results. Remind students that in the examinations, they must show all their working in these types of questions.

Functional maths and problem-solving help
It is important for students to start with question 1 of Exercise 12C, in order to consolidate the process taught in this lesson before meeting it in context in the functional maths (FM) questions, 3c and d, 4–6 and 7c. Emphasise the need to be careful about the group boundaries and of rounding their answers to give a sensible answer. The problem-solving (PS) question, 8, requires students to manipulate the numbers in the table to find the answer. They need to read the question carefully.

Starter
- Students often require revision of inequality symbols. Go through $<, >, \leq$ and $\geq$. Discuss a set of intervals given in terms of inequalities. Ask what the following mean:
$0 \leq x < 10, 10 \leq x < 15$. Explain terms such as 'lower class boundary' and 'upper class boundary' for each interval.
- Explain why it is best to use inequalities when describing continuous intervals, rather than simply expressing a range such as 0–10, 10–15.

© HarperCollinsPublishers 2012

Lesson 12.3 Grouped data

Main lesson activity
- Review the previous lesson when the class looked at data in tabular form and were able to find the mode, median and mean. Explain that when there is a lot of data, it is often provided in ranges in frequency tables.
- Say that if you were to ask how much each student spends on lunch, they would all name different amounts; it would not be practical to set up a table with a row for each amount. A better solution would be to use ranges.
- Ask how many students spend a pound or less on their lunch, and how many spend between one and two pounds. Then ask if they can see the problem with asking this question. Lead students to understand that £1 falls into both groups. Each piece of data must fall in only one group in any table.
- Either build a table from data within the class or use Example 3 to illustrate this issue. Write the table on the board but only complete the first two columns. Talk about the mode – for grouped data, this is called the 'modal group', or sometimes 'modal class'. Make sure students are familiar with both terms.
- Explain that it is especially easy to find the modal group as it is simply the group with the largest frequency.
- Ask students why it is not possible to find the mode. (The precise numbers within each group are not known.)
- Ask students how to find the mean from this data. Lead them to recognise the similarity to the table from the previous lesson, except that now the individual numbers in each group are unknown. Explain that it is customary to assume that all the data values in a group take the middle value for the group. This is because, statistically, a full range of values would be expected within each group, with as many values above the halfway value as below. The bigger the frequency, the more likely this is. If necessary, for **less able** students, demonstrate with examples of data in a range.
- Now complete the third column by adding the two class boundaries and halving for each group. Tell **more able** students that using the midway values can make calculating the mean quite difficult. In this case, the values can be rounded, e.g. to one decimal place, as the calculation will only be an estimate.
- Compare this routine to that of the previous lesson. Find the total amount of money for all students; estimate for each group by multiplying the midway value by the frequency, then total that column. Complete the fourth column of the table, and then complete the calculation to find an estimate of the mean. Divide the total by the total frequency and round suitably.
- Now students can do Exercise 12C.

Plenary
- Discuss the reasons why class widths of a certain size might be chosen before collecting a sample. Must they all be the same width? Under what circumstances might it make sense to choose different widths?
- Examine the change in estimate of the men if different class widths are chosen. For example, look at question 2 from Homework Book Exercise 3.3, but use these three class intervals: $0 \leq x < 4$, $4 \leq x < 8$ and $8 \leq x < 12$.

Answers
Answers to Exercise 12C can be found at the end of this chapter.

© HarperCollins*Publishers* 2012

12.4 Frequency diagrams

Curriculum references

Functional Skills standards
R3 Select a range of mathematics to find solutions.

PLTS
Independent enquirers, Creative thinkers, Reflective learners, Team workers, Self-managers, Effective participators

APP
Handling data L7.4

Collins references
- Student Book pages 278–286
- Interactive Book: 10 Quick Questions

Learning objectives
- Draw frequency polygons for discrete and continuous data. [Grade D–C]
- Draw histograms for continuous data with equal intervals. [Grade D–C]

Draw pie charts. [Grade D]Learning outcomes
- Students **must** be able to draw a frequency polygon for continuous data. [Grade C]
- Students **should** be able to draw a histogram [Grade C]
- Students **could** be able to compare two distributions using frequency diagrams. [Grade C]

Key words
- frequency density
- continuous data
- grouped data
- sector
- discrete data
- histogram
- pie chart
- frequency polygon

Prior knowledge
Students should have covered Lessons 12.1 to 12.3. This lesson is a natural continuation.

Common mistakes and remediation
The most common mistake students make when drawing frequency polygons is using grouped labels on data axes rather than continuous scales, e.g. labelling a whole section of the axes '5 to 10' or '5–10' instead of a clear 5 at one mark and 10 at the next mark, with equal-sized gaps between marks. Although in this section all class intervals involving histograms will be of equal size and, hence, bar widths will be the same size' this should not be assumed, as students will soon be introduced to histogram problems involving unequal class intervals. So students should check calculated bar widths. Help students to avoid making these errors by pointing out these things regularly.

Useful tips
When students are drawing histograms, they will find that calculating the bar widths is straightforward if they use inequalities to describe class intervals, as these give the upper and lower class boundaries clearly. However, if they do not use inequalities to define class intervals, they must take care to calculate the bar widths correctly. For example, if the times of an event are noted to the nearest second, and the intervals are given as 6–10s, 11–15s, 16–20s, etc., then the first bar width should be 5 (10.5–5.5) and not 4 (10–6).

© HarperCollins*Publishers* 2012

Lesson 12.4 Frequency diagrams

Functional maths and problem-solving help
Question 4 of Exercise 12D is functional maths (FM). Help students to interpret the scales and boundaries of the groups. Help them to see the link between using the midway value of the tables in the previous lesson and the need to use the midway values in frequency polygons. There is not much work on histograms here as most of the work is covered in the next lesson (unequal intervals). For the problem-solving (PS) question, 5, students need to remember the link between frequency polygons and bar charts or histograms. They should then be able to calculate the total number of students and hours.

Starter
- Statistical data is often presented in pictorial or graphical form. Ask students to name the types of presentations they may have already encountered.
- Ask them to suggest any advantages one format might have over another. For example, a pie chart, which illustrates frequency as an area, could arguably be more 'visually correct' than a bar chart, in which the bar widths may be misleading to the eye.

Main lesson activity
  Pie charts and frequency polygons
- Students should be familiar with pie charts, so this section will provide practice. Work through the text under 'Pie charts' and Example 4.
- Use Example 5 to illustrate drawing a frequency polygon from 'discrete data'. Use this expression as it will help students to distinguish the different types of data as they occur. Point out that drawing frequency polygons with ungrouped data, as in this Example, is grade D work, but when they are drawn from grouped data (Example 6) this moves the topic into grade C.
- Show the data table from Example 6 on the board and ask what will be different about a frequency polygon drawn from this. Guide students to recognise that the group boundaries need to be placed on the scale; discuss where to draw the height of the polygon. Many will want to draw the frequency height above the higher boundary, so discuss why a truer representation is to use the midway mark, as consistent with the previous lesson's estimation of the mean.
- Display the completed frequency diagram on the board; discuss why the polygon ends where it does and is not taken down to the horizontal axis.
- Ask students to complete Exercise 12D.
  Bar charts and histograms, Using your calculator
- Look at the bar chart, and then work through Example 7. Display the table on Student Book pg 292 on the board, omitting the frequency density row. Ask how to illustrate the data. A bar chart could be used but, as the data is continuous, there is a better way. Explain that in a histogram the area of the bar – not just the height – represents the frequency; there are no gaps between the bars bar, unless the frequency of one value happens to be 0.
- Study the histogram in the Student Book. Explain that the numbers on the vertical axis show the frequency density, found by dividing the frequency for the class interval by the class width. To read a histogram, multiply the frequency density by the width of the bar to find the frequency (the area of the bar). Say that histograms are most commonly used with data that has unequal interval widths, which they will learn more about in the next lesson.
- If students have graphics calculators, demonstrate how to use them to drawn histograms.
- Now students can complete Exercise 12E.

Plenary
- Refer **more able** students to Example 9. What coordinates should be plotted to obtain a frequency polygon from the data of the seedlings' heights? Ask students to draw a frequency polygon and histogram of this data, and to compare and contrast the results.

Answers
Answers to Exercise 12D and 12E can be found at the end of this chapter.

© HarperCollins*Publishers* 2012

12.5 Histograms with bars of unequal width

Curriculum references

Functional Skills standards
I2 Draw conclusions and provide mathematical justifications

PLTS
Independent enquirers, Creative thinkers, Reflective learners, Team workers, Self-managers, Effective participators

APP
Handling data L8.2

Collins references
- Student Book pages 287–293
- Interactive Book: Worked Exam Questions

Learning objectives
- Draw and read histograms where the bars are of unequal width. [Grade A–A*]
- Find the median, quartiles and interquartile range from a histogram. [Grade A–A*]

Learning outcomes
- Students **must** be able to draw a histogram with unequal column widths. [Grade A]
- Students **should** be able to interpret a histogram with unequal column widths. [Grade A]
- Students **could** be able to estimate the median and the interquartile range from a histogram. [Grade A*]

Key words
- class interval
- median
- interquartile range
- upper quartile
- lower quartile

Prior knowledge
Students should understand the work covered in Lessons 12.3 and 12.4.

Common mistakes and remediation
Students often make mistakes in finding the quartiles because of the difficulties involved in finding their position in the class and then estimating their values. Practice and careful thought will enable students to remember and understand this routine.

Useful tips
Explain that quartiles are simply a measure that can be used to compare different distributions. This adds to the information gained from comparing the medians.

Functional maths and problem-solving help
In Exercise 12F, questions 1 and 4–6 provide students with practice in the skill of creating and reading histograms. Questions 2, 10d and 11 are functional maths (FM) as they are set in context. To complete it, students must use their knowledge of histograms and quartiles. For problem-solving (PS) question, 11, students need to remember the definition of frequency density.
Note: In a histogram some groups are combined to give a bigger picture rather than using equal bars that might mask a tendency.

Starter
- Draw a normal distribution (bell-shaped) curve on the board and explain that it is a statistical diagram, used in many real-life situations, e.g. people's heights. These distributions occur normally – and so are called normal distributions – in which the frequencies are greatest for the middle values and smallest at the extremes.

© HarperCollins*Publishers* 2012

Lesson 12.5 Histograms with bars of unequal width

- Discuss quartiles, by which the population is divided into four equal parts: the first quarter point gives the lower quartile, the second point the median (second quartile) and the top quarter point the upper quartile.
- Show these on the normal distribution curve and point out that the area under the curve at each division is one-quarter of the whole frequency.

Main lesson activity
- Recap what a histogram is. Some students may think it is a bar chart with no gaps, but this is not accurate. Explain that it is a bar chart in which the area of the bars, not the height, represents the frequency and the vertical axis is the frequency density, not frequency.
- Copy the horizontal table from Example 8 onto the board; ask what students notice about the class intervals (different sizes). This may be the first time students have met unequal class widths. Explain that to draw a histogram you need to calculate the heights for each bar to plot the frequency density. You may need to remind **less able** students of the definition of frequency density (frequency divided by class interval width).
- Rewrite the table vertically, adding two columns. List the class widths (difference between class boundaries) in the third column. Divide each frequency by class width to complete the frequency density column. Begin to draw the histogram, in which the frequency density goes up to 47, so mark the scale to 50 as shown. Complete it.
- Ask students how to find the median (they will find it from the table, not the histogram).
- Add all the frequencies (262), then halve this (131). As this is a grouped frequency table, the median will only be an estimate, so take the height of the 131st girl (in order) rather than the mean for the 131st and the 132nd girls. **More able** students may challenge this; briefly explain that, in practice, after rounding they will get the same value unless the frequency is quite small.
- Find the 131st girl by looking down the table and counting which height group she falls into. Adding frequencies shows that she is in the third group, $154 \leq h < 155$, which contains the 108th to 154th girls. Estimate the height of the 131st girl: (131 - 108) ÷ 47 = 0.49 along the division from 154 to 155 on the scale. This division is 1 cm, so the median is 154.49 cm. (This rounds to 154.5 but avoid in this example, in case students assume it is always halfway.) Carefully explain this a few times to **less able** students. Draw the median mark on the histogram as a vertical line at that point.
- Show that the lower quartile can be found similarly. Divide 262 by 4 (65.5) and find the 65.5th girl (estimate).
  This is just into the second group, which includes the 65th to 107th girls. Work out (65.5 - 65) ÷ 43 = 0.01; then the lower quartile is at 153 + 0.01 = 153.01 cm. Here, it is sensible to round to 153 cm. Draw the lower quartile on the histogram as a vertical line here. Say that in this case, since it is an estimate and is just into this next group, 153 cm could have been estimated without calculation, but most questions will need a calculation.
- For completeness, ask students to calculate the third (upper) quartile. Find the (65.5 x 3) = 196.5th girl, who is in the fourth group. (196.5 - 155) ÷ 96 = 0.43, so the upper quartile is 155.43 cm. Make sure all students understand how to do this; draw the upper quartile on the histogram.
- Use Example 9 for more practice. Now students can do Exercise 12F.

Plenary
- Students have considered two ways to display continuous grouped data with unequal class intervals: the frequency polygon and the histogram. Ask which they think is more useful, or does the 'usefulness' depend on the type of data being analysed?

Answers to Exercise 12F can be found at the end of this chapter.

© HarperCollins*Publishers* 2012

12.6 Surveys

Curriculum references

Functional Skills standards
R2 Identify the situation or problems and identify the mathematical methods needed to solve them.

PLTS
Independent enquirers, Creative thinkers, Reflective learners, Team workers, Self-managers, Effective participators

APP
Handling data L6.1

Collins references
- Student Book pages 294–296
- Interactive Book: Student Demo

Learning objectives
- Conduct surveys. [Grade D–C]
- Ask good questions in order to collect reliable and valid data. [Grade D–C]

Learning outcomes
- Students **must** be able to design and use a data collection sheet. [Grade D]
- Students **should** be able to design a two-way table to collect data. [Grade C]
- Students **could** be able to identify criteria to use when designing a data collection sheet. [Grade C]

Key words
- data collection sheet
- survey
- hypothesis

Prior knowledge
Students must be able to construct and use tables. They should also know how to use a tallying procedure accurately and effectively.

Common mistakes and remediation
Students make mistakes in the design of response options. They often forget (or do not handle well) the options to respond at either extreme. When presented with data collection sheets or survey questions, **less able** students also often find it difficult to find the faults. To help students avoid these errors, explain their mistakes clearly.

Useful tips
Encourage students to study the options in their data collection sheets carefully and check that nobody could tick more than one box for any question.

Functional maths and problem-solving help
Almost all the questions in Exercise 12G are functional maths (FM); students are being asked to engage in real-life mathematical activities testing hypotheses and seeking opinions. It is vital that they keep in mind the impartiality needed to avoid any bias.
There are no problem-solving (PS) questions in this exercise.

Lesson 12.6 Surveys

Starter
- Ask students to count how much money they have on them, but not to tell you. Give them a (deliberately poor) choice of responses, e.g. i between £0 and £1, ii between £1 and £5 or iii between £5 and £10.
- Collect students' responses by tallying them on the board.
- Ask for comments. Lead the discussion towards these issues.
- Why not just ask exactly how much everybody has?
- What do you do if you have, e.g. exactly £5?
- What do you do if you have more than £10?

Main lesson activity
- Ask, "If you could choose a trip to go on, where would you go?" After a short discussion, ask how they could find out where everyone wants to go. This should lead to 'asking people and taking a survey'.
- Show the simple data collection sheet in the Student Book, with options of Blackpool, Alton Towers, etc. and discuss the merits of this type of survey as opposed to a more open-ended survey. Both are possible but the first is easier to use. Emphasise that, in any survey, the questions are important and should be clear and unambiguous.
- Discuss another survey that could be done in the school on the number of hours per week that students spend doing their homework. Ask students to look at the table in Example 10 and ask if they can see any problems. If necessary, lead them to comment on the overlapping times, e.g. 5 hours is in two groups, the 0–5 and the 0–10, and 10 hours is in both 0–10 and 10–20. Point out that when choices of answers are provided people must be able to tick only one box. Go through the changes made to the second table in Example 10.
- Students can now do Exercise 12G. Work through question 1 with **less able** students before they attempt the remaining questions. **More able** students should not need this help.
- In this exercise, questions 1, 2, 4, 5, 6 and 8 are FM. Question 8 assesses understanding (AU).

Plenary
Ask students to suggest response options for topics such as:
- how often people use the internet
- how many days off work or school people have had during the last year.

Look for a sensible range of response options with no overlaps and realistically catered-for extremes.

Answers
Answers to Exercise 12G can be found at the end of this chapter.

© HarperCollins*Publishers* 2012

12.7 Questionnaires

Curriculum references

Functional Skills standards
R2 Identify the situation or problems and identify the mathematical methods needed to solve them.

PLTS
Independent enquirers, Creative thinkers, Reflective learners, Team workers, Self-managers, Effective participators

APP
Handling data L6.1

Collins references
- Student Book pages 297–299

Learning objectives
- Ask good questions in order to collect reliable and valid data. [Grade D–C]

Learning outcomes
- Students **must** be able to judge the quality of questions in a questionnaire. [Grade D]
- Students **should** be able to design a questionnaire to test a hypothesis. [Grade C]
- Students **could** be able to trial a questionnaire and suggest improvements. [Grade C]

Key words
- data collection sheet
- questionnaire
- hypothesis
- survey
- leading question

Prior knowledge
Students should have worked through at least some of Exercise 12G in the previous lesson.

Common mistakes and remediation
Students often use overlapping groups or omit a value from the answer sets. To create awareness of these errors, encourage students to check each other's questionnaires.
They should also check that the options do cover all possible responses. The use of bullet points encourages students to list ideas briefly and succinctly.

Useful tips
Encourage students to study the options in their data collection sheets carefully and check that nobody could tick more than one box for any question.

Functional maths and problem-solving help
In Exercise 12H, questions 3–6 are functional maths (FM). Encourage students to discuss their questionnaires with one another. Ensure that you also talk about their questions and the reasons for their choices.
All of the problem-solving (PS) questions (3–6 and 8) require students to put into practice the guidance they have learned in this lesson.

© HarperCollins*Publishers* 2012

Lesson 12.7 Questionnaires

Starter
- Ask students to review the previous lesson and list, in bullet points, what they need to remember when conducting surveys.

Main lesson activity
- Explain to students that this lesson is about the questions used in questionnaires. Say that you want to find out what type of music students like to listen to. Say, "Reggae music is brilliant, do you like it?" Discuss whether this is a good or bad question and why.
- Lead the class to see that this is a leading question as it suggests that reggae music is good, which might encourage some people to say 'yes' when they might otherwise have said 'no'. **More able** students should see this as an example of bias. **Less able** students may need more examples before they fully understand what a leading question is.
- Ask, "What do you weigh?" but stop before anyone answers or gets embarrassed. Ask why this is a bad (or biased) question. (It is very personal and people may not wish to talk about their weight.) Tell students to try to avoid questions like this.
- Ask, "How many times have you visited … (local place)?" Write 'tick' box answers on the board, e.g. 'Twice', 'Five times', 'More than 10 times'. Discuss the choices of answers, pointing out that not every choice is available, including 'Never'. Explain to **more able** students that, ideally, the number of choices should be six at the most.
- Now students can do Exercise 12H. They can do questions 1 and 2 in small groups in order to discuss their answers, but ensure that each student writes up his or her own answers. They can do questions 3 to 8 individually, but **less able** students may find it easier to work in pairs and discuss their questionnaires before writing them up on their own.
- In this exercise, questions 3–6 are FM. Questions 3–6 and 8 are PS and 9 is an assessing understanding (AU) question.

Plenary
- Ask students to comment on a question such as, "Do girls really not like football?" Choose a question that is suitable for the class.
- Also ask students to come up with their own questions.

Answers
Answers to Exercise 12H can be found at the end of this chapter.

© HarperCollins*Publishers* 2012

12.8 The data-handling cycle

Curriculum references

Functional Skills standards
R2 Identify the situation or problems and identify the mathematical methods needed to solve them.

PLTS
Independent enquirers, Creative thinkers, Reflective learners, Team workers, Self-managers, Effective participators

APP
Handling data L6.1

Collins references
- Student Book pages 300–301

Learning objectives
- Use the data-handling cycle. [Grade C]

Learning outcomes
- Students **must** be able to plan how to collect data to test a hypothesis. [Grade C]
- Students **should** be able to design and refine a data collection sheet. [Grade C]
- Students **could** be able to demonstrate how to use each part of the data handling cycle. [Grade C]

Key words
- hypothesis
- secondary data
- primary data

Prior knowledge
Students need to understand what a good survey is. Students should also be familiar with the previous lessons in this chapter.

Common mistakes and remediation
Students often forget to use all four parts of the cycle in their descriptions. Encourage them to make a mental note to remember always to add the four parts of the data-handling cycle.

Functional maths and problem-solving help
Question 2 of Exercise 12I is functional maths (FM) and all of them are based on problem solving (PS). Students need to understand how in real life, this cycle would be applied by different people with different needs. For example, a large company would run a much larger sample than would a school (possibly because they would have better resources or income at their disposal).

Starter
- Divide the class into groups of four, and ask, "What is the average amount of pocket money for your group?"
- Allow time for each group to calculate these and give you their answers, and write these on the board.
- Then ask what type of average everyone has given. Is it the mode, median or mean? Ask students why they gave that type of average and not one of the others.
- If necessary, discuss the fact that in most cases the mean would give the best average.

Main lesson activity
- Tell the class that you have heard it said that 'Boys have more pocket money than girls.' Discuss this statement as a hypothesis, i.e. what someone thinks may be true. Say that you want to test this, and ask the nearest boy and girl to say what pocket money they receive. Then ask: Is the hypothesis true?
- Ask if we could improve on our investigation of this hypothesis. Hopefully you will get the answer: Yes, by asking more students. Then do this; i.e. ask students to calculate their average pocket money as two groups, collate the results, and then display them.
- Discuss with the class whether the hypothesis is true or not.

© HarperCollins*Publishers* 2012

Lesson 12.8 The data-handling cycle

- Ask if we can we improve on this investigation. The answer you want to illicit from the class is: Yes, by asking different ages, etc. Encourage **more able** students to suggest other improvements. **Less able** students may find that just one improvement is enough to see how the cycle is repeated.
- Now introduce the data-handling cycle as per the text in the Student Book. Explain that the idea is to start with a given problem, which is often a hypothesis, as we did above. Then we make a plan as to how we will test this, as we also did. Then we gather the data for our plan and look at it to see if it proves the hypothesis. Finally, we go back to the beginning and decide if we need to modify our plan. This cycle is known as the data-handling cycle.
- Talk through how we might apply the data-handling cycle to the hypothesis of 'Teachers are quicker at mental arithmetic than students.' Remind students that they are not doing the actual investigation; they are simply seeing how the data handling cycle is applied to this problem.
- Once you are confident that all students understand the cycle, let them complete Exercise 12I, in which question 2 is FM and 3 is an assessing understanding question (AU).

Plenary
- Give the class another hypothesis, 'The quicker you clear up, the quicker you can leave the classroom.'
- Ask them how the data-handling cycle would help them to test this hypothesis.

Answers
Answers to Exercise 12I can be found at the end of this chapter.

12.9 Other uses of statistics

Curriculum references

Functional Skills standards
I2 Draw conclusions and provide mathematical justifications.

PLTS
Independent enquirers, Creative thinkers, Reflective learners, Team workers, Self-managers, Effective participators

APP
Not in assessment criteria.

Collins references
- Student Book pages 302–304

Learning objectives
- Apply statistics in everyday situations. [Grade D–B]

Learning outcomes
- Students **must** be able to use indices to calculate values. [Grade D]
- Students **should** be able to understand a range of social statistics. [Grade C]
- Students **could** be able to interpret a time series. [Grade B]

Key words
- margin of error
- Retail Price Index
- national census
- social statistics
- polls
- time series

Prior knowledge
Students should be able to interpret information from various forms of frequency polygons, histograms, graphs and pie charts.

Common mistakes and remediation
When using the Retail Price Index (RPI) students often try to increase the amounts from year to year instead of returning to the base year. Pointing out this error should help to minimise it.

Functional maths and problem-solving help
All the questions in Exercise 12J are functional maths (FM), and lead students to areas where, in the future, they could well find themselves asking such questions. The topics covered will be new to most students and will introduce them to concepts they may not have encountered before. If necessary, take time to discuss the questions with students, giving them an idea of how to begin. There are no problem-solving (PS) questions in Exercise 12J.

© HarperCollins*Publishers* 2012

Lesson 12.9 Other uses of statistics

Starter
- Ask students if they know how many people live in Britain (was about 60 million as at July 2008). Discuss how we know this. Ask, "What is the exact figure? Can we find the exact figure?"
- Mention the national census and discuss what sort of data this collects (number of people per house, ages, gender, religion, ethnicity, education, and so on), how often it is taken (every 10 years) and if it is compulsory (yes).

Main lesson activity
General Index of Retail Prices, Time Series, National census
- Ask if anyone knows what the Retail Price Index (RPI) is. Few will know or even have heard of it. Explain that it is simply a means of keeping some kind of check on prices, year on year, as the cost of living changes.
- Discuss how pocket money has increased over time. Students' parents probably received about 50p. What do they get? Explain that relative values could be about the same – their parents could probably buy with 50p the same as students can with the average pocket money today.
- Go through question 1 of Exercise 12J carefully, explaining a 'base year' and the 'price index'. Say that the 2001 price is an increase of 3% (where did that 3 come from?). You will probably have to remind **less able** students how to calculate 3% of 78p so they can add this to the 78p price in 2000. Ensure that **more able** students use the multiplier 1.03 to find 103% of the 78p (1.03 × 78p).
- Talk through the 2002 change, increasing by 8% from 2000. Then ask students to complete the question.
- Now students can complete Exercise 12J, in which all questions are FM, and 5 is also an assessing understanding (AU) question.

Plenary
- On the board, draw this graph, which shows a rough distribution of ages in Britain in 1950 and 2000.
- Ask students how such information might be useful to a government. Explain that the blip in the 2000 graph is the post-war baby boom. Discuss how an ageing population could mean pension problems, and fewer young people could mean that fewer school places will be needed in five years' time.

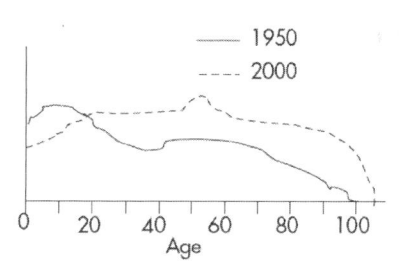

Answers
Answers to Exercise 12J can be found at the end of this chapter.

© HarperCollins*Publishers* 2012

12.10 Sampling

Curriculum references

Functional Skills standards
R2 Identify the situation or problems and identify the mathematical methods needed to solve them.

PLTS
Independent enquirers, Creative thinkers, Reflective learners, Team workers, Self-managers, Effective participators

APP
Not in assessment criteria.

Collins references
- Student Book pages 304–308
- Interactive Book: 10 Quick Questions

Learning objectives
- Understand different methods of sampling. [Grade D–C]
- Collect unbiased reliable data. [Grade D–C]

Learning outcomes
- Students **must** be able to identify reasons why a sample may be biased. [Grade C]
- Students **should** be able to design a stratified sample. [Grade C]
- Students **could** be able to plan how to answer a real life problem using a stratified sample. [Grade C]

Key words
- population
- stratified
- random
- unbiased
- sample

Prior knowledge
There is no direct prerequisite for studying sampling methods, although this lesson naturally complements Lessons 12.6 on surveys and 12.7 on questionnaires.

Common mistakes and remediation
Students often get confused when working with the stratified sampling and their proportions. Encourage them to use a table of all the groups and try to ensure that this results in the final sample size aimed for.

Useful tips
The activity Using the internet, after Exercise 12K suggests some hypotheses that students could test if time permits.

Functional maths and problem-solving help
This lesson is about context and asking questions in real life, and questions 1–4 and 8 of Exercise 12K are functional maths (FM).
Question 6 is a problem-solving (PS) question and requires students to apply what they have learnt in this lesson, working 'back' to an answer.

© HarperCollins*Publishers* 2012

Lesson 12.10 Sampling

Starter
- Tell students that now you will do a quick survey on their television-watching habits. Say that you will choose five people randomly. Pick five girls (or boys) and ask silly questions about programmes that were shown on television last night.
- After asking the questions, ask the open question, "What was wrong with my sample?" Students may argue it was not completely random, but you can argue otherwise!
- Ask the class how to choose a more appropriate sample. Must the sample be exactly half boys and half girls?
- In a single-sex class, you could question the five oldest students. Age and gender often play a part in determining viewing habits, especially if the sample contains people from a certain age range.

Main lesson activity
  Sampling methods
- Discuss the different ways of sampling. In random sampling, people are chosen at random. For example, they could be the first 10 to walk into the room, or the first 50 to walk out of the supermarket. Choosing a random sample will very often give random results.
- Stratified sampling takes into consideration the subgroups (or strata) of the population to be surveyed. The survey includes people from each stratum or layer of the population, in the same proportion as that stratum is to the whole population. This means that the sample has the same make-up, proportionally, as the whole population. Most polls currently aim for stratified sampling.
- Explain that students should always choose stratified sampling over purely random sampling, especially if they suspect that subgroups may have significant differences in characteristics to be measured. For example, views on sports between men and women are often quite different, so a sports-based survey on a sample of a club with membership of 25% men and 75% women should be based on a sample that includes three times as many women as men.
- Go through Example 12, showing students how to find the proportions of each group in the school and then multiply that by the sample size, rounding the data at the end.
- Explain to **more able** students that this type of rounding can lead to a very slightly different sample size from that chosen, but it would always be no more than 1 out and so would not really affect the results. If these students need convincing, ask them to experiment with some figures to demonstrate this.

  Sample size
- Discuss sample size with the class. Ideally, a survey would include everyone. Although this would show what people think, it is not usually practical and can be expensive, so a sample is taken – but how many people should be included? The larger the sample, the more accurate the findings, but at GCSE level there is no 'right' sample size.
- Work through Example 13. This will stimulate discussion and allow students to explore the idea of sampling. Then students can do Exercise 12K, in which questions 1–4 and 8 are FM. Question 6 is PS and 7 is an assessing understanding (AU) question.

Plenary
- Another method of sampling is systematic sampling. Ask students to suggest what this might be and whether it has any advantages over the other two types of sampling discussed so far. Students could try the activity, Using the internet.

Answers
Answers to Exercise 12K can be found at the end of this chapter.

© HarperCollins*Publishers* 2012

Functional maths
Fishing competition on the Avon

Curriculum references

Functional Skills standards
R3 Select a range of mathematics to find solutions.
A1 Apply a range of mathematics to find solutions.
I1 Interpret and communicate solutions to multistage practical problems in familiar and unfamiliar contexts and situations.

PLTS
Independent enquirers, Reflective learners

APP
Using and applying mathematics L7.3

Collins references
- Student Book pages 318–319

Learning objectives
- Identify the mode, median, mean and range.
- Calculate a mean from a grouped frequency table.
- Understand the advantages and disadvantages of each type of average.

Key words
- estimated mean
- grouped data
- mean
- median
- mode
- range

Functional maths help
'Getting started' will help students understand that data must be organised in order to find an average, and that every number has a relative value. To help **less able** students, work through this set of questions to ensure full understanding.

- Write '5' on the board and ask for a number larger than this. Students should be able to name many numbers.
- Now ask for the smallest number, larger than five. Students may say six, but the number should be smaller. Encourage them to name ever-smaller decimals. For example, if students say '5.1', ask for a smaller number (5.01). Continue this as they work their way downwards, from 5.01 to 5.001, and so on.
- Some students may realise that the very smallest number larger than five must have very many zeros, followed by a 1. **More able** students will realise that it is impossible to write down the smallest number larger than five. Explain that in the inequality $5 < X$, X could be any number they have suggested. Refer students to the 'Handy hints' box.
- Now ask for a number smaller than five. Again, students should give a wide variety of correct answers.
- Ask students to give the largest number that is smaller than five. As before, this should lead to a set of decreasing decimals (4.9, 4.99, 4.99, 4.99999, and so on). Students should recognise that the largest number smaller than five is a 4 followed by an indefinite number of 9s.
- Explain that in the inequality $Y > 5$, Y could be any of the numbers that students have identified. They should now recognise that they have established a continuous line of values. This is the starting point for finding out any average.

Lesson plans
Build
Begin by referring to 'Functional maths help' (above).

Activity 1
Ask students to draw line graphs to show the data for Weeks 1–4 and to comment on what their graphs show. (Ensure that they leave room for the data from Week 5.)

© HarperCollinsPublishers 2012

Functional maths Fishing competition on the Avon

Activity 2
Ask students to calculate an estimated mean for the following for Week 5.
- The number of fish caught.
- The time spent fishing.
- The weight of fish caught.
- The longest fish caught. (20 cm)

They should then add these averages to their graph from Activity 1.
To help **less able** students, work through the 'number of fish caught' in Week 5 as a class. Then ask them to calculate an estimate for the mean from the remaining information. If necessary, remind students of the definition of each average and how to calculate them. Writing these key principles on the board at the start of the lesson will help less able students.
Ask students to determine if Week 5 was a better week than the previous four. They must justify their answer, including a statement (using mathematical terms) of how they define 'better'.

Activity 3
Ask students to use the information gathered during Activities 1 and 2 to describe the 'average angler'. There are several possible descriptions and most will be valid as long as they are supported by a set of clear criteria defining the word 'average' in this context.

Activity 4
In a class discussion, decide which of students' 'average anglers' is truly the most average.

Activity 5
Ask students to investigate their own data about a subject that interests them. They should calculate averages and spreads for the data and draw conclusions. Then they should produce a short report to present to the class.

Apply
After working through the 'Getting started' section, hold a class discussion on which type of graph best represents each average. Get students to decide how they can make comparisons using their graphs.
Discuss how best to describe the average angler, including which averages to use and how to calculate them. If students are 'building', provide them with new data. **More able** students can use this activity as an opportunity to discuss sampling techniques. Point out that these activities are ideal opportunities for students to show mastery.

Master
In order to demonstrate mastery of the learning objectives, students should be able to:
- Use an appropriate graph to represent the data.
- Make at least three appropriate statements commenting on the results.
- Use several averages and possibly the range to describe the average angler.
- Consider why the data varied over the weeks.
- Collect and display their own data, using this to support conclusions in their final report.
- Discuss sampling techniques.

Plenary
After students' presentations, ask the class to provide feedback on one good aspect of the presentation and one aspect that could be improved.

© HarperCollins*Publishers* 2012

Answers Lessons 12.1 – 12.10

Quick check
1 a 7 b 6 c 8 d 6

12.1 Averages
Exercise 12A
1 Mode
2 Three possible answers: 12, 14, 14, 16, 18, 20, 24;
or 12, 14, 14, 16, 18, 22, 24;
or 12, 14, 14, 16, 20, 22, 24
3 53
4 a Median (Mean could be unduly influenced by results of very able and/or very poor candidates.)
b Median (Mean could be unduly influenced by pocket money of students with very rich or generous parents.)
c Mode (Numerical value of shoe sizes irrelevant, just want most common size.)
d Median (Mean could be distorted by one or two extremely short or tall performers.)
e Mode (The only way to get an 'average' of non-numerical values.)
f Median (Mean could be unduly distorted by very low weights of premature babies.)
5 The mean is 31.5, which rounds up to 32, so the statement is correct (though mode and median are 31).
6 a i £20 000 ii £28 000 iii £34 000
b A 6% rise would increase the mean salary to £35 060, a £1500 pay increase would produce a mean of £35 500.
7 a Median b Mode c Mean
8 Tom – mean, David – median, Mohamed – mode
9 a 9 b 7
10 a $6x$ b 8
11 $2x - 4$
12 $2x + 3$
13 11.6
14 42.7 kg
15 24

16 a Possible answer: 1, 6, 6, 6, 6
b Possible answer: 2, 5, 5, 6, 7
17 Boss chose the mean while worker chose the mode.
18 $5x = 17 + x + y$, $4x = 17 + y$, $y = 4x - 17$
19 a $a = 2y - x$

b $b = \dfrac{x + 3y}{4} = \dfrac{x}{4} + \dfrac{3y}{4}$

20 B = 66, C = 69, D = 93, E = 69

12.2 Frequency tables
Exercise 12B
1 a i 7 ii 6 iii 6.4
b i 8 ii 8.5 iii 8.2
2 a 668 b 1.9 c 0 d 328
3 a 2.2, 1.7, 1.3
b Better dental care
4 a 50 b 2 c 2.8
5 a Roger 5, Brian 4
b Roger 3, Brian 8
c Roger 5, Brian 4
d Roger 5.4, Brian 4.5
e Roger, smaller range
f Brian, better mean
6 a 40 b 7 c 3 d 2 e 2.5 f 2.5 g 2.4
7 a 2 b 1.9 c 49%
8 5
9 The total frequency could be an even number where the two middle numbers have an odd difference.
10 a 34
b $x + 80 + 3y + 104 = 266$, so $x + 3y = 266 - 184 = 82$
c $x = 10$, $y = 24$
d 2.5
11 The mean for 2009 = 396 ÷ 12 = 33, Range = 38
12 a 7 b 4 c 1.75
13 Any of (1, 8), (3, 7), (5, 6), (7, 5), (9, 4), (11, 3), (13, 2), (15, 1)

© HarperCollins*Publishers* 2012

12.3 Grouped data
Exercise 12C

1. **a** i $30 < x \leq 40$ ii 29.5
   b i $0 < y \leq 100$ ii 158.3
   c i $5 < z \leq 10$ ii 9.43
   d i 7–9 ii 8.41

2. **a** $100\text{ g} < w \leq 120\text{ g}$ **b** 10.86 kg
   c 108.6 g

3. **a** $175 < h \leq 200$
   b 31%
   c 193.3 hours
   d No, the mean was under 200.

4. **a** Yes, average distance is 11.7 miles per day.
   b Because shorter runs will be done faster which will affect the average.
   c Yes, because the shortest could be 1 mile, the longest 25 miles.

5. 24

6. Soundbuy; average increases are Soundbuy 17.7p, Springfields 18.7p, Setco 18.2p

7. **a** 160 **b** 52.6 minutes
   c Modal group **d** 65%

8. The first 5 and the 10 are the wrong way round.

9. Find the midpoint of each group, multiply that by the frequency and add those products. Divide that total by the total frequency.

10. **a** Yes, as total in first two columns is 50, so median is between 39 and 40.
    b He could be correct, as the biggest possible range is 69 – 20 - 49, and the lowest is 60 - 29 = 31.

12.4 Frequency diagrams
Exercise 12D

1. **a** 36
   b Pie charts with these angles: 50°, 50°, 80°, 60°, 60°, 40°, 20°
   c Check students' bar charts.
   d Bar chart, because easier to make comparisons from it.

2. **a** Pie charts with these angles: 124°, 132°, 76°, 28°
   b Split of total data can be seen at a glance.

3. **a** 55° **b** 22 **c** 33%

4. **a** Pie charts with these angles:
   Strings: 36°, 118°, 126°, 72°, 8°
   Brass: 82°, 118°, 98°, 39°, 23°
   b Overall, the brass candidates did better, as a smaller proportion got the lowest grade. A higher proportion of strings candidates scored the top two grades.

5. The sector for 'Don't know' has angle 360° − (80° + 90° + 150°) = 40°.
   As a fraction of the whole circle, the 'Don't know' sector is $\frac{40}{360} = \frac{1}{9}$ or, as a percentage, 11%.

6. Identify the possible ways in which students might come to school and, on the morning in question, use a tally chart to record how each one arrives.

Exercise 12E

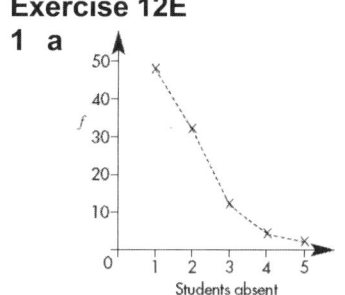

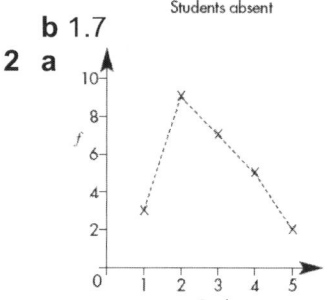

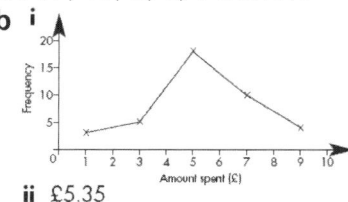

1. **a** [frequency diagram: Students absent vs f]
   b 1.7

2. **a** [frequency diagram: Goals vs f]
   b 2.8

3. **a** i 17, 13, 6, 3, 1 ii £1.45
   b i [frequency diagram: Amount spent (£) vs Frequency]
   ii £5.35
   c Much higher mean. Early morning, people just want a paper or a few sweets; later, people are buying food for the day.

Answers Lessons 12.1 – 12.10

4 a

[graph: frequency polygon, Height vs f, peaking near 135]

b

[graph: histogram, fd vs Height 120–170]

c 140.4 cm

5 a

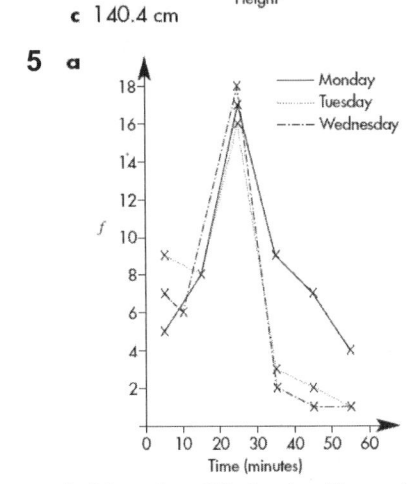

[graph: frequency polygons for Monday, Tuesday, Wednesday; Time (minutes) vs f]

b Monday 28.4 min, Tuesday 20.9 min, Wednesday 21.3 min
c There are more patients on a Monday, and so longer waiting times, as the surgery is closed during the weekend.

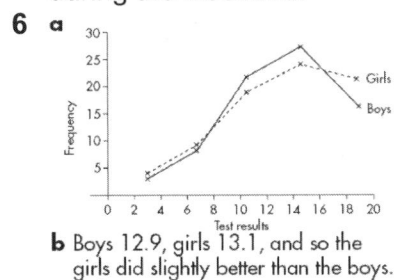

6 a

[graph: Frequency vs Test results, Girls and Boys]

b Boys 12.9, girls 13.1, and so the girls did slightly better than the boys.

7 2.17 hours
8 That is the middle value of the time group 0 to 1 minute. It would be very unusual for most of them to be exactly in the middle at 30 seconds.

12.5 Histograms with bars of unequal width
Exercise 12F

1 The respective frequency densities on which each histogram should be based are:
 a 2.5, 6.5, 6, 2, 1, 1.5
 b 4, 27, 15, 3
 c 17, 18, 12, 6.67
 d 0.4, 1.2, 2.8, 1
 e 9, 21, 13.5, 9

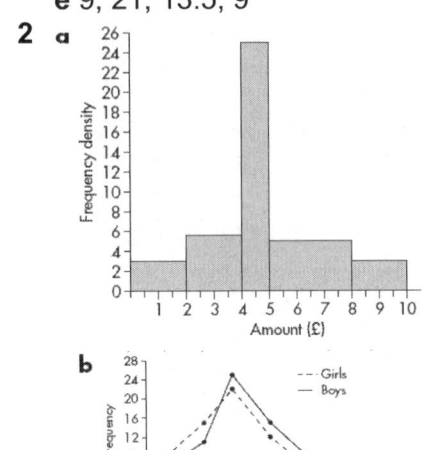

2 a

[histogram: Frequency density vs Amount (£)]

b

[graph: Frequency vs Pocket money (£), Girls and Boys]

c Girls £4.36, boys £4.81. Boys get more pocket money than girls do.

3

[histogram: Frequency density (000) vs Year 1940–2001]

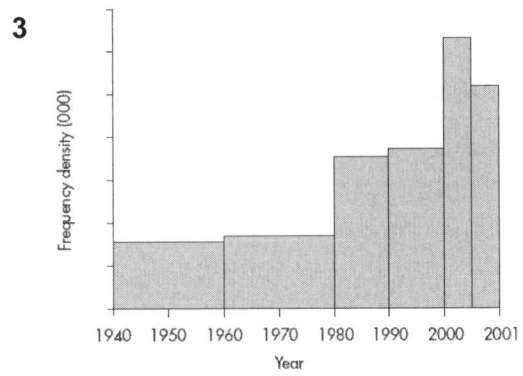

4 a 775 **b** 400

5 Divide the frequency of the class interval by the width of the class interval.

© HarperCollins*Publishers* 2012

Answers Lessons 12.1 – 12.10

6 a i

Age, y (years)	$9 < y \leq 10$	$10 < y \leq 12$	$12 < y \leq 14$	$14 < y \leq 17$	$17 < y \leq 19$	$19 < y \leq 20$
Frequency	4	12	8	9	5	1

  ii 10–12 **iii** 13 **iv** 11, 16, 5 **v** 13.4

b i

Temperature, t (°C)	$10 < t \leq 11$	$11 < t \leq 12$	$12 < t \leq 14$	$14 < t \leq 16$	$16 < t \leq 19$	$19 < t \leq 21$
Frequency	15	15	50	40	45	15

  ii 12–14°C **iii** 14.5°C **iv** 12°C, 17°C, 5°C **v** 14.8°C

c i

Weight, w (kg)	$50 < w \leq 70$	$70 < w \leq 90$	$90 < w \leq 100$	$100 < w \leq 120$	$120 < w \leq 170$
Frequency	160	200	120	120	200

  ii 70–90 kg and 120–170 kg **iii** 93.33 kg **iv** 74 kg, 120 kg, 46 kg **v** 99.0 kg

7 a 7.33 hours **b** 8.44 hours **c** 7 hours

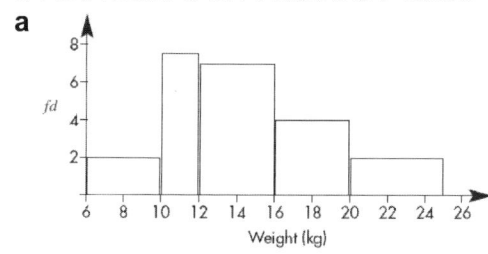

8 a

[Histogram: frequency density vs Weight (kg), bars approximately at 6–10: 2, 10–12: 8, 12–16: 7, 16–20: 4, 20–26: 2]

b 14.2 kg **c** 14.7 kg **d** 33 plants

9 a

Speed, v (mph)	$0 < v \leq 40$	$40 < v \leq 50$	$50 < v \leq 60$	$60 < v \leq 70$	$70 < v \leq 80$	$80 < v \leq 100$
Frequency	80	10	40	110	60	60

  b 360 **c** 64.5 mph **d** 59.2 mph

10 a 102 **b** 35 **c** 104 **d** 75

11 0.45

12.6 Surveys
Exercise 12G

1–5 Check students' answers and designs, which will vary.

6 a Possible answer:
Question – Which of the following foods would you normally eat for your main meal of the day?

Name	Sex	Chips	Beef burgers	Vegetables	Pizza	Fish

  b Yes, as a greater proportion of girls ate healthy food.

7 Possible answer:
Question – What kind of tariff do you use on your mobile phone?

Name	Pay as you go		Contract	
	200 or over free texts	Under 200 free texts	200 or over free texts	Under 200 free texts

(Any sheet in which choices that can distinguish one from the other have to be made will be accepted.)

8 Possible answers: shop names, year of student, tally space, frequency.

12.7 Questionnaires
Exercise 12H

1 a It is a leading question, and no option to disagree with the statement.
  b Unbiased, and the responses do not overlap.

2 a Responses overlap.
  b Provide options:
  up to £2, more than £2 and up to £5, more than £5 and up to £10, more than £10.

3–6 Check students' questionnaires.

7 a This is a leading question with no possibility of showing disagreement.
  b This is a clear direct question that has an answer, and good responses as only one selection can be made.
  c Check students' questions.

© HarperCollins*Publishers* 2012

Answers Lessons 12.1 – 12.10

8 Possible questionnaire:
Do you have a back problem?
☐ Yes ☐ No
Tick the diagram/text that best illustrates/describes how you sit.
☐ shoulders back awkwardly, curved spine
☐ slumped, straining lower back
☐ caved chest, pressure on spine
☐ balanced, head and spine aligned

9 The groups overlap, and the 'less than £15' is also in the 'less than £25'.

12.8 The data-handling cycle
Exercise 12I
1 a secondary data
  b primary data
  c primary or secondary
  d primary or secondary
  e primary data
  f primary data
2 Students' answers will vary.
3 For example, Kath may carry out a survey among her friends or class-mates.

12.9 Other uses of statistics
Exercise 12J
1 Price 78p, 80.3p, 84.2p, 85p, 87.4p, 93.6p
2 a £1 ≈ $1.88
  b Greatest drop was from June to July.
  c There is no trend in the data.
3 a 9.7 million   b 4.5 years
  c 12 million   d 10 million
4 £74.73
5 General cost of living in 2009 dropped to 98% of the costs in 2008.
6 a Holiday month
  b i 140 thousand
    ii 207 thousand (an answer of 200–210 thousand over the 3 months is acceptable).

12.10 Sampling
Exercise 12K
1–4 Check students' answers as they will vary.
5 a How many times, on average, do you visit a fast food outlet in a week?
    ☐ Never   ☐ 1 or 2 times
    ☐ 3 or 4 times ☐ More than 4 times

  b
	Boys	Girls
Y9	11	8
Y10	10	11
Y11	10	10

6 555
7 Find the approximate proportion of men and women, girls and boys, then decide on a sample size. Work out the proportion of men in the whole group and find that proportion of the sample size to give the number of men in the sample. Similarly work out the number of women, boys and girls.
8 a There are many possible correct answers. Below are two examples.
  How far from Meadowhall do you live?
  ☐ Less than 5 miles
  ☐ Between 5 and 10 miles inclusive
  ☐ More than 10 miles
  When you visit Meadowhall, approximately how much do you usually spend?
  ☐ £50 or less
  ☐ Between £50 and £100
  ☐ £100 or more

  b $Y7 = 143 \times \frac{100}{670} = 21$ $Y10 = 131 \times \frac{100}{670} = 20$
     $Y8 = 132 \times \frac{100}{670} = 20$ $Y11 = 108 \times \frac{100}{670} = 16$
     $Y9 = 156 \times \frac{100}{670} = 23$

9 a There are many more girls than boys.
  b 20

Answers Lessons 12.1 – 12.10

Examination questions

1. **a i** 8 **ii** 23 **iii** 19
   b There is no space for 0 hours and 6 hours appears in two groups.
2. **a** 1.7
   b Which days of the week are you prepared to car share?
   ☐ None ☐ Mon ☐ Tues ☐ Wed ☐ Thu ☐ Fri
3. 28.2 min
4. **a** 4060 ÷ 100 **b i** 125 **ii** 140.6
   c i

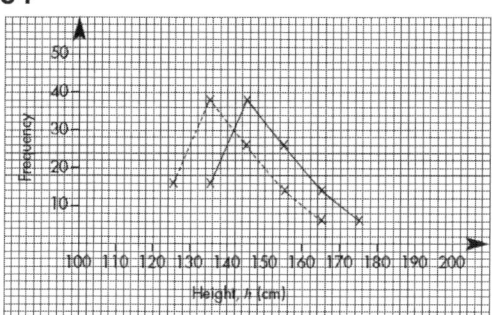

   ii On average, the boys are about 10 cm taller than the girls, the range of the heights in both groups is the same, 40 cm.
5. **a**

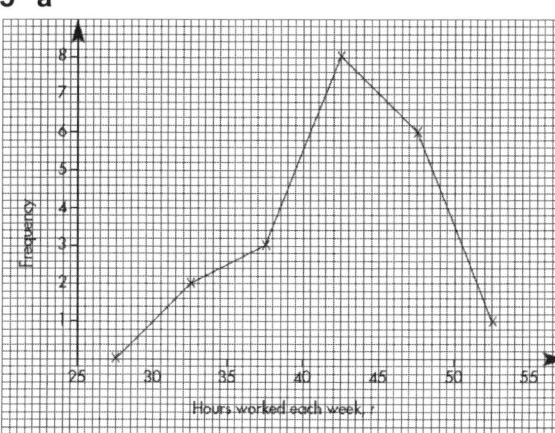

   b 60
   c It is what they thought and not what they measured.

6. **a** Level 5 **b** 3.2
   c Because of all the zeros at the lower grades in German
7. Because over half the students have more than £10 pocket money, so the mean must be more than £10.
8. **a** 24.6
9. **a** Polygon from (5, 6), (15, 10), (25, 20), (35, 8), (45, 6)
   b Mean for girls is lower. Girls' times are more consistent.
10. **a** 25 **b** 50 **c** 20 years
11. 70
12. **a**

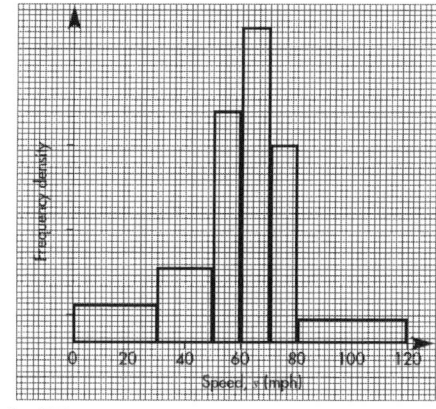

    b 57
13. **a**

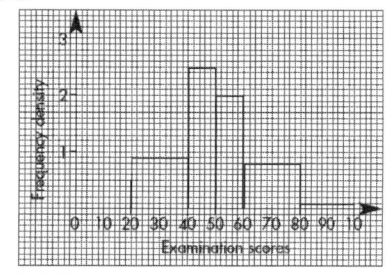

    b 38
14. NUT 1040, ATL 680, NATFHE 280
15. 12 − 9 = 3 more
16. Single 4, Couple 11, Family 15

© HarperCollins*Publishers* 2012

Chapter 13 Algebra: Real-life graphs

Overview

| 13.1 Straight-line distance–time graphs | 13.2 Other types of graphs |

Lesson 13.1 is an introduction to distance–time graphs, and how to interpret them. Lesson 13.2 includes a variety of other graphs, including the 'filling a container' graph that is so popular with examiners.

Context
Lesson 13.1 covers distance–time graphs and the interpretation of these, including average speed. Lesson 13.2 covers more unusual graphs, such as those that help to calculate tax and mortgage payments – both relevant to many people.

Curriculum references
KS4 Programme of Study references
1.1a Applying suitable mathematics accurately within the classroom and beyond.
1.3b Understanding that mathematics is used as a tool in a wide range of contexts.
2.1c Simplify the situation or problem in order to represent it mathematically, using appropriate variables, symbols, diagrams and models.
2.2k Make accurate mathematical diagrams, graphs and constructions on paper and on screen.
2.4b Engage in mathematical discussion of results.
4d Work on problems that arise in contexts beyond the school.

Linear specification
N6.12 Discuss, plot and interpret graphs (which may be non-linear) modelling real situations.

Functional Skills standards
R2 Identify the situation or problems and identify the mathematical methods needed to solve them.
I1 Interpret and communicate solutions to multistage practical problems in familiar and unfamiliar contexts and situations.

PLTS
Independent enquirers identify questions to answer and problems to resolve; explore issues, events or problems from different perspectives. **Reflective learners** assess themselves and others, identifying opportunities and achievements. **Team workers** collaborate with others to work towards common goals. **Self-managers** work towards goals, showing initiative, commitment and perseverance.

APP
Algebra L6.5 Construct functions arising from real-life problems and plot their corresponding graphs; interpret graphs arising from real situations. **Using and applying mathematics L8.2** Reflect on lines of enquiry when exploring mathematical tasks.

Route mapping

Exercise	Grades			
	D	C	B	A
A	1–6	7		
B	1	2–6		
C		1	2	3

© HarperCollins*Publishers* 2012

Chapter 13 Algebra: Real-life graphs

Overview test
Question 1 is grade E; question 2 is grade D.
1 This is a graph showing how far Cedric the slug travels across his lawn.

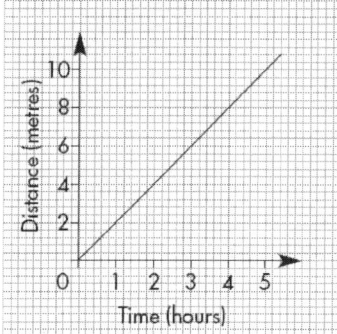

   a How far does Cedric travel in 1 hour?
   b How long does it take Cedric to travel 6 metres?
   c What is Cedric's speed in metres per hour (m/h)?
2 The following shapes are containers that will be filled with water from a hosepipe.
   The water flows at a steady rate all the time.

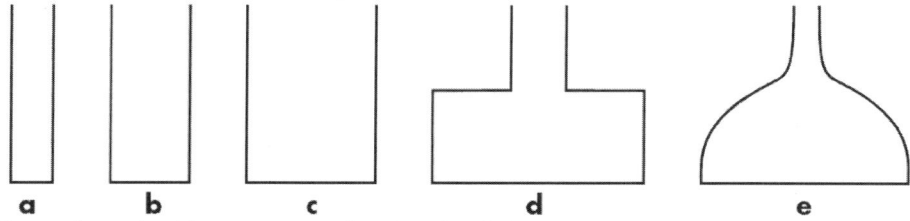

   Describe what happens to the depth of the water in each container as it fills.
   Does the depth of the water increase quickly, slowly, at a constant rate, or at a changing rate?

Answers to overview test
1 **a** 2 metres **b** 3 hours **c** 2 m/h
2 **a** Fills very fast, at an even rate.
   b Fills at an even rate, fairly fast, but not as fast as **a**.
   c Fills slowly, at an even rate.
   d Fills slowly at first at an even rate, then very fast at an even rate.
   e Fills slowly initially, but the rate changes, getting faster and faster – not even.

Why this chapter matters
Cross-curricular: This work has links with the sciences, Geography, History, Design and Technology.
Introduction: A line graph offers a simple and clear way of illustrating a link between two variables. These graphs can be used to represent simple direct proportion, as in currency conversion graphs, or more complicated relationships, as in distance–time graphs. An interesting application is used in question 2 of the Overview test, above.
Discussion points: Why are graphs useful for showing data? How are they different from data listed in tables? What does a currency conversion graph tell you? Why is it useful? How are graphs useful in measuring performance in sports? How are they useful in finance? What does a horizontal straight line on a distance–time graph mean? Why can there not be a vertical straight line on such as graph?
Plenary: As well as those shown in the Student Book, have a display of graphs collected from newspapers, magazines and the internet for discussion.

Answers to the quick check test can be found at the end of the chapter.

© HarperCollins*Publishers* 2012

13.1 Straight-line distance–time graphs

Curriculum references

Functional Skills standards
R2 Identify the situation or problems and identify the mathematical methods needed to solve them.

PLTS
Independent enquirers, Creative thinkers, Reflective learners, Team workers, Self-managers, Effective participators

APP
Algebra L6.5

Collins references
- Student Book pages 322–329
- Interactive Book: Real Life Video

Learning objectives
- Interpret distance–time graphs. [Grade D–C]

Learning outcomes
- Students **must** be able to find average speeds from a distance-time graph. [Grade D]
- Students **should** be able to interpret a distance–time graph. [Grade C]
- Students **could** be able to find the gradient of a straight line graph. [Grade C]

Key words
- average speed
- gradient
- distance
- speed
- distance–time graphs
- time

Prior knowledge
Students need to be confident in plotting points on coordinate axes, and they should be able to read scales on axes. Students should also know that:
distance = speed × time
time = distance ÷ speed
speed = distance ÷ time

Common mistakes and remediation
Students often interpret the minutes as a decimal, e.g. entering 2 hours 30 minutes as 2.3 on a calculator instead of 2.5. If this problem persists, provide some common conversions on the board or on an information sheet.
Students often misread scales and thus make mistakes, so remind them to pay special attention to this and give them extra practice in reading scales of different types, if necessary.

Useful tips
Time is always plotted as the horizontal axis. The gradient of the line is the speed.
Parts of journeys are always shown as straight lines, showing a steady average speed. However, in reality, speed varies due to traffic conditions.

Functional maths and problem-solving help
Distance–time graphs provide functional mathematics (FM) questions. Students should cope well with these, but may need help with average speed. The problem-solving (PS) questions require students to be able to interpret graphs. **Less able** students may benefit from working through a question, with your guidance.

Lesson 13.1 Straight-line distance–time graphs

Starter
- Ask students to convert the following hours, minutes and decimals.
  - From minutes, to hours and minutes: 230 mins (3h 50), 165 mins (2h 45), 84 mins (1h 24)
  - From hours and minutes, to minutes only: 1h 55 (115 mins), 2h 15 (135 mins), 3h 16 (196 mins)
  - From hours and minutes, to decimals: 1h 30 (1.5h), 2h 45 (2.75), 3h 20 (3.3333…)
  - From decimals in hours, to hours and minutes: 2.666… (2h 40), 3.25 (3h 15), 1.1 (1h 6)
- Make sure students are familiar with the decimal equivalents of 15 minutes, 30 minutes, etc.

Main lesson activity
  On the board, sketch a graph as shown. Mark the axes as time and distance. Ask what sort of journey this could represent e.g. a form of public transport. (Students should recognise this as being a bus journey.)
- Ask, "Why are the sloping lines straight?" Explain that the graph models the real situation; the sloping lines show average or steady speed. In reality, speed varies and the line would not be straight. Ask what the horizontal lines represent. Make sure they understand that horizontal lines represent zero speed.
- Now sketch a graph as shown. Discuss the differences, compared to the previous graph. (It returns to the original point, so is an out-and-back graph.) Discuss the slopes of the lines. Ask which part shows the fastest part of the journey. Make sure they understand that the steeper the line, the higher the speed. Then ask which is the second fastest part of the graph. This may cause confusion, as the return journey is the second fastest. Make it clear that the slope direction is not relevant.
- Ask students if they can recall the connection between speed, distance and time. Then draw the speed–distance–time 'road sign'.
- Work through Example 1.
- **Less able** students may have difficulty in working out average speed. Make sure they understand that it is the total distance travelled divided by the total time taken.
  Students can now do Exercise 13A, in which questions 1–3 and 5 are FM, 4-6 are PS and 7 is an assessing understanding (AU) question.
  Gradient of straight-line distance–time graphs
- Work through this section which demonstrates how to calculate the gradient. Stress the following points.
  - When drawing a triangle to measure the gradient, use grid points so that values are integers.
  - Always divide distance in the vertical direction by distance in the horizontal direction (which will be time).
  - Lines that slope from top left top bottom right have a negative gradient, which on a distance–time graph means that the person or vehicle is on the way back.
  - The gradient, whether positive or negative, is the speed.
- **More able** students should understand the concept of velocity. It is a vector, so has a value and a direction.
- Students can now do Exercise 13B, in which questions 1 and 6 are FM, 5 is AU and 6 is PS.

Plenary
- Ask some quick-fire questions about calculating speed: How fast would you be travelling if you covered 60 miles in 2 hours? 3 kilometres in 30 minutes? 10 kilometres in 15 minutes? 210 miles in 3 hours?
- Ensure students can convert to distance per hour (questions above: 30 mph, 6 km/h, 40 km/h, 70 mph).

Answers to Exercises 13A and 13B can be found at the end of this chapter.

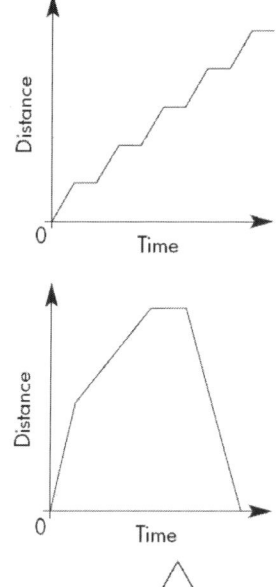

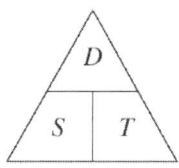

© HarperCollins*Publishers* 2012

13.2 Other types of graphs

Curriculum references

Functional Skills standards
I1 Interpret and communicate solutions to multistage practical problems in familiar and unfamiliar contexts and situations.

PLTS
Independent enquirers

APP
Algebra L6.5

Collins references
- Student Book pages 329–330
- Interactive Book: Matching Pairs

Learning objectives
- Identify and draw some of the more unusual types of real-life graphs. [Grade C–A]

Learning outcomes
- Students **must** be able to match a graph to a particular real life situation. [Grade C]
- Students **should** be able to sketch a graph to match a particular situation. [Grade B]
- Students **could** be able to draw a graph to show particular information. [Grade A]

Key words
- curved
- gradient
- depth
- measure

Prior knowledge
Students should know how to work out 10% of a quantity, and then extend this to 20%.

Common mistakes and remediation
Students may not realise that some change in depth is gradual, which would make the graph a curve, and will use straight-line graphs instead. More examples of both curved and straight-line graphs would help to remediate this.

Useful tips
Remind students that in graphs involving time, this is always plotted on the horizontal axis. There can be no vertical parts in a graph involving time as it never stands still.

Functional maths and problem-solving help
This whole chapter gives plenty of opportunity for functional mathematics (FM). The question in the exercise is based on income tax, which is something everyone needs to understand.
Two questions in Exercise 13C, one of them a problem-solving (PS) task, are based on filling containers at a steady rate. Students will encounter the same problems as they did previously, so offer support as needed. Encourage students to be methodical when setting out the working on the question about tax.
Note: This section is best introduced using dynamic imagery. The student CD provides this.

Lesson 13.2 Other types of graphs

Starter
- Display a target board, as shown below but without the answers in brackets, and ask students to work out 10% of each quantity.

£75 (£7.50)	£20 (£2)	£5 (£0.50)	£35 (£3.50)
£120 (£12)	£45 (£4.50)	£32 (£3.20)	£10 (£1)
£100 (£10)	£16 (£1.60)	£50 (£5)	£250 (£25)
£400 (£40)	£8 (£0.80)	£150 (£15)	£500 (£50)

- Then ask students to work out 5% of each quantity (this is half of each amount in brackets in the target board, above), and 20% (double each amount in brackets).

Main lesson activity
- If the CD is not available, work through the following plan. Obtain a variety of containers, e.g. conical flasks (the science department may be a good source).
  Explain that you are going to plot the depth of liquid in the flask as liquid is poured in at a steady rate. Start with a cylinder and assume that 10 ml is added every second. Add 10 ml of liquid and measure the depth in the cylinder. Plot this on a graph. Add another 10 ml and measure depth and plot the points. Repeat until the shape of the graph is established. (This will be a straight-line graph.) Now repeat with a conical flask, but first ask students if they can predict any differences between this graph and the previous graph. Now add 10 ml, in stages, and plot depths. This graph will increase in gradient. Repeat with another differently-shaped container.
- Now go through this section in the Student Book, which shows more examples, including filling a flask with a round body but a cylindrical neck.
- Now students can do Exercise 13C, in which question 1 is an assessing understanding (AU) question, 2 is PS and question 3 is FM.
- **Less able** students will find the concept of how the depth changes over time difficult. Discussing the change in volume, and how this relates to the cross-sectional area, may help, but it is not easy for some students to visualise 3D concepts. Encourage **more able** students to work with those who are having difficulty.

Plenary
- Draw differently-shaped containers and ask students to draw graphs on the board. Or, ask students to test one another by drawing their own container on the board.
- Draw a graph and ask students to work out the shape of the container.

Answers
Answers to Exercises 13C can be found at the end of this chapter.

© HarperCollins*Publishers* 2012

Functional maths
Planning a motorbike trip to France

Curriculum references

Functional Skills standards
R2 Identify the situation or problems and identify the mathematical methods needed to solve them.
I1 Interpret and communicate solutions to multistage practical problems in familiar and unfamiliar contexts and situations.

PLTS
Reflective learners, Team workers, Self-managers

APP
Using and applying mathematics L8.2

Collins references
- Student Book pages 336–337

Learning objectives
- Communicate findings effectively using correct mathematical terminology, and accurate diagrams and graphs.
- Apply mathematical accuracy.
- Identify the mathematical aspect of a situation or problem.
- Explore the effects of varying values.

Key words
- conversion
- convert
- estimate
- investigate

Functional maths help
Refer students to the conversions in the 'Handy hints' box in the Student Book. For the benefit of **less able** students, you may wish to supply conversion charts to aid them in working out the questions.

Lesson plans
Activity 1
Ask students to work out the answers to the following questions.
- How many kilometres are there in 10 miles (16 km), 15 miles (24 km), 40 miles (64 km) and so on? (Provide other distances for students to work out.)
- How many litres are there in a tank of fuel? For this question and the next, students should refer to the text towards the end of the 'Handy hints' box. (12 litres)
- How much would it cost for one tank of fuel? (€14.40)
- How much would it cost to travel 150 miles? (€48)

Activity 2
Refer students to the 'Getting started' text. Then ask them to draw a conversion graph of British pounds to Euros. For the benefit of **less able** students, draw a set of axes on the board to prompt them. Let students think about and work through the remaining points in the text.

Activity 3
Students should use the map to work out at least two routes from Boulogne to Perpignan. For each route, they should calculate the total distance and approximate driving times.
More able students could also work out an average time for their routes. You could provide criteria to help guide and challenge them. For example, you take a 20-minute break for every two hours travelled.
These activities should give students some firm direction in approaching the task. They can then begin to think about further information that they will need and to create their own graphs to represent this information.

Activity 4
Ask students to carry out the same investigation and produce different routes for the friend's journey.

© HarperCollins*Publishers* 2012

Functional maths Planning a motorbike trip to France

Activity 5
Ask students to use the internet to investigate what would happen if they chose the Channel Tunnel instead of the ferry. They should answer the following questions.
- What effect would this have on the cost?
- How would miles travelled differ?

Activity 6
Ask students to work out the following problem that the travellers encounter.
Just before travelling, the pound drops against the euro and the exchange rate is now: 1 euro is 98p.
How will this alter the price of the holiday?

Activity 7
Students should be able to price the journey through the UK to the ferry port. Ask them to imagine that they will begin the journey from their home town. Then they should answer these questions.
- What will it cost to travel to the south coast, or ferry port?
- How much of the cost would it be in pounds and how much in Euros?
- If you combine this information with a change in the euro exchange rate, what is the overall percentage change in the group's travelling costs?

Ask **more able** students to produce a plan of costs and make contingency recommendations.

Apply
Leaving the task more open, give students minimal guidance; this could include the following points.
- Using the information given to work out at least two routes. For each route, calculate the distance travelled, time taken and cost implications.
- Use the information given on the page to provide other data. This could include converting a number of mileages (or mph) to kilometres per hour (kph) in order to ensure that they stick to the speed limits in France.

Note: Speed limits in France might be approximately 50 kph in towns, 90 kph on open roads, 110 kph on dual carriageways and 130 kph on motorways.

For Activities 4–7, provide support to **less able** students as needed. Encourage students to ask and answer their own questions. **More able** students should be able to evaluate their questions. For example, were they useful, and did I approach the tasks in the best way?

Master
Students should be able to ask mathematical questions of their own and should be expected to take the task in a number of different directions.
In order to demonstrate mastery of the learning objectives, students will need to take their investigation down a path that will achieve the following success criteria:
- Drawing at least one conversion graph.
- Reading and using information from a conversion graph.
- Using **all** the data supplied to make statements about the journey through France, such as identifying at least one route, and producing different routes for the friend's journey.
- Using a range of values in calculations.
- Approaching the problem in a systematic way.
- Explaining and writing findings accurately and appropriately.

Plenary
Share students' work and ask them to highlight the mathematics that they have used in the task.
As a class, decide if the information that each group has gathered would be useful if they were to undertake such a journey and if there is anything else that they would need to find out.

© HarperCollins*Publishers* 2012

Answers Lessons 13.1 – 13.2

Quick check
1 A(3, 0), B(1, 4), C(4, 5)

13.1 Straight-line distance–time graphs
Exercise 13A
1 **a i** 2 h **ii** 3 h **iii** 5 h
   b i 40 km/h **ii** 120 km/h **iii** 40 km/h
   c 6.30 am
2 **a i** 125 km **ii** 125 km/h
   b i Between 2 pm and 3 pm
   ii About $12\frac{1}{2}$ km/h
3 **a** 30 km **b** 40 km **c** 100 km/h
4 **a i** 263 m/min (3 sf) **ii** 15.8 km/h (3 sf) **b** 500 m/min **c** Paul by 1 minute
5 **a** Patrick ran quickly at first, then had a slow middle section but he won the race with a final sprint. Araf ran steadily all the way and came second. Sean set off the slowest, speeded up towards the end but still came third.
   b i 1.67 m/s **ii** 6 km/h
6 There are three methods for doing this question. This table shows the first, which is writing down the distances covered each hour.

Time	9am	9:30	10:00	10:30	11:00	11:30	12:00	12:30
Walker	0	3	6	9	12	15	18	21
Cyclist	0	0	0	0	7.5	15	22.5	30

The second method is algebra. Walker takes T hours until overtaken, so $T = \frac{D}{6}$; Cyclist takes $T - 1.5$ to overtake, so $T - 1.5 = \frac{D}{15}$.

Rearranging gives $15T - 22.5 = 6T$, $9T = 22.5$, $T = 2.5$.
The third method is a graph.

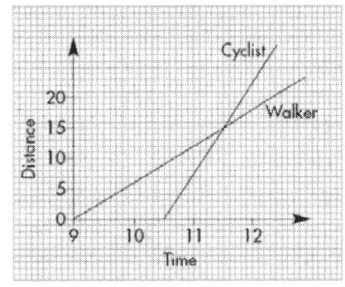

2 **a** 18 **b** 265

All methods give the same answer of 11.30 when the cyclist overtakes the walker.

7 **a i** Because she stopped several times.
   ii Ravinder
   b Ravinder at 3.58 pm, Sue at 4.20 pm, Michael at 4.35 pm
   c i 24 km/h **ii** 20.5 km/h **iii** 5

Exercise 13B
1 **a** 39.2°C
   b Day 4 and 5, steepest line
   c Day 8 and 9, steepest line
   d i Day 5 **ii** 4
   e 37° C
2 **a** $\frac{15}{2}$ **b** $\frac{25}{8}$ **c** $\frac{15}{16}$ **d** $\frac{2}{25}$ **e** $\frac{6}{35}$
   f $\frac{1}{2}$ **g** $-\frac{4}{5}$
3 **a** 2.5 km/h **b** 3.75 m/s **c** 2.5 km/h
4 **a** AB: 30 km/h, BC: 6 km/h, CD: 0 km/h, DE: 36 km/h (in opposite direction)
   b FG: 4 m/s, GH: 16 m/s, HI: 2 m/s (in opposite direction), IJ: 16 m/s (in opposite direction)
5 Rob has misread the scales. The gradient is actually 2.
The line is $y = 2x + 2$, when $x = 10$, $y = 22$.
6 **a** 4 m **b** 1 m
   c i $\frac{4}{3}$ m **ii** 3 m

13.2 Other types of graphs
Exercise 13C
1 a Graph C
  b Any container with a regular horizontal cross-section

2 a, b, c, d, e, f (depth vs time graphs)

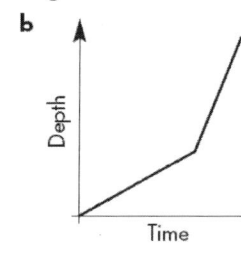

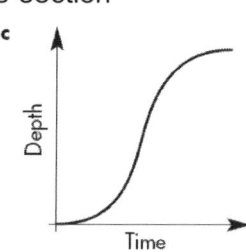

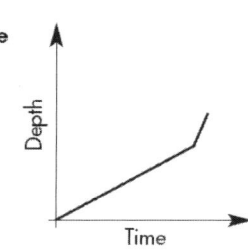

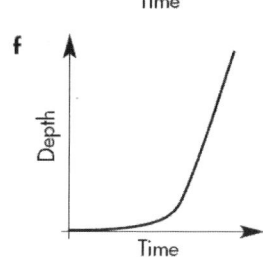

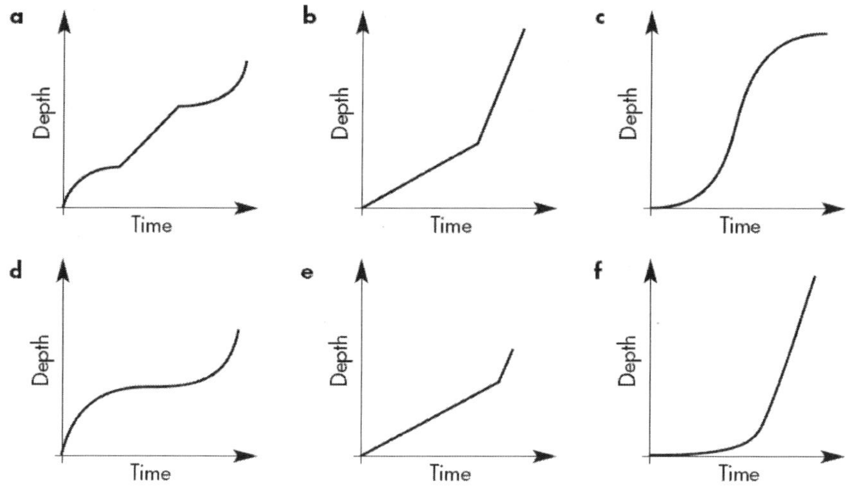

3 (graph of Tax paid (£1000s) vs Income (£1000s))

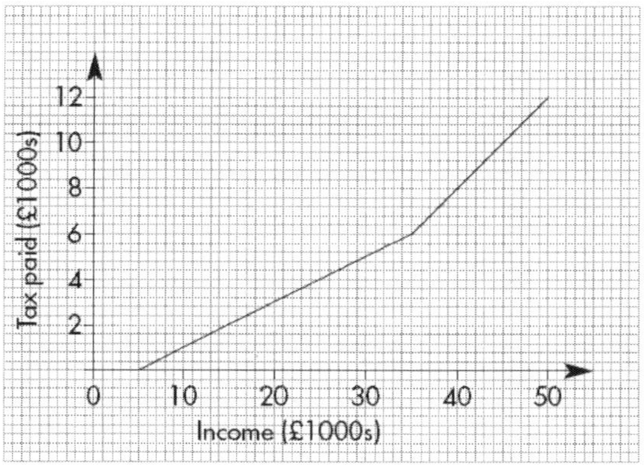

Examination questions
1 a (distance from home vs time graph)
  b 16 km/h
2 a 15 minutes **b** 3 km
  c D to E, line is steepest.
3 a 150 miles
  b 10 minutes
  c 50 mph
4 68 mph
5 20 kilometres per litre
6 a Grant
  b 93 seconds
  c 65 m/min
  d i Mark **ii** Steeper line
7 a (distance vs time graph)
  b 25 km/h
8 a High **b** 30 in
  c Line from (18, 18) to (40, 40) and labelled.

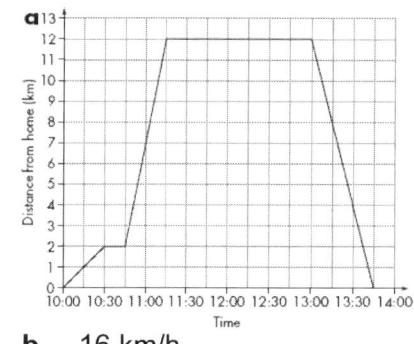

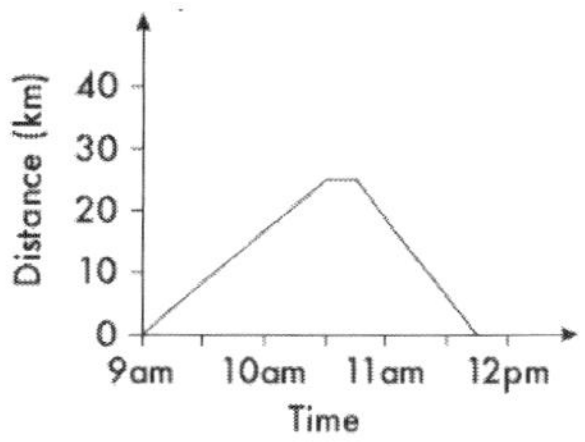

© HarperCollins*Publishers* 2012

221

Chapter 14 Statistics: Statistical representation

Overview

| 14.1 Line graphs | 14.3 Scatter diagrams |
| 14.2 Stem-and-leaf diagrams | |

This chapter covers more of the statistical topics required at this level.

Context
Line graphs have various uses, but here they indicate how data changes over time, and show trends. This idea may need to be explained. Stem-and-leaf diagrams offer a quick way to summarise and display data presented as a list of numbers. Students should do this in two stages, first simply listing the 'leaves' as they appear in the raw form, then refining the diagram to produce an ordered stem-and-leaf diagram. Scatter diagrams as used to compare to sets of data that may or may not be related, such as height and shoe size, or eye-colour and handspan. Students may think that some sets of data are clearly related while others are not, but scatter diagrams are an interesting way of checking.

Curriculum references
KS4 Programme of Study references
1.1a Applying suitable mathematics in the classroom and beyond.
1.3b Understand that mathematics is used as a tool in a wide range of contexts.
2.3d Look at data to find patterns and exceptions.
3.3b Presentation and analysis of large sets of grouped and ungrouped data, including box plots and histograms, lines of best fit and their interpretation.
3.3c Measures of central tendency and spread.

Linear specification
S1 Understand and use the statistical problem-solving process which involves:
- specifying the problem and planning
- collecting data
- processing and presenting the data
- interpreting and discussing the results.

S3.2h Histograms with unequal class intervals, box plots, cumulative frequency diagrams, relative frequency diagrams.
S3.3h Calculate median, mean, range, quartiles and inter-quartile range, mode and modal class.
S4.1 Interpret a wide range of graphs and diagrams and draw conclusions.
S4.3 Recognise correlation and draw and/or use lines of best fit by eye, understanding what these represent.
S4.4 Compare distributions and make inferences.

Functional Skills standards
R2 Identify the situation or problems and identify the mathematical methods needed to solve them.
A2 Use appropriate checking procedures and evaluate their effectiveness at each stage.
I2 Draw conclusions and provide mathematical justifications.

PLTS
Independent enquirers identify questions to answer and problems to resolve; analyse and evaluate information, judging its relevance and value; support conclusions, using reasoned arguments and evidence. **Creative thinkers** question their own and others' assumptions; adapt ideas as circumstances change. **Reflective learners** review progress, acting on the outcomes; invite feedback and deal positively with praise, setbacks and criticism; communicate their learning in relevant ways for different audiences. **Team workers** collaborate with others to work towards common goals; reach agreements, managing discussions to achieve results. **Self-managers** work towards goals, showing initiative and perseverance; organise time and resources; seeking advice and support when needed.

© HarperCollinsPublishers 2012

Chapter 14 Statistics: Statistical representation

APP
Handling data L6.2 Select, construct and modify on paper and using ICT: pie charts for categorical data, bar charts and frequency diagrams for discrete and continuous data, simple time graphs for time series, scatter graphs, and identify which are most useful in the context of the problem. **L7.6** Examine critically the results of a statistical enquiry, and justify the choice of statistical representation in written presentation. **Using and applying mathematics L7.2** Give reasons for choice of presentation, explaining selected features and showing insight into the problems structure.

Route mapping

Exercise	Grades D	Grades C
A	all	
B	all	
C	1	2–8

Overview test
Question 1 is grade D and questions 2 and 3 are grade C.
1. Draw a set of coordinate axes, marking each axis (x and y) from –4 to +4.
   The coordinates (3.5, 1.5), (3.5, –1.5) and (–2.5, 1.5) are three vertices of a rectangle.
   Plot them and state the coordinates of the missing vertex.
2. What can you say about the gradients of each of these lines?
   a $y = x$ **b** $y = -x$ **c** $y = 3$ **d** $x = 3$
3. The table shows the lengths of some marrows grown on Lucy's allotment.
   a State the modal group.
   b Estimate the mean length of a marrow.
   c In which class interval must the median lie?

Length L (cm)	Frequency (f)
$0 \leq L < 20$	4
$20 \leq L < 30$	16
$30 \leq L < 35$	12
$35 \leq L < 40$	10
$40 \leq L < 50$	6
$50 \leq L < 60$	2

Answers to overview test
1. (–2.5, –1.5)
2. **a** Gradient is positive (= 1). **b** Gradient is negative (= –1). **c** Gradient is 0. **d** Gradient is infinite.
3. **a** $20 \leq L < 30$ **b** 31.7 cm **c** $30 \leq L < 35$

Why this chapter matters
Cross-curricular: This work links to Geography, Citizenship, Science and any other subject that involves statistical information.
Introduction: What is a normal distribution? It really is as straightforward as it sounds. It shows how values of a variable are distributed over a range of data, with most occurring near the middle and fewer at the boundaries, to give the classic bell-shaped curve. At this level, most of the work on statistics that is covered is not difficult, once students know what the various terms mean.
Discussion points: More people live in urban areas than in the countryside. How do we know this? Do tall people need bigger hats and shoes? How can we find out? Do people who read their horoscopes live longer? Is there any connection between eye colour and height? Or sight defects?
Plenary: Carry out a quick survey of arm length and shoe size, or hair colour and favourite subjects. Discuss how to show the results.

Answers to the quick check test can be found at the end of the chapter.

© HarperCollinsPublishers 2012

14.1 Line graphs

Curriculum references

Functional Skills standards
I2 Draw conclusions and provide mathematical justifications.

PLTS
Independent enquirers, Creative thinkers, Reflective learners, Team workers, Self-managers, Effective participators

APP
Handling data L6.2

Collins references
- Student Book pages 340–341
- Interactive Book: Worked Exam Questions

Learning objectives
- Draw a line graph to show trends in data. [Grade D]

Learning outcomes
- Students **must** be able to draw a line graph. [Grade D]
- Students **should** be able to compare two line graphs. [Grade D]
- Students **could** be able to draw conclusions from a line graph. [Grade D]

Key words
- line graphs
- trends

Prior knowledge
Students should be able to understand data presented in tabular form. They need to be able to plan and read scales on axes in a variety of situations.

Common mistakes and remediation
The most common mistake students make here is the poor choice, or incorrect use, of scales to represent times. Students also often make intermediate or future estimates without giving realistic thought to the data. Point out these errors to help students avoid making them.

Useful tips
Remind students that they should always think carefully about the data they are dealing with.

Functional maths and problem-solving help
The functional maths (FM) questions in Exercise 14A are 1d and 3c, and use real-life scenarios. Question 1 of the exercise requires students to recognise a trend. Encourage them to take note of the differences from year to year in order to grasp that though these differences keep changing, the trend is still that the number of tourists is increasing yearly. This could lead to a discussion on the 'green' issue of the use of aeroplanes and the demand for foreign holidays, which increases emissions of greenhouse gases. In the problem-solving (PS) question, 4, students are asked to determine weight according to the data in the table.

© HarperCollins*Publishers* 2012

Lesson 14.1 Line graphs

Starter
- Draw a straight line segment divided by markers into four or five unequal sections. Put an appropriate number at each marker. Indicate points on the line between markers and ask students to say what values they represent. Repeat with different lines and values.
- Repeat the exercise with times along the line segments. (A variety of time units would be useful, e.g. minutes, years and dates.)

Main lesson activity
- Draw a bar chart on the board, using the data from Example 1. Tell the class that this bar chart represents outside temperatures at various times.
- Ask, "Why is this type of bar chart not as helpful as it could be?" Lead students to respond that it does not tell you anything about the temperatures in between the marked points in time.
- Put a cross on the middle of the top of each bar in the chart and join the crosses. Students should agree that this is better. (If necessary, demonstrate to **less able** students the reason why.) Point to a position along the line and ask what time it represents. Then ask for the temperature at that point.
- Explain that line graphs are often used instead of bar charts when time is involved. Mention to **more able** students that the correct terminology is a 'time series' although they do not need to use the term. With a line graph estimates between the given values can be made.
- Look at Example 1 again and explain that although the temperature was taken only at certain times, it is still a time series, as time is continuous.
- Now students can complete Exercise 14A, in which questions 1d and 3c are FM. Question 4 is PS and 5 is an assessing understanding (AU) question. **Less able** students can work in pairs if you feel this will help them.

Plenary
- Have a classroom discussion about the value of line graphs, by asking, e.g. when they can be used, why they can be useful, and so on.

Answers
Answers to Exercise 14A can be found at the end of this chapter.

14.2 Stem-and-leaf diagrams

Curriculum references

Functional Skills standards
A2 Use appropriate checking procedures and evaluate their effectiveness at each stage.

PLTS
Independent enquirers, Creative thinkers, Reflective learners, Team workers, Self-managers, Effective participators

APP
Not in assessment criteria.

Collins references
- Student Book pages 342–344
- Interactive Book: Multiple Choice Questions

Learning objectives
- Draw and read information from an ordered stem-and-leaf diagram. [Grade D]

Learning outcomes
- Students **must** be able to construct a stem-and-leaf table. [Grade D]
- Students **should** be able to find the median and the mode from a stem-and-leaf table. [Grade D]
- Students **could** be able to compare distributions from stem-and-leaf tables. [Grade D]

Key words
- discrete data
- raw data
- ordered
- unordered

Prior knowledge
Students should be able to order a set of numbers. They should also have a clear understanding of place value in order to classify the digits of a set of numbers into the same category.

Common mistakes and remediation
The most common mistake is in misreading stem-and-leaf diagrams. Students often forget to recombine the two digits of the original data value, or do not do so consistently. They may also forget to use the key, and so make mistakes with the size of the numbers.
When creating a stem-and-leaf diagram, students may forget to provide a key. They may also be careless with spacing the digits, which reduces the effectiveness of the diagram. For remediation, refer to 'Useful tips' and also provide plenty of practice.

Useful tips
Emphasise the need for neatness, which will help to reduce careless errors.

Functional maths and problem-solving help
Although this lesson is functional mathematics (FM), there are no FM questions in Exercise 14B. Students are shown that they can represent a situation by using mathematics in a novel way. Tell them that stem-and-leaf diagrams are a relatively new way of showing data and that their parents are unlikely to have come across them unless they use them at work.
Problem-solving (PS) question 4 of Exercise 14B shows a different way to use a stem-and-leaf diagram. Again, students are required to make some interpretation from the table and to identify the key. They may need help in interpreting this diagram, but encourage them to make their own observations first.

Lesson 14.2 Stem-and-leaf diagrams

Starter
- Show the class a set of numbers, and to arrange them in order, from lowest to highest. Repeat for sets of numbers of different sizes, e.g. a set of all two-digit integers, all two-digit decimal numbers, all numbers with one unit digit and one decimal digit, and so on.
- Show the class this set of numbers; 23, 54, 20, 48, 26, 58, 41. Ask them to suggest ways to select groups of numbers from this collection that have something in common. Examples include 54 and 48 (both multiples of 6); 23 and 41 (both odd numbers), etc. Someone may suggest groups with the same tens digit (20, 23 and 26; 41 and 48; and 54 and 58).

Main lesson activity
Raw data
- Put a stem-and-leaf diagram on the board. It is irrelevant what the data represents as long as it looks something like the diagram in the Student Book. Be sure to include the key.
- Explain that this is a stem-and-leaf diagram, and that it represents numbers. Explain that the key just shows how the diagram represents numbers. It must show a number that is in the range of the data. This may not be clear to **less able** students, who will need a few examples using the key before they understand the notation.
- Explain how the stem is almost always on the left-hand side and typically represents tens, whereas each leaf is the unit digit of a number and the leaves appear on each branch, on the right.
- You may wish to demonstrate an alternative that they will encounter in Exercise 14B question 4, where the stem is central, as it is used for two sets of data and the leaves appear on both sides. Part of the problem is to identify the key. Explain that, as long as the key is clearly defined, this should not cause confusion. This example may confuse **less able** students, so although it is important to explain this type to the class before they attempt the exercise, ensure that they have had practice in using the usual type of stem and leaf.
- Explain that this display is an easy way to represent a set of numbers without having to write them all down, and that it allows you to put the numbers into order, ready for analysis. Work through Example 2.
- Challenge **more able** students with collections of numbers that have decimal numbers or numbers in the thousands.
- Now students can complete Exercise 14B, in which question 4 is PS and question 5 is an assessing understanding (AU) question. **Less able** students can complete the exercise in pairs.

Plenary
- Ask students to discuss the advantages or disadvantages of stem-and-leaf diagrams over grouped frequency tables.
- Ask students to bring in one of their own CDs and to present the CD track times in the form of a stem-and-leaf diagram.

Answers
Answers to Exercise 14B can be found at the end of this chapter.

© HarperCollins*Publishers* 2012

14.3 Scatter diagrams

Curriculum references

Functional Skills standards
I2 Draw conclusions and provide mathematical justifications.

PLTS
Independent enquirers, Creative thinkers, Reflective learners, Team workers, Self-managers, Effective participators

APP
Handling data L6.2

Collins references
- Student Book pages 344–348
- Interactive Book: Common Misconceptions

Learning objectives
- Draw, interpret and use scatter diagrams. [Grade D–C]

Learning outcomes
- Students **must** be able to describe the type of correlation in a scatter diagram. [Grade D]
- Students **should** be able to draw a line of best fit on a scatter diagram. [Grade C]
- Students **could** be able to estimate missing values in a scatter diagram. [Grade C]

Key words
- line of best fit
- positive correlation
- negative correlation
- scatter diagram
- no correlation
- variable

Prior knowledge
In terms of statistical work, no prior knowledge is needed for studying the topic in this lesson. Students should, of course, be able to use appropriate scales on axes and be able to interpret intermediate positions within the scales.

Common mistakes and remediation
Students often draw lines of best fit inappropriately. Remind them that the line need not necessarily go through the origin, nor does it actually have to go through any plotted points.

Useful tips
The line of best fit should have as many values above it as below it.
No correlation indicates that there is no clear link between the two sets of data, which is sometimes useful to know.

Functional maths and problem-solving help
Most of Exercise 14C involves functional maths (FM). Questions 3 and 5 require students to make justifications, and every question requires them to recognise some trend, and then use it to estimate further data that might not actually be recorded.
Problem-solving (PS) question 7 requires students to manipulate the data for a value that does not appear in the table. They must read the question carefully and take care with the times, which are given as decimal numbers of hours.

© HarperCollins*Publishers* 2012

Lesson 14.3 Scatter diagrams

Starter
- Draw a straight-line graph on a coordinate grid. Give the value of one variable and ask students to find the corresponding value of the other variable.

Main lesson activity
Correlation
- Ask, "If you get a good mark in a mathematics test, are you likely also to get a good mark in an English test?" Discuss trends and the fact that some people 'buck the trend'.
- Refer to the scatter diagram in the Student Book, showing mathematics and English results. Mention that these points have been plotted from data and that the pattern they show clearly demonstrates that, for this group of students, there is a clear link between a good English mark and a good mathematics mark – the higher their English mark, the higher is likely to be their mathematics mark.
- Introduce the word 'correlation' and explain that, in this case, the diagram indicates positive correlation – the bigger one variable gets the bigger the other gets.
- Explain that this is not always the case. Refer to the three diagrams that have positive, negative and no correlation. Ensure that students grasp the clear difference between positive and negative correlation. Look at Example 3 with students.
- Extend this discussion with **more able** students to include strong positive and weak positive correlation, and the equivalent negative correlations. Some patterns show a slight correlation which can be described as weak, or a very clear link, hence strong. Show some examples on the board.

Line of best fit
- Ask students to look at the diagrams illustrating lines of best fit and ask them to describe the lines. If necessary, explain them in more detail, mentioning that there should be just one straight line that shows the trend – it does not need to go through any of the given points, nor through the origin – it should simply be as close to them all as possible, with as many above it as below.
- Explain 'outliers' to **more able** students: these are points that clearly buck the trend. This will sometimes happen and, as long as only one or two are in the data set, they can be ignored. Mention to **more able** students that a line of best fit can be a curve, but again it must be only one smooth curve and not an attempt to join every point together.
- Ask questions about missing marks from the set, e.g. "What English mark might you expect from the girl who gained 75 marks in the mathematics test?" (73) Ensure that students know how to use the line of best fit to answer such a question and, if necessary, go through the method carefully, while asking further questions. Emphasise to **more able** students that this is only an approximation.
- Now students can complete Exercise 14C, in which questions 2d, 3d and 4–6 are FM, question 7 is PS and 8 is an assessing understanding (AU) question. **Less able** students can work in pairs to complete the questions. Offer help as needed.

Plenary
- Ask students for examples of variables that would give the different types of correlation. They should provide examples to show: positive, negative and no correlation.

Answers
Answers to Exercise 14C can be found at the end of this chapter.

© HarperCollins*Publishers* 2012

Functional maths
Reporting the weather

Curriculum references

Functional Skills standards
R2 Identify the situation or problems and identify the mathematical methods needed to solve them.
I2 Draw conclusions and provide mathematical justifications.

PLTS
Self-managers, Creative thinkers

APP
Handling data L7.6; Using and applying mathematics L7.2

Collins references
- Student Book pages 354–355

Learning objectives
- Read data in tables.
- Represent data in charts.
- Work with averages.

Key words
- averages
- bar chart
- box-and-whisker diagrams
- cumulative frequency
- frequency
- frequency table
- stem-and-leaf diagrams

Functional maths help
Students should already have an understanding of frequency diagrams and bar charts. However, they may misread the chart or they may mix up the maximum and minimum temperatures. To prevent students from making these kinds of errors, make sure they understand the meaning of all the columns in the chart.

Lesson plans
Build
This task is very open, but should still be accessible to students as there are many ways to represent the raw data. However, if you want to build students' functional skills, you may wish to close the task by directing students to complete the questions in the following activities, which offer full structure and allow students to 'answer' questions while still producing a variety of charts.

Activity 1
Ask students to create a frequency table for the different types of weather on Friday 21st April 2010 (as shown in the chart). Then, from their frequency table, ask them to draw a bar chart to illustrate the data. They should use their bar chart to describe the general weather in the UK on that day.

Activity 2
Ask students to create a frequency table for the amount of sun and a frequency table for the amount of rain on Friday 21st April 2010 (as shown in the chart). As in Activity 1, students should draw a bar chart for each frequency table and use these to describe the general weather in the UK for that day.

Activity 3
Ask students to create a frequency table for the maximum and minimum temperatures in the towns in Scotland (Aberdeen, Edinburgh, Glasgow), Northern Ireland (Belfast) and Wales (Cardiff, Holyhead and Rhyl). Then ask them to draw a dual bar chart for each country to show the maximum and the minimum temperatures of each of these towns. Ensure that students use the same scale for each bar chart. Finally, ask them to use the bar charts to compare the temperatures in the three countries for that day.

© HarperCollins*Publishers* 2012

Functional maths Reporting the weather

Activity 4
Ask students to look at the map that shows the weather for Saturday 22nd April 2010. During a class discussion, elicit a comparison of the weather on the two days.

Activity 5
Ask pairs or groups of three students to select a few towns from the chart and create a poster to recommend one town to visit for a weekend. Posters should include every aspect of the weather shown in the chart, a comparison of the weather in other towns, and a justification of the recommendation.
Extend the task by asking students to supplement the data with findings from their own research. For example, they could add extra cities, look at data for another date for the same cities, compare seasons by bringing in two sets of their own data, or decide whether the weather for the day is 'average'.

Apply
Lead a class discussion about which diagrams are the most appropriate to display the data provided in the chart and map. During the discussion, encourage **more able** students to think about why we present the weather forecast and summary in the format of a map and chart in a newspaper, and what we cannot learn from these diagrams. (Students should recognise that the chart and map do not allow us to analyse or compare data.) Following the class discussion, allow students to represent the data in the ways they think are most suitable.
Less able students may need guidance on how to collate the data before they can begin constructing the diagrams. Activities 1–4 may help to provide this guidance.
Then the class can complete Activity 5 in pairs or groups of three.
Ask students to extend the task as described. **More able** students should be able to do this with ease. Provide **less able** students with support, if necessary.

Master
This set of activities can be left as an open-ended task that students can work on over a series of lessons.
In order to demonstrate mastery of the learning objectives, students should be able to:
- Show competence in deciding when to use certain graphs.
- Show the ability to decide on what each diagram should show.
- Show an understanding of which averages to calculate.
- Show competence in summarising all the data provided.

Plenary
Students should present their diagrams to the class. (Each student could present to a subset of the group.)
They should use this presentation to discuss the strategies they used and to explain the mathematical rationale behind their choices.

© HarperCollins*Publishers* 2012

Answers Lessons 14.1–14.3

Quick check
1 29.0

14.1 Line graphs
Exercise 14A
1 a

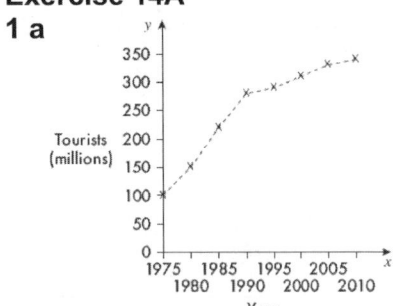

b About 328 million
c Between 1980 and 1985
d Rising living standards

2 a

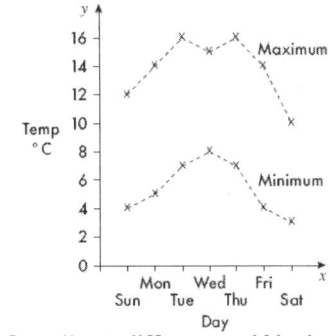

b Smallest difference Wednesday and Saturday (7°), greatest difference Friday (10°)

3 a

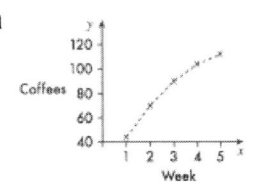

b 119
c The same people keep coming back and tell others, but new customers each week become more difficult to find (the graph becomes less steep).

4 Use a graph to estimate about 1040–1050 g.

5 All the temperatures were presumably higher than 20 °C.

14.2 Stem-and-leaf diagrams
Exercise 14B
1 a 2 | 8 9
    3 | 4 5 6 8 8 9
    4 | 1 1 3 3 3 8 8
b 43 cm **c** 39 cm **d** 20 cm

2 a 0 | 2 8 9 9 9
    1 | 2 3 7 7 8
    2 | 0 1 2 3
b 9 messages   **c** 15 messages

3 a 0 | 7 8 9 9
    1 | 0 2 3 4 5 8 8 9 9 key 2 | 3 = 23
    2 | 0 3 4 4 6 8
    3 | 1
b 18   **c** 24

4 The girls' heights are on the right, 15 | 3 means 153 cm tall.
The boys' heights are on the left, 6 | 15 means 156 cm tall.

5 All the data items start with a 5 and there are only two digits.

14.3 Scatter diagrams
Exercise 14C
1 a Positive correlation, reaction time increases with amount of alcohol drunk
b Negative correlation, you drink less alcohol as you get older
c No correlation, speed of cars on M1 is not related to the temperature
d Weak, positive correlation, older people generally have more money saved in the bank

2 a and b

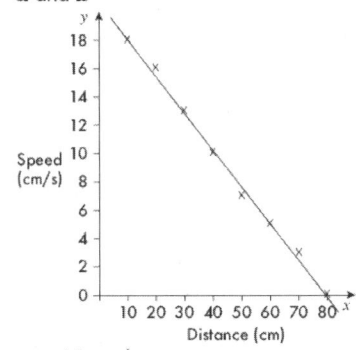

c ≈ 19 cm/s
d ≈ 34 cm

© HarperCollins*Publishers* 2012

3 a and **b**

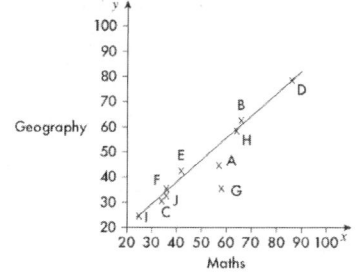

(scatter graph: Geography vs Maths, points labelled A–J and D)

c Greta **d** ≈ 70 **e** ≈ 70

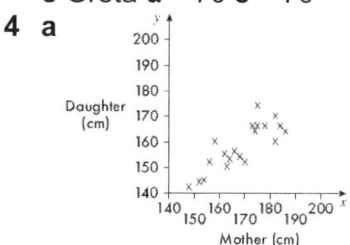

4 a

(scatter graph: Daughter (cm) vs Mother (cm))

b Yes, usually good correlation.

5 a

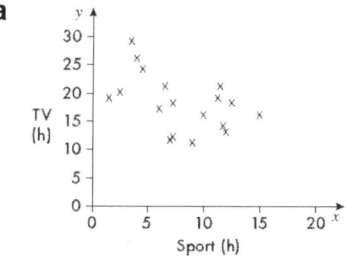

(scatter graph: TV (h) vs Sport (h))

b No. Little correlation, so cannot draw a line of best fit.

6 a and **b**

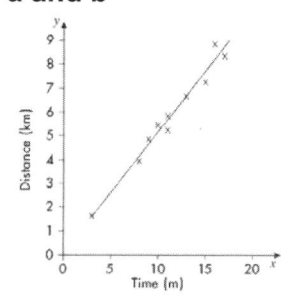

(scatter graph: Distance (km) vs Time (m))

c About 2.4 km
d About 9 minutes

7 About 23 mph

8 Points showing a line of best fit sloping down, from top left to bottom right.

Answers Lessons 14.1–14.3

Examination questions

1
```
6 | 5 7
7 | 0 0 2 6      6|7 represents 67 minutes
8 | 0 2 4 5 7
9 | 1
```

2 a Negative correlation
  b i Check students' lines of best fit.
   ii 7
  c He was a good typist.
  d No, as the graph goes no further than 10 hours and the correlation could change beyond that point.

3 a and **c**

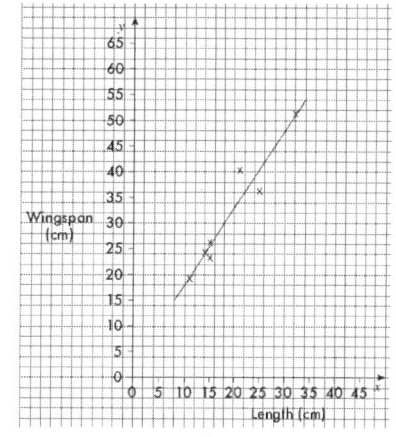

(scatter graph: Wingspan (cm) vs Length (cm))

b Strong positive **d** 32 cm
e No data given is higher than a length of 32 cm and the correlation might not hold at 41 cm.

4 a and **b** Check students' scatter diagrams and lines of best fit.
  c £7.20
  d 820 pages

5 a 3 **b** 28 **c** 88

6 a 17|7 9,
     18|1 4 6 6 7 8
     19|0 4
  b 186
  c There will be 11 members and the middle number will be the 6th in order. This will still be 186.

7 a Check correct values plotted.
  b Strong negative correlation
  c Check students' lines of best fit.
  d 75, or whatever value is read from the line of best fit
  e The data is only valid up to 40 births per thousand.

8 £29

Chapter 15 Probability: Calculating probabilities

Overview

15.1	Experimental probability	15.4	Two-way tables
15.2	Mutually exclusive and exhaustive events	15.5	Addition rule for events
15.3	Expectation	15.6	Combined events

This chapter covers more of the probability material in the Higher syllabus.

Context
Students can link probability to their own experiences, such as games involving chance. It is helpful throughout the study of probability to use the language of chance and prediction, to remind students that probabilities do not predict results in an exact way, only the likelihood of particular results.

Curriculum references
KS4 Programme of Study references
1.3b Understand that mathematics is used as a tool in a wide range of contexts.
1.4a Knowing that mathematics is essentially abstract and can be used to model, interpret or represent situations.
3.3d Experimental and theoretical probabilities of single and combined events.

Linear specification
S5.1 Understand and use the vocabulary of probability and the probability scale.
S5.2 Understand and use estimates and measures of probability from theoretical models (including equally likely outcomes), or from relative frequency.
S5.3 List all outcomes for single events, and for two successive events, in a systematic way and derive related probabilities.
S5.4 Identify different mutually exclusive outcomes and know that the sum of the probabilities of all these outcomes is 1.
S5.5h Know when to add or multiply two probabilities: if A and B are mutually exclusive, then the probability of A or B occurring is P(A) + P(B) whereas, if A and B are independent events, the probability of A and B occurring is P(A) × P(B).
S5.7 Compare experimental data and theoretical probabilities.
S5.8 Understand that if an experiment is repeated, this may result in different outcomes.
S5.9 Understand that increasing sample size generally leads to better estimates of probability and population characteristics.

Functional Skills standards
R2 Identify the situation or problems and identify the mathematical methods needed to solve them.
A1 Apply a range of mathematics to find solutions.
I2 Draw conclusions and provide mathematical justifications.

PLTS
Independent enquirers analyse and evaluate information, judging its relevance and value; support conclusions, using reasoned arguments and evidence. **Creative thinkers** question their own and others' assumptions; adapt ideas as circumstances change. **Reflective learners** review progress, acting on the outcomes. **Team workers** collaborate with others to work towards common goals; reach agreements, managing discussions to achieve results. **Self-managers** work towards goals, showing initiative, commitment and perseverance; organise time and resources, prioritising actions; respond positively to change, seeking advice and support when needed.

APP
Handling data L5.2 In probability, select methods based on equally likely outcomes and experimental evidence as appropriate. **L6.3** Find and record all possible mutually exclusive outcomes for single events and two successive events in a systematic way. **L7.5** Understand relative frequency as an estimate of probability and use this to compare outcomes of an experiment.

© HarperCollins*Publishers* 2012

Chapter 15 Probability: Calculating probabilities

L8.3 Know when to add or multiply two probabilities. **Using and applying mathematics L7.3** Justify generalisations, arguments or solutions.

Route mapping

Exercise	Grades			
	D	C	B	A
A		all		
B		1–8	9–12	
C		1–9	10–12	
D	1–5	6–11	12	
E	1–5	6–12	13	
F	1–5	6–12	13	14

Overview test
All these questions are grade D.
1 For each of these percentages, work out how much must be added to reach 100%.
   a 70%   b 23%   c 6%   d 91%

2 For each fraction or decimal, work out how much must be added to reach 1.
   a $\frac{2}{3}$ b $\frac{3}{8}$ c 0.4   d 0.18   e $\frac{5}{12}$

3 Work out each of these, simplifying your fraction answers when possible.
   a $\frac{1}{8} + \frac{5}{8} =$ d $0.32 + 0.59 =$
   b $\frac{3}{10} + \frac{2}{5} =$ e $0.28 + 0.4 =$
   c $\frac{2}{3} + \frac{1}{6} =$ f $0.3 + 0.09 =$

4 The travellers on a ferry were asked the reason for their journeys. The results are in this table.
   a What fraction of the travellers were visiting family?
   b What percentage were on business?
   c What fraction were on holiday?
   d If 2500 people use the ferry in a month, how many, based on these results, would you expect to be travelling on business?

Reason for journey	Holiday	Business	Shopping	Family visit	Other
Frequency	40	25	20	18	22

5 Work out each of these calculations.
   a $\frac{1}{4}$ of 96   b $\frac{2}{3}$ of 75   c $\frac{3}{5}$ of 225   d $\frac{7}{8}$ of 128

Answers to overview test
1 a 30%   b 77%   c 94%   d 9%
2 a $\frac{1}{3}$ b $\frac{5}{8}$ c 0.6   d 0.82   e $\frac{7}{12}$
3 a $\frac{3}{4}$ b $\frac{7}{10}$ c $\frac{5}{6}$ d 0.91   e 0.68   f 0.39
4 a $\frac{18}{125}$ b 20%   c $\frac{8}{25}$ d 500
5 a 24   b 50   c 135   d 112

Why this chapter matters
Cross-curricular: This work links to the sciences and Geography.
Introduction: Probability has been an integral part of life for a very long time, as outlined in the Student Book. It extends further than, for example, weather forecasts, horse-racing bets and the National Lottery. In all these cases, mathematical principles apply to the outcome, although we may not have enough information about the variables to make an accurate forecast.
Discussion points: Can you predict accurately the chance of rain tomorrow? Why not? What is the chance of winning the lottery? The chance of scoring a six on a dice? In this case, we can understand the chance.
Plenary: Ask for the possible results when a dice is thrown, List the possibilities, then note how many chances there are of a six not being thrown. Do a similar trial with a bag of 10 numbered counters. Compare the results from each and try to elicit a rule determining the chance of a particular number being thrown or drawn.
Answers to the quick check test can be found at the end of the chapter.

© HarperCollinsPublishers 2012

15.1 Experimental probability

Curriculum references

Functional Skills standards
R2 Identify the situation or problems and identify the mathematical methods needed to solve them.

PLTS
Independent enquirers, Creative thinkers, Reflective learners, Team workers, Self-managers, Effective participators

APP
Handling data L5.2

Collins references
- Student Book pages 359–365
- Interactive Book: Paper Animation

Learning objectives
- Calculate experimental probabilities and relative frequencies. [Grade C]
- Estimate probabilities from experiments. [Grade C]
- Use different methods to estimate probabilities. [Grade C]

Learning outcomes
- Students **must** be able to calculate experimental probability. [Grade C]
- Students **should** be able to choose an appropriate method to find a probability. [Grade C]
- Students **could** be able to compare relative frequencies with expected frequencies and draw conclusions. [Grade C]

Key words
- experimental probability
- trials
- relative frequency

Prior knowledge
Students should know how to find the probability of an event, using the definition:
$P(A) = \dfrac{\text{number of ways A can happen}}{\text{number of possible outcomes}}$, where A is an event.

Common mistakes and remediation
When data is presented in tabular form, students often have difficulty in identifying the correct information. Make it clear to them that they should use 'number of correct results' and 'number of trials' in all cases, but they need to be able to interpret a table correctly to find these values. Some tables will be cumulative frequencies of the required outcome and others will be frequencies of all the results. Stress to students that they should read the tables carefully.

Useful tips
Remind students that the more an experiment is repeated (i.e. the more trials are carried out), the closer the probability gets to the expected value (theoretical probability). Calculators are essential for this lesson for students to convert fractions to decimals.

Functional maths and problem-solving help
None of the questions in Exercise 15A are marked as functional maths (FM), but the questions require students to give reasons. It will help students if they can discuss their reasons before writing them down, as articulating these will help them to formulate a reasonable answer.
Problem-solving (PS) question 12 may sound familiar to many students. The question requires them to look carefully at the table and use what they have been learning to find the answer.

© HarperCollins*Publishers* 2012

Lesson 15.1 Experimental probability

Starter
- Prepare a bag or box containing a number of different coloured balls or counters, or similar.
- Ask students to select one without looking and then put it back. While this is happening, ask one student to collect the data as it is generated.
- At different stages, e.g. after four selections and again after 10, ask for suggestions as to what is in the bag. Ensure that students give reasons for their suggestions.

Main lesson activity
- Read through the preamble and Example 1 in the Student Book.
- Ask, "What is the probability of tossing a coin and getting a head?" (or 0.5) Most students will know this.
- Say that this is the theoretical probability – but does it actually work out in practice?
- Ask all students to toss a small-value coin and see how many heads there are. Make a note of this on the board, in a table like the one below. Show how to find the probability and explain the need to express it as a decimal for comparison as more coins are tossed. Remind **less able** students how to do this, by dividing the numerator by the denominator, and explain that no more than four decimal places are needed.

Number of throws	Number of heads	Probability
30	18	$\frac{18}{30} = 0.6$

- Ask the class to toss their coins again, count the number of heads and add the results to the table.

Number of throws	Number of heads	Probability
30	18	$\frac{18}{30} = 0.6$
30 + 30 = 60	18 + 14 = 32	$\frac{32}{60} = 0.5333$

- Continue to add to the table. You should notice that the probability gets closer and closer to 0.5.
- Ensure that **more able** students realise that decimal notation is nearly always the best way to compare fractions.
- Explain that this is experimental probability and that this is often the only way to find a probability, e.g. finding the probability of dropping a slice of jam-spread bread on the floor and it ending up jam side down!
- Explain that historical data is used in order to find probabilities of events where experiments are impossible, e.g. asteroids hitting the earth, volcanoes erupting, etc.
- Discuss Example 2 in the Student Book.

Finding probabilities
- Work through this section and Example 3.
- Now students can complete Exercise 15A, in which there are no FM questions, 12 is PS and 11, 13 and 14 are assessing understanding (AU) questions. **Less able** students should be able to complete the exercise but suggest they work with a partner if necessary.

Plenary
- Ask students to share their results for question 6, Exercise 15A. Collating these results should show students they can obtain more accurate experimental probabilities.

Answers
Answers to Exercise 15A can be found at the end of this chapter.

© HarperCollins*Publishers* 2012

15.2 Mutually exclusive and exhaustive events

Curriculum references

Functional Skills standards
I2 Draw conclusions and provide mathematical justifications.

PLTS
Independent enquirers, Creative thinkers, Reflective learners, Team workers, Self-managers, Effective participators

APP
Handling data L6.3

Collins references
- Student Book pages 366–370
- Interactive Book: 10 Quick Questions

Learning objectives
- Recognise mutually exclusive, complementary and exhaustive events. [Grade C–B]

Learning outcomes
- Students **must** be able to calculate probabilities for mutually exclusive events. [Grade C]
- Students **should** be able to use different methods to find probabilities of combined events. [Grade C]
- Students **could** be able to draw conclusions based on probabilities. [Grade B]

Key words
- complementary
- mutually exclusive
- exhaustive events

Prior knowledge
Remind students of the make-up of a pack of cards. While this may have been common knowledge at one time, it cannot be assumed that all students will be familiar with them today. There is also the question of jokers in the pack. Say that, unless stated otherwise, students should assume a pack consists of 52 cards, with no jokers.

Common mistakes and remediation
Mistakes or misunderstandings inevitably occur within the vicinity of fractions. For questions involving mutual exclusivity, ensure that students are familiar with calculations that require the application of $P(B) = 1 - P(A)$.
Run through a few examples, and get students to evaluate subtractions such as:
$1 - \frac{7}{12}$ and $1 - \frac{13}{25}$ $\left(\frac{5}{12}, \frac{12}{25}\right)$

Useful tips
Mutually exclusive events cannot happen at the same time – they are separate and independent.

Functional maths and problem-solving help
Functional maths (FM) questions 10 and 11 bring a real-life context to the ideas of mutually exclusive events. These terms are not used in daily life, but the concepts are familiar. Students should understand the concepts; the vocabulary is useful to **more able** students who may need it if they move on to the more advanced statistics.
Problem-solving (PS) question 8 requires students to manipulate fractions, decimals and percentages to get them into a form in which they can be compared and combined.

© HarperCollins*Publishers* 2012

Lesson 15.2 Mutually exclusive and exhaustive events

Starter
- Ask students simple probability questions based on a pack of cards. For example, suppose a single card is taken from a pack. What is the probability that it is:
  a the 2 of clubs ($\frac{1}{52}$) **c** red card ($\frac{26}{52} = \frac{1}{2}$) **e** an Ace ($\frac{4}{52} = \frac{1}{13}$)
  b a spade ($\frac{13}{52} = \frac{1}{4}$) **d** a royal card ($\frac{12}{52} = \frac{3}{13}$) **f** an odd-numbered card? ($\frac{26}{52} = \frac{1}{2}$)
- Ask for all probabilities to be given as simplified fractions.

Main lesson activity
- Ask students if a card can be both a club and a King. (the King of clubs)
- Now ask if a card can be a club and a heart at the same time. (no) You may need to point out that it could be either a club or a heart. Explain that this means the two outcomes – picking a club and picking a heart – are mutually exclusive. Emphasise the importance of this concept and write the expression on the board so that it is visible throughout the lesson.
- Ask the class to suggest other events that are mutually exclusive. (any other suits, or numbers, at the same time, a head and a tail at the same time after tossing a coin, etc.)
- Now ask students to suggest events that are not mutually exclusive. (a card that is a six and a diamond, rolling a dice and getting a total higher than 1 and an even number, selecting at random a student from the class who is male and has brothers, etc.)
- Ask: "What is the probability of rolling a 3 on a dice and the probability of not rolling a 3? Are they mutually exclusive?" (yes, as they cannot happen at the same time) Explain that these are complementary events – one event happening and another event not happening. Add this expression to the list on the board.
- Ask for the probability of rolling a dice and getting 3, and of rolling a dice and not getting 3. ($\frac{1}{6}$ and $\frac{5}{6}$) Write the fractions on the board next to each other.
- Ask for the probability of cutting a pack of cards and getting an Ace, then of cutting a pack of cards and not getting an Ace. ($\frac{4}{52}$ and $\frac{48}{52}$) Write these fractions next to each other.
- Now ask students to add together the fractions in each pair – they add to 1 in each case. Say that the probability of an event and its complement will always total 1.
- Discuss Example 4, which also introduces exhaustive events – events that cover all possibilities. Add this expression to the list.
- Now work through Example 5, which includes all the terms introduced during the lesson. You may need to spend some time revising fractions with **less able** students.
  Complementary events
- Read through this section in the Student Book. Ensure that **more able** students learn the correct terminology, while **less able** students should at least understand the concepts.
- Now students can complete Exercise 15B. The FM questions are 10 and 11, 8 is PS and 12 is an assessing understanding (AU) question. **Less able** students can complete questions 1–8 then work with a partner to do 9–12.

Plenary
- Ask students to pick a single card from a pack, at random. Ask for descriptions of two events that are:
  a mutually exclusive and exhaustive
  b mutually exclusive and not exhaustive
  c not mutually exclusive.

Answers
Answers to Exercise 15B can be found at the end of this chapter.

© HarperCollins*Publishers* 2012

15.3 Expectation

Curriculum references

Functional Skills standards
R2 Identify the situation or problems and identify the mathematical methods needed to solve them.

PLTS
Independent enquirers, Creative thinkers, Reflective learners, Team workers, Self-managers, Effective participators

APP
Handling data L7.5

Collins references
- Student Book pages 370–372
- Interactive Book: Multiple Choice Questions

Learning objectives
- Predict the likely number of successful events, given the number of trials and the probability of any one event. [Grade C–B]

Learning outcomes
- Students **must** be able to use probabilities to find expected values. [Grade C]
- Students **should** be able to explain the difference between expected values and actual values. [Grade C]
- Students **could** be able to find expected frequencies in practical situations. [Grade C]

Key words
- expectation

Prior knowledge
Students should know how to find the probability of an event, and also how to calculate a fraction of a quantity.

Common mistakes and remediation
Mistakes here are likely to be confined to one of two kinds. Some students may work out the required probability incorrectly, while others work out the fraction of the number of trials incorrectly. Provide plenty of practice and go through any errors with them individually.

Useful tips
Students will need calculators for some questions.
Expectation is the probability multiplied by the number of times the experiment is carried out.

Functional maths and problem-solving help
None of the questions in Exercise 15C are marked as functional maths (FM), but they do include the use of decimals. These may cause problems for **less able** students, who may need help, particularly with questions 9 and 11.
Problem-solving (PS) question 11 requires students to know the structure of an ordinary pack of cards and to manipulate probabilities associated with cards.

© HarperCollins*Publishers* 2012

Lesson 15.3 Expectation

Starter
- Recap, or ask a student to explain, how to find a fraction of a quantity.
- Display the number 240 and ask students to choose a fraction to find of this number. It may be more appropriate to limit the choice of fractions to halves, thirds, quarters, fifths, sixths, eighths, tenths, twelfths.

Main lesson activity
- Ask, "If you roll a dice 600 times, how many times would you expect to roll the number 5?" Discuss this with the class and ensure they would expect 100 of each of the six numbers, so 100 fives.
- Ask, "If you cut a pack of cards 100 times, how many times would you expect to cut the Ace of diamonds?"
- This is more awkward as there are 52 cards in a pack and not 50, so you will need to lead the class to the expected value close to 2, since if you cut the pack 104 times, you might expect to have each card twice.
- Explain to **more able** students that, although it is common to refer to an expected number, in practice this exact number rarely occurs. The result is, however, usually close to it. Expected numbers are always an estimate.
- Show the class, in the dice example, they can find the expected number by multiplying the theoretical probability by the number of rolls, i.e. $\frac{1}{6} \times 600 = 100$. So, for the cards example, multiply the probability of 16 cutting the Ace of diamonds ($\frac{1}{52}$) by 100, which, using a calculator, is 1.923. You would expect to get the card 1.9 times, which rounds to 2.
- Ensure that students understand the term 'expectation'.
- Talk through Examples 6 and 7 with the class.
- Now students can complete Exercise 15C, in which no questions are marked as FM, 11 is PS and 12 is an assessing understanding (AU) question. **Less able** students can complete questions 1–9, but may need support to work through 10–12, which are grade B. Working with a partner may help these students.

Plenary
- Display a table of the total scores from the roll of two dice. Ask students how many times they would expect to get different scores (such as 4, 12, 5, less than 4), from 360 rolls of the dice. Which score would they expect to get most often? Ask why you picked 360 rolls.

Answers
Answers to Exercise 15C can be found at the end of this chapter.

15.4 Two-way tables

Curriculum references

Functional Skills standards
A1 Apply a range of mathematics to find solutions.

PLTS
Independent enquirers, Creative thinkers, Reflective learners, Team workers, Self-managers, Effective participators

APP
Handling data L6.3

Collins references
- Student Book pages 373–377
- Interactive Book: Multiple Choice Questions

Learning objectives
- Read two-way tables and use them to work out probabilities and interpret data. [Grade D–C]

Learning outcomes
- Students **must** be able to interpret values in a two-way table. [Grade D]
- Students **should** be able to estimate probabilities from a two-way table. [Grade D]
- Students **could** be able to construct two-way tables to calculate probabilities. [Grade C]

Key words
- two-way table

Prior knowledge
Students should be able to retrieve information presented in tabular form. They should also know how to find the probability of an event. In addition, they should be able to find the expected number of outcomes from the total number of trials and the probability of the outcome in question.

Common mistakes and remediation
Most mistakes will come from misunderstandings in how to read the data from a two-way table. To assist students' understanding of how to interpret these tables, go through many examples, with as much discussion as possible.

Useful tips
It will help students if they are able to change fractions to percentages and cancel fractions, as these occur in the first few questions of Exercise 15D.

Functional maths and problem-solving help
The functional maths (FM) question 7 is structured and leads students to apply their knowledge to build up the answer.
For problem-solving (PS) question 10, students need to start by finding all the combinations that will give a product of more than 40.
Less able students may need extra help with questions 6, 7 and 8, which require them to provide reasons. Encourage students, even the **more able**, to articulate their reasons before writing them down.

Lesson 15.4 Two-way tables

Starter
- Collect some data from the class that can be tallied into a two-way table. Ask students, e.g. to name their favourite pop group, football team or sport from a short list drawn up before the lesson. Use the results to construct a two-way table with column headings 'Male' and 'Female', and up to four row headings. (In a single-sex class, find some other way to divide the class into two groups.)
- Ask the class to provide the probabilities of selecting different classes of people (e.g. a girl who supports Man Utd, or a boy whose favourite sport is golf). Choose a relatively simple multiple of the number of students in the class, and ask how many of this total they would expect to be in each category in the table.

Main lesson activity
- Explain that two-way tables are used to display information about groups, and can be used to calculate probabilities as well as to estimate numbers in larger groups.
- Go through Example 8, especially with **less able** students, then go through Example 9 with the class.
- If necessary, revise changing fractions to percentages and cancelling fractions, as these occur in the first few questions. Then students can complete Exercise 15D, in which question 7 is FM, 10 is PS and 6 and 8 are assessing understanding (AU) questions. **Less able** students should be able to complete questions 1–5, which are grade D, on their own, and then work in pairs to complete questions 6–10, which are grade C.
- The variety of questions will allow **less able** students plenty of practice at earlier questions and **more able** students to move swiftly on to the more challenging questions.

Plenary
- Revisit the information in the starter. Ask students to suggest how to calculate the percentages of the total that each represents.

Answers
Answers to Exercise 15D can be found at the end of this chapter.

15.5 Addition rule for events

Curriculum references

Functional Skills standards
A1 Apply a range of mathematics to find solutions.

PLTS
Independent enquirers, Creative thinkers, Reflective learners, Team workers, Self-managers, Effective participators

APP
Handling data L8.3

Collins references
- Student Book pages 378–380
- Interactive Book: Matching Pairs

Learning objectives
- Work out the probability of two events such as P(A) or P(B). [Grade D–B]

Learning outcomes
- Students **must** be able to calculate the probability of one of two events occurring. [Grade C]
- Students **should** be able to use the addition rule for probabilities. [Grade C]
- Students **could** be able to solve practical problems involving probabilities. [Grade C]

Key words
- either

Prior knowledge
Students should know how to find the probability of an event. They should also know how to express one number as a fraction of another, and how to simplify, add and subtract fractions.

Common mistakes and remediation
Students often make mistakes when adding probability fractions. They often learn how to add probabilities, but do not always understand when it is appropriate to do so. Question 8 of Exercise 15E illustrates well an occasion when it is not appropriate to do so.

Useful tips
Tell students that they should remember to discuss their reasons for their explanations, so that their arguments are clear. They should remember that for independent events they should multiply the probabilities, and for exclusive events, they should add the probabilities.

Functional maths and problem-solving help
The functional maths (FM) and problem-solving (PS) questions are in real-life contexts. They rely on students' understanding of when they can add the probability fractions and when not.
Question 10, which is an assessing understanding (AU) question, requires an explanation.

Starter
- Display the fractions $\frac{1}{2}, \frac{1}{3}, \frac{1}{4}, \frac{1}{6}$ and $\frac{1}{12}$.
- Ask students to choose a pair to add. Then ask them to choose any three to add, then any four, and finally to add all five.

Main lesson activity
- Ask students to explain what mutually exclusive events are, as a useful reminder of terminology. They are events that cannot happen at the same time.
- Draw on the board a bag containing 2 white balls, 3 yellow balls and 5 black balls. Ask students for the probabilities of picking each colour. (white $\frac{2}{10}$, yellow $\frac{3}{10}$ and black $\frac{5}{10}$) Write these fractions on the board.

© HarperCollins*Publishers* 2012

Lesson 15.5 Addition rule for events

- Now ask for the probability of choosing a black or a white ball. They should give the answer $\frac{7}{10}$ simply from counting the possible balls in the bag, but point out that the two relevant probability fractions ($\frac{2}{5}$ and $\frac{5}{10}$) also add to $\frac{7}{10}$.
- Repeat the previous point, choosing a yellow or a white ball ($\frac{5}{10}$) and say that the probability of this, too, is the sum of the probabilities of each separate event. Explain that to find a probability of event A or event B happening, and they are mutually exclusive, then just add their probabilities.
- Give another example: Jake is playing a game. The probability that he wins the game is 0.45, the probability that he draws the game is 0.2; what is the probability that he loses the game? (0.35) Write the figures on the board. Discuss students' suggestions and ask for explanations. If necessary, explain that as these are exhaustive events (there are no other possibilities) the probabilities add to 1, so 1 − (0.45 + 0.2) = 0.35.
- Emphasise that you can add the two probabilities, as the events are mutually exclusive and that the chance of losing is the complement of the sum of the other two.
- Discuss Example 10. Explain that it is not always helpful to cancel the probability fractions immediately, as you may need to add them together later; then you often have the required common denominator. **More able** students should always cancel but should also remain aware that the original fractions can be used for addition.
- Discuss, with **more able** students, the point made after Example 10 – the probability of neither green nor blue is not the sum of the probability of not green or not blue. Make sure they see that these are not mutually exclusive.
- Now students can complete Exercise 15E, in which questions 6 and 11 are FM, 12 is PS and 8e, 10 and 13 are AU questions. **Less able** students can complete questions 1–5, but may need support in order to complete questions 6–13.

Plenary
- Display a target board with whole numbers on it. Ask for probabilities of events such as 'picking a multiple of 4 or 6', 'picking a prime number or a factor of 100', 'picking a number which is a factor of 40 and 60'. The choice of questions will depend on the numbers displayed.

Answers
Answers to Exercise 15E can be found at the end of this chapter.

© HarperCollins*Publishers* 2012

15.6 Combined events

Curriculum references

Functional Skills standards
I2 Draw conclusions and provide mathematical justifications.

PLTS
Independent enquirers, Creative thinkers, Reflective learners, Team workers, Self-managers, Effective participators

APP
Handling data L8.3

Collins references
- Student Book pages 381–384
- Interactive Book: Worked Exam Questions

Learning objectives
- Work out the probability of two events occurring at the same time. [Grade D–C]

Learning outcomes
- Students **must** be able to complete and use a sample space diagram. [Grade D]
- Students **should** be able to be able to calculate probabilities of combined events. [Grade C]
- Students **could** be able to solve practical problems by designing and using a sample space diagram. [Grade C]

Key words
- probability space diagram
- sample space diagram

Prior knowledge
Students should know how to find the probability of an event.

Common mistakes and remediation
Students are often careless about reading instructions. Make it clear that a sample space diagram can either record two separate results in each cell or the outcome of both results, such as adding them or finding the difference between them, or whatever is required by the question. Students should make their own decisions about what to record.

Useful tips
Tell students that it is important for them to read questions thoroughly, before they begin to work on them.

Functional maths and problem-solving help
Functional maths (FM) question 10 in Exercise 15F reflects a familiar situation. It also reveals that the likelihood of winning competitions based on chance can be small.
Problem-solving (PS) question 11 presents a practical situation that **less able** students can try for themselves.
Explain that although it is not always essential to draw a sample space diagram, it can help to show all the possibilities more clearly. Simple diagrams can be made by using crosses to indicate the intersection of rows and columns without the need to write anything in the cells. Numbers are needed if the space demands finding totals, but often, the known numbers are all that are needed.

Lesson 15.6 Combined events

Starter
- Divide the class into pairs, providing each pair with two dice. Allowing a two-minute time limit, ask each pair to roll the dice, add the scores on the dice on each roll, and record each total score.
- While they do this, prepare a tally chart on the board. Collate the results from all pairs when they have finished. Ask for the relative frequencies, or experimental probabilities, for each score.

Main lesson activity
Throwing two dice, Throwing coins, Dice and coins
- Ask students which total they got most often with two dice. Each group may have a different total, but the answer 7 would be expected to appear most frequently.
- Now ask why some numbers seem to occur more than others. For example, the totals 12 and 11 should occur very rarely, as there is only one way to get a 12 (6, 6) and two ways to get 11 (5, 6 and 6, 5), The total 1 is impossible and 2 is only achieved by (1, 1). There are several ways of getting some other numbers.
- Ask the class how many ways there are of getting each number. Can they think of a way to find out? Discuss this and then, on the board, show them a sample space diagram for two dice. Show that there are six ways to get a total of 7, but only two ways to get 11 as a total, so they can expect to get more 7s than 11s.
- Discuss the examples in the Student Book and the different ways in which to draw sample spaces: a list of possibilities for tossing two coins and a grid for the coin and the dice.
- Show **less able** students that each point represents one equally likely event. So the diagram gives the total number of possibilities as well as the number of the required event.
- Now students can complete Exercise 15F, in which question 10 is FM, 11 is PS and 12 is an assessing understanding (AU) question. **Less able** students can complete questions 1–5 and move on to 6–12 if they are confident to do so. Offer help, as needed.

Plenary
- Ask if students can notice any difference between their theoretical probabilities to question 1 of Exercise 15F and the experimental probabilities in the starter. Should they expect any differences?
- Emphasise that the more often they repeat an experiment, the closer the experimental probabilities will be to the theoretical probabilities.

Answers
Answers to Exercise 15F can be found at the end of this chapter.

© HarperCollins*Publishers* 2012

Functional maths
Fairground games

Curriculum references

Functional Skills standards
R2 Identify the situation or problems and identify the mathematical methods needed to solve them.
A1 Apply a range of mathematics to find solutions.

PLTS
Independent enquirers, Team workers

APP
Handling data L8.4; Using and applying mathematics L7.3

Collins references
- Student Book pages 390–391
- Interactive Book: Paper Animation

Learning objectives
- Produce a sample space diagram showing different outcomes.
- Calculate the probability of different outcomes.

Key words
- cost
- planning
- schedule

Functional maths help
Remind the class about expectation and explain that fairground stallholders tend to have a good grasp of what they expect their customers will achieve on their stalls. Any stallholder with a new game must test it and consider the experimental probability before opening to the public. They want to offer big prizes to entice people to their stall, but need to make a profit, too.

Lesson plans
Build
Guide students through the 'Getting started' task. They should find the following answers.
- Probability of spinning:
  * a one ($\frac{1}{5}$, 0.2) * a five ($\frac{3}{5}$, 0.6) * a two ($\frac{1}{5}$, 0.2) * a number other than five ($\frac{2}{5}$, 0.4)
- If the spinner is spun 20 times, probability of spinning:
  * a five (12 times)   * not a five (8 times)   * an odd number (16 times)
- If the spinner is spun 100 times, probability of spinning:
  * a two (20 times)   * not a two (80 times) * a prime number (80 times)

To give more structure for those building their functional skills, ask students to complete these activities rather than writing a report.

Activity 1
Tell students to assume that Joe wants to charge 30p for two balls to roll down and give a prize of 50p for totals over 7.
First ask students to think about the probabilities connected with Joe's game. Discuss as a class how you could go about finding these out.
More able students should realise that they can create a sample space table of the different totals that can be made with the two balls.

	1	3	5	3	1
1	2	4	6	4	2
3	4	6	8	6	4
5	6	8	10	8	6
3	4	6	8	6	4
1	2	4	6	4	2

For Joe's game, the probability of getting a 1 is $\frac{2}{5}$ or 40%; the probability of getting a 3 is $\frac{2}{5}$ or 40%; and the probability of getting a 5 is $\frac{1}{5}$ or 20%. If the class is **more able**, ask them to work out the probability of scoring different totals of their own choosing, using two balls. **Less able** students may need more structure. Provide the following questions for them to answer:

1 What is the probability of rolling two balls down this board and getting a total of:
  a 2 ($\frac{4}{25}$)   **b** 4 ($\frac{8}{25}$)   **c** 6 ($\frac{8}{25}$)   **d** 8 ($\frac{4}{25}$)   **e** 10 ($\frac{1}{25}$)

© HarperCollinsPublishers 2012

2 What is the probability that you will score over 7 with your two balls? ($\frac{5}{25}$ or $\frac{1}{5}$)
3 If 100 people have a go at this game, how many of them would you expect to win a prize? (20)

Activity 2
Now students should think about how Joe calculates his profit, after working out the probability of his customers winning. Set up the following scenarios and ask students to work out the profit in each case.
1 100 people play the game with the expected number of winners. (20 possible winners = 30p × 100 = £30 taken, minus £10 (20 × 50p possible prize money) = £20 profit.)
2 60 people play the game with the expected number of winners. (60 players with 12 possible winners = £18 taken, minus £6 prize money = £12 profit.)
3 Joe changes the prize for getting a total of 10 to £1. He keeps the other prizes at 50p. (If the prize for a score of 10 is £1 and other prize money remains 50p, and 100 people play (£4 + £8 prize money) = £18 profit (£30 − £12). If 60 people play, (£3 + £6 prize money) = £9 profit (£18 − £9).)
4 Joe changes the prize for getting a total over 7 to 30p. (If the prize money for scoring over 7 is changed to 30p, Joe would have to pay 20 × 30p = £6 if 100 people play, and would make £24 profit (£30 − £6). If 60 people play, he will pay £3.60 and make a profit of £14.40 (£18 − £3.60).)
More able students could find solutions to their own scenarios.

Activity 3
Tell students that Joe wants over 300 people to take part in the game and to make at least £100 profit.
They must find the charge and prize money that they feel would attract enough people to play and give Joe the profit he wanted. (Students should work out the entrance fee, and then deduct the possible prize money. A possible example is to charge 40p per go with a £1 prize for a total of 10 only, giving a profit on 300 people of £108. (There will be many different correct answers.)

Activity 4
Students can go on to create their own games. If the students are **less able** they can work in groups of two or three. Encourage them to ask and answer similar questions about their games.

Apply
Students working at this level should write a report to combine all their findings. However, they may need help with the layout of the report and what you want to be included. The activities above could be used to help students structure their report into 'calculating probability', 'finding the profit' and 'finding the best set-up'.
They may also need to be given specific values to begin with. You could give them the following:
    30p to roll the two balls, 100 people playing, a 50p prize for a score over 7
Now students can start to introduce their own values. They should investigate, using their own questions, what the predicted profit is if a specific number of people play and how many people must play if you have a specific desired profit in mind. **Less able** students may need to be told that they must first draw a tree diagram to show the different outcomes before they can methodically work out the probability of each outcome occurring and the probability of winning the game.

Master
Students should be able to:
- Draw a tree diagram to show the different outcomes.
- Calculate the probability of these outcomes.
- Use probability to write a report that looks at maximising profit for a school fair game taking into account how different variables alter the outcome.
- Design a school fair game that will make a profit.

© HarperCollins*Publishers* 2012

Answers Lessons 15.1 – 15.6

Quick check
1 a Perhaps around 0.6
   b Very close to 0.1
   c Very close to 0
   d 1
   e 1

15.1 Experimental probability
Exercise 15A
1 a $\frac{1}{5}, \frac{2}{25}, \frac{1}{10}, \frac{21}{200}, \frac{37}{250}, \frac{163}{1000}, \frac{329}{2000}$
   b 6   c 1
   d $\frac{1}{6}$ e 1000
2 a $\frac{19}{200}, \frac{27}{200}, \frac{4}{25}, \frac{53}{200}, \frac{69}{200}$
   b 40
   c No, it is weighted towards the side with numbers 4 and 5.
3 a 32 is too high, unlikely that 20 of the 50 throws between 50 and 100 were 5.
   b Yes, all frequencies fairly close to 100.
4 a $\frac{1}{5}, \frac{1}{4}, \frac{38}{100}, \frac{21}{50}, \frac{77}{200}, \frac{1987}{5000}$
   b 8
5 a 0.346, 0.326, 0.294, 0.305, 0.303, 0.306
   b 0.231, 0.168, 0.190, 0.16, 0.202, 0.201
   c Red 0.5, white 0.3, blue 0.2   d 1
   e Red 10, white 6, blue 4
6 a Students' answers will vary. b 20
   c Answer depends on students' results.
   d Answer depends on answer to c.
7 a 6
   b and c Answer depends on students' results.
8 a Caryl, most throws
   b 0.43, 0.31, 0.17, 0.14
   c Yes, it is more likely to give a 1 or 2
9 a Method B b B c C d A e B
   f A g B h B
10 a Not likely b Impossible
   c Not likely d Certain e Impossible
   f 50–50 chance g 50–50 chance
   h Certain i Quite likely
11 Thursday
12 The missing top numbers are 4 and 5; the two bottom numbers are likely to be close to 20.
13 Although you would expect the probability to be close to $\frac{1}{2}$, hence 500 heads, it is more likely that the number of heads is close to 500 rather than actually 500.
14 Roxy is correct, as the expected numbers are: 50, 12.5, 25, 12.5. Sam has not taken into account the fact that there are four red sectors.

15.2 Mutually exclusive and exhaustive events
Exercise 15B
1 a Yes b Yes c No d Yes
   e Yes f Yes
2 Events **a** and **f**
3 $\frac{3}{5}$
4 a i $\frac{3}{10}$ ii $\frac{3}{10}$ iii $\frac{3}{10}$
   b All except iii c Event iv
5 a Jane/John, Jane/Jack, Jane/Anne, Jane/Dave, Dave/John, Dave/Jack, Dave/Anne, Anne/John, Anne/Jack, Jack/John
   b i $\frac{1}{10}$ ii $\frac{3}{10}$
      iii $\frac{3}{10}$ iv $\frac{7}{10}$
   c All except iii d Event ii
6 a $\frac{3}{8}$ b $\frac{1}{8}$
   c All except ii d Outcomes overlap
7 $\frac{3}{20}$
8 $\frac{1}{75}$
9 Not mutually exclusive events
10 a i 0.25 ii 0.4 iii 0.7
   b Events not mutually exclusive
   c Man/woman, American man/American woman
   d Man/woman
11 a i 0.95 ii 0.9 (assuming person chooses one or other) iii 0.3
   b Events not mutually exclusive
   c Possible answer: pork and vegetarian
12 These are not mutually exclusive events.

© HarperCollins*Publishers* 2012

Answers Lessons 15.1 – 15.6

15.3 Expectation
Exercise 15C
1. 25
2. 1000
3. a 260 b 40 c 130 d 10
4. 5
5. a 150 b 100 c 250 d 0
6. a 167 b 833
7. 1050
8. a Each score expected 10 times
   b 3.5
   c Find the average of the scores, which is 21 (1 + 2 + 3 + 4 + 5 + 6) divided by 6.
9. a 0.111 b 40
10. 281 days
11. Multiply the number of tomato plants by 0.003
12. 400

15.4 Two-way tables
Exercise 15D
1. a 23 b 20% c $\frac{4}{25}$ d 480
2. a 10 b 7 c 14% d 15%
3. a

	1	2	3	4
5	6	7	8	9
6	7	8	9	10
7	8	9	10	11
8	9	10	11	12

   b 4
   c i $\frac{1}{4}$ ii $\frac{3}{16}$ iii $\frac{1}{4}$

4. a 16 b 16 c 73 d $\frac{51}{73}$
5. a

	1	2	3	4	5	6
1	2	3	4	5	6
2	4	6	8	10	12

   b 3 c $\frac{1}{4}$

6. a The greenhouse sunflowers are bigger on average.
   b The garden sunflowers have a more consistent size (smaller range).
7. a 40% b 45%
   c No, as you don't know how much the people who get over £350 actually earn.
8. Either Reyki because she had bigger tomatoes or Daniel because he had more tomatoes.
9.

	Men	Women	Children
Left footed	4	3	5
Right footed	21	12	15

10. $\frac{22}{36} = \frac{11}{18}$
11. a

Score of second spinner						
10	10	11	13	15	17	19
8	8	9	11	13	15	17
6	6	7	9	11	13	15
4	4	5	7	9	11	13
2	2	3	5	7	9	11
0	0	1	3	5	7	9
	0	1	3	5	7	9

Score of first spinner

   b 9 or 11 c 0
   d $\frac{15}{36} = \frac{5}{12}$ e $\frac{30}{36} = \frac{5}{6}$

12. a 200 b $\frac{75}{200} = \frac{3}{8}$
    c $\frac{45}{105} = \frac{3}{7}$ d $\frac{25}{75} = \frac{1}{3}$

15.5 Addition rules for events
Exercise 15E
1. a $\frac{1}{6}$ b $\frac{1}{6}$ c $\frac{1}{3}$
2. a $\frac{1}{4}$ b $\frac{1}{4}$ c $\frac{1}{2}$
3. a $\frac{1}{13}$ b $\frac{1}{13}$ c $\frac{2}{13}$
4. a $\frac{2}{11}$ b $\frac{4}{11}$ c $\frac{6}{11}$
5. a $\frac{1}{3}$ b $\frac{2}{5}$ c $\frac{11}{15}$
   d $\frac{11}{15}$ e $\frac{1}{3}$
6. a 0.6 b 120
7. a 0.8 b 0.2
8. a 0.75 b 0.6 c 0.5 d 0.6
   e i Cannot add P(red) and P(1) as events are not mutually exclusive
   ii 0.75 (= 1 − P(blue))
9. a $\frac{17}{20}$ b $\frac{2}{5}$ c $\frac{3}{4}$
10. Probability cannot exceed 1, and probabilities cannot be summed in this way as events are not mutually exclusive.
11. a i 0.4 ii 0.5 iii 0.9
    b 0.45 c 2 hours 12 minutes
12. $\frac{5}{52}$ or 0.096 to 3 decimal places
13. a $\frac{13}{20}$ as it cannot be square rooted
    b $\frac{1}{9}$ as this gives a ratio of red to blue of 1 : 2

© HarperCollinsPublishers 2012

Answers Lessons 15.1 – 15.6

15.6 Combined events
Exercise 15F
1 **a** 7
 b 2, 12
 c $P(2) = \frac{1}{36}$, $P(3) = \frac{1}{18}$, $P(4) = \frac{1}{12}$,
  $P(5) = \frac{1}{9}$, $P(6) = \frac{5}{36}$, $P(7) = \frac{1}{6}$,
  $P(8) = \frac{5}{36}$, $P(9) = \frac{1}{9}$, $P(10) = \frac{1}{12}$,
  $P(11) = \frac{1}{18}$, $P(12) = \frac{1}{36}$
 d i $\frac{1}{12}$ **ii** $\frac{5}{9}$ **iii** $\frac{1}{2}$
   iv $\frac{7}{36}$ **v** $\frac{5}{12}$ **vi** $\frac{5}{18}$

2 **a** $\frac{1}{12}$ **b** $\frac{11}{36}$ **c** $\frac{1}{6}$ **d** $\frac{5}{9}$

3 **a** $\frac{1}{36}$ **b** $\frac{11}{36}$ **c** $\frac{5}{18}$

4 **a** $\frac{5}{18}$ **b** $\frac{1}{6}$ **c** $\frac{1}{9}$
 d 0 **e** $\frac{1}{2}$

5 **a** $\frac{1}{4}$ **b** $\frac{1}{2}$ **c** $\frac{3}{4}$ **d** $\frac{1}{4}$

6 **a** 6
 b i $\frac{4}{25}$ **ii** $\frac{13}{25}$
   iii $\frac{1}{5}$ **iv** $\frac{3}{5}$

7 **a** $\frac{1}{8}$ **b** $\frac{3}{8}$ **c** $\frac{7}{8}$ **d** $\frac{1}{8}$

8 **a** 16 **b** 32 **c** 1024 **d** 2^n

9 **a** $\frac{1}{12}$ **b** $\frac{1}{4}$ **c** $\frac{1}{6}$

10 **a**

	1	2	3	4	5	6
1	2	3	4	5	6	7
2	3	4	5	6	7	8
3	4	5	6	7	8	9
4	5	6	7	8	9	10
5	6	7	8	9	10	11
6	7	8	9	10	11	12

 b $\frac{1}{18}$ **c** 18 **d** Twice

11 0.5

12 You would need a 3D diagram or there would be too many different events to list.

13 $\frac{19}{100}$

14 **a** If he takes black, striped and spotted with his first three picks he must get one of these with his next pick to make a pair.
 b There are only 19 socks left, of which only 3 are black
 c The events are not independent. The second probability depends on what was taken out the first time.
 d $\frac{132}{380} = \frac{33}{95}$

Examination questions

1. Vegetarian row: 4, 6
   Non-vegetarian row: 18, 12
   Total row: 22, 18
2. He is more likely to win, as there are 13 ways to score 5, 6 or 7. This is a probability of $\frac{13}{25}$, which is just over 0.5.
3. **a i** 0.9 **ii** 12
   b Yes. She should throw 20 if the dice is fair.
4. The two possibilities, off or not, are not necessarily equally likely chances.
5. **a** $\frac{4}{10} = \frac{2}{5}$ **b** 140
6. **a** 0.126
   b All values should be about 250 but 1 and 3 are nowhere near this.
   c 100
7. Throw both dice at least 100 times. Record the results in a table. Compare the results with the expected values which, for a fair six-sided dice, would be about 16 or 17 times for each number for 100 throws. The dice with results that are not in line with that will be the biased dice.
8. **a** 0.15 **b** 0.65
9. Once

Chapter 16 Algebra: Algebraic methods

Overview

16.1 Number sequences	**16.3** Special sequences
16.2 Finding the *n*th term of a linear sequence	**16.4** General rules from given patterns

This chapter provides an in-depth study of sequences and patterns.

Context
This work is a prequel to manipulation of algebraic expressions and changing the subject of a formula, which are very important skills that apply to many other areas of mathematics, as well as physics and other subjects.

Curriculum references
KS4 Programme of Study references
1.3a Knowing that mathematics is a rigorous, coherent discipline.
2.2m Manipulate numbers, algebraic expressions and equations and apply routine algorithms.
2.2n Use accurate notation, including correct syntax when using ICT.
3.1e Linear, quadratic and other expressions and equations.
4a Develop confidence in an increasing range of methods and techniques.
4b Work on sequences of tasks that involve using the same mathematics in increasingly difficult or unfamiliar contexts, or increasingly demanding mathematics in similar contexts.

Linear specification
N5.3h Simplify rational expressions.
N5.4h Solve simultaneous equations in two unknowns.
N5.6 Change the subject of a formula.
N6.1 Generate terms of a sequence using term-to-term and position-to-term definitions of the sequence.
N6.2 Use linear expressions to describe the *n*th term of an arithmetic sequence.

Functional Skills standards
R2 Identify the situation or problems and identify the mathematical methods needed to solve them.
R3 Select a range of mathematics to find solutions.
A1 Apply a range of mathematics to find solutions.
A2 Use appropriate checking procedures and evaluate their effectiveness at each stage.
I2 Draw conclusions and provide mathematical justifications.

PLTS
Independent enquirers explore issues, events or problems from different perspectives. **Effective participators** present a persuasive case for action. **Self-managers** work towards goals, showing initiative, commitment and perseverance. **Creative thinkers** generate ideas and explore possibilities.

APP
Algebra L6.3 Generate terms of a sequence using term-to-term and position-to-term definitions of the sequence, on paper and using ICT; write an expression to describe the *n*th term of an arithmetic sequence. **L6.4** plot the graphs of linear functions, where *y* is given explicitly in terms of *x*; recognise that equations of the form $y = mx + c$ correspond to straight-line graphs. **L8.3** Derive and use more complex formulae and change the subject of a formula. **Using and applying mathematics L6.3** Precise a concise, reasoned argument, using symbols, diagrams, graphs and related explanatory texts.

© HarperCollins*Publishers* 2012

Chapter 16 Algebra: Algebraic methods

Route mapping

Exercise	Grades D	Grades C	Grades B
A	1–3	4–11	
B		1–6	7–10
C		all	
D		all	

Overview test

Questions 1–3 are grade D; questions 4–8 are grade E.

1 $\dfrac{2}{3} \times \dfrac{1}{2} =$
2 $\dfrac{2}{3} \div \dfrac{3}{4} =$
3 $\dfrac{2}{3} + \dfrac{3}{4} =$

4 Find the value of p in each of these equations.
 a $p = 4 + 3 \times 2$
 b $4 = p + 3 \times 2$
 c $4 = 3 + 2 \times p$

5 Solve: $4x - 7 = 13$.

6 Write the next two terms of each sequence.
 a 2, 4, 6, 8, …, …
 b 3, 7, 11, 15, …, …
 c 3, 4, 6, 9, 13, …, …
 d 30, 27, 24, 21, …, …
 d 1, 4, 9, 16, 25, …, …
 f 1, 10, 100, 1000, …, …

7 Work out the values for the expression $2n$, when:
 a $n = 1$
 b $n = 2$
 c $n = 4$

8 Work out the values for the expression $4n - 3$, when:
 a $n = 1$
 b $n = 2$
 c $n = 3$

Answers to overview test

1 $\dfrac{2}{6}$ or $\dfrac{1}{3}$

2 $\dfrac{8}{9}$

3 $\dfrac{17}{12}$ or $1\dfrac{5}{12}$

4 a 10 b −2 c $\dfrac{1}{2}$

5 $x = 5$

6 a 10, 12 b 19, 23 c 18, 24 d 18, 15
   e 36, 49 f 10 000, 100 000

7 a 2 b 4 c 8

8 a 1 b 5 c 9

Why this chapter matters
Cross-curricular: This work links with the sciences, Design and Technology and Art and Crafts.
Introduction: Any sequence that is built upon a rule can be expressed algebraically. Once the formula has been found, a shortcut exists for finding any term without tedious calculation and the attendant risk of error.
Discussion points: How do numerical sequences determine some of the shapes we see in the natural world, such as those shown in the Student Book? Do sequences determine the shapes, or do the shapes just happen to fit the sequences? Do numerical rules govern other things in nature, apart from shapes?
Plenary: Set this task.
Make up a sequence, then write, in words, the rule that governs it. Test the rule to ensure that it works for all situations. Now try to describe the rule that you have created, algebraically. Can it be done? If not, why not? What are the advantages of an algebraic formula?

Answers to the quick check test can be found at the end of the chapter.

© HarperCollinsPublishers 2012

16.1 Number sequences

Curriculum references

Functional Skills standards
I2 Draw conclusions and provide mathematical justifications.

PLTS
Independent enquirers, Effective participators

APP
Algebra L6.3

Collins references
- Student Book pages 394–397
- Interactive Book: Worked Exam Questions

Learning objectives
- Recognise how number sequences are building up. [Grade D–C]
- Generate sequences, give the nth term. [Grade D–C]

Learning outcomes
- Students **must** be able to extend a number sequence and identify patterns. [Grade D]
- Students **should** be able to use a formula for the nth term of a sequence. [Grade D]
- Students **could** be able to identify connections between number sequences. [Grade C]

Key words
- coefficient
- sequence
- consecutive
- term
- difference
- term-to-term
- nth term

Prior knowledge
Students need to know how to substitute into simple expressions.

Common mistakes and remediation
Students find sequences accessible and usually do well. They make errors mainly with methods, which more practice will iron out. Often, when asked for the 10th term, **less able** students will double the 5th term, so point out that this will not work. They may also look only at the first two numbers and assume that the rest follow this pattern, so stress the importance of using all the information provided and using this to check if their sequences work.

Useful tips
Point out that finding the difference between terms is the first thing to do with a pattern they do not recognise. If this fails to get a result, they should look for another system, e.g. repeated multiplication if the numbers increase quickly.

Functional maths and problem-solving help
There are no functional maths (FM) questions in Exercise 16A. Questions 7 and 9 are problem solving (PS). Question 7 requires some thought. **Less able** students may try to list all the terms; **more able** students may devise a different method. Students are required to break down the problem into manageable steps. Refer students to the hint, provided to help them work out how many letters there are in the sequence and then to work out how many sequences are needed.

© HarperCollins*Publishers* 2012

Lesson 16.1 Number sequences

Starter
- Write the numbers 2 and 4 on the board. Ask students what the next two numbers in the pattern could be. The most likely answer is 6 and 8. Write this on the board and ask how the pattern is building up. (+ 2 each time)
- Now write 2, 4, 8, 16, … on the board and ask how this is building up. (× 2)
- Write 2, 4, 7, 11, … on the board and ask students how this is building up. (+ 2, + 3, + 4, …)
- Repeat with 1, 5, …. Ask for other series starting 1, 5, …, and a description of how each one builds up.

Main lesson activity
- Ask students to suggest some sequences and to describe how they build up.
- First ask for series that increase by a fixed amount, such as + 3.
- Then ask for series that decrease by a fixed amount, such as − 4. Then ask for series that increase by a different amount each time, such as + 1, + 2, + 3, … or + 2, + 4, + 6.
- Next, ask for sequences that multiply by a fixed number each time, such as × 2. However, as these get large very quickly, keep the starting numbers simple or allow calculators.
- Now ask for sequences that divide by a fixed number each time, such as ÷ 10. These soon get to decimals, often recurring, so stick to ÷ 2, ÷ 5 or ÷ 10.
- Most students can usually see intuitively how these patterns build up.
- **Differences**
- Go through the introductory text and Example 1, which demonstrates how to use differences. This may confuse **less able** students, so be prepared to go through it several times, until they get the idea.
- **Generalising to find the rule**
- Go through Examples 2 and 3. **Less able** students may find the concept of the general term difficult. In Example 2 it is, essentially, adding a term that is itself increasing by a rule. Help students to see it as 4, 4 + 1 × 3, 4 + 2 × 3, etc. Example 3 reinforces the idea.
- Make sure students are familiar with the highlighted words in the text and that they understand that how the series builds up is called the 'term-to-term rule'.
- Now students can do Exercise 16A, in which questions 7 and 9 are PS and 8, 10 and 11 are assessing understanding (AU) questions. **Less able** students can complete questions 1–5, which are grade D; then they can move on to questions 6–11, which are grade C.

Plenary
- Write some 'silly' sequences on the board and ask for the next term. For example:
    - 31, 28, 31, 30, 31, 30, … (31: days in the months)
    - O, T, T, F, F, … (S: initial letters of the numbers, one, two, three, etc,)
    - 1, 2, 5, 10, 20, … (50: British coins).
- Ask students to think of or find some silly or unusual sequences for a future lesson.

Answers
Answers to Exercise 16A can be found at the end of this chapter.

© HarperCollins*Publishers* 2012

16.2 Finding the *n*th term of a linear sequence

Curriculum references

Functional Skills standards
A2 Use appropriate checking procedures and evaluate their effectiveness at each stage.

PLTS
Independent enquirers, Effective participators

APP
Algebra L6.3

Collins references
- Student Book pages 398–401

Learning objectives
- Find the *n*th term of a linear sequence. [Grade C]

Learning outcomes
- Students **must** be able to find the *n*th term of a linear sequence. [Grade C]
- Students **should** be able to find the *n*th term of simple non-linear sequences. [Grade C]
- Students **could** be able to use number sequences in real life applications. [Grade C]

Key words
- *n*th term

Prior knowledge
Students should know how to recognise number patterns and how to find differences between consecutive terms.

Common mistakes and remediation
Students often miscalculate the constant term. To avoid this, refer to the 'Useful tips'.

Useful tips
Students should look carefully at all the numbers provided before determining the constant.

Functional maths and problem-solving help
The first functional maths (FM) question in Exercise 16B uses different formulae for different ranges, which is a common pricing practice. The question is structured so students should find it accessible. Problem-solving (PS) question 8 is also structured, but the concept of the limit as *n* tends to infinity may be difficult, especially for **less able** students. Keying very large numbers into the calculator will show that the decimal is 0.666 666 6, which all students should recognise as two-thirds.
The second FM and PS questions (questions 9 and 10 respectively) are reverse concepts of those in previous work, and are structured, so should be accessible to most students.

Lesson 16.2 Finding the *n*th term of a linear sequence

Starter
- Give students a starting number and an answer, e.g. 3 and 16. Ask for a rule of the form 'multiply by a number then add or subtract a number' to get from 3 to 16, such as: × 3 + 7. Write this as a linear expression in *n*, e.g. 3*n* + 7.
- Ask students for similar rules and to write them as linear expressions. This should produce rules such as: 5*n* + 1, 2*n* + 10, 4*n* + 4, 6*n* − 2, 7*n* − 5.

Main lesson activity
- Point out that the rules written down are linear expressions, as they contain a single term in the variable *n* and no powers.
- Ask a student to give a similar linear expression in *n*. For example, if the student says 3*n* + 2, ask the class to substitute *n* = 1, 2, 3, etc. into this expression and write the results in a list, i.e. 5, 8, 11, 14, 17, …
- Ask if they can see any connections with previous work. They should see that the expression generates a series with a constant difference. Repeat with another linear expression.
- Now write the expression n^2 on the board. Ask students how this is different from other expressions. Explain that this is a non-linear expression as it has a power in it.
- Ask students what sequence is generated by substituting *n* = 1, 2, 3. etc. It gives 1, 4, 9, 16, 25, …, which students should recognise as square numbers.
- Taking any of the sequences used at the beginning of the lesson, write the linear expression and the series on the board and ask students if they can see any connections.
- The difference and the coefficient of *n* is the most obvious. Students may also spot that the first term of the sequence is the difference plus or minus the constant term.
- Work through the text and Examples 4, 5, and 6 in the Student Book.
- Now students can do Exercise 16B, in which questions 7 and 9 are FM, questions 8 and 10 are PS. **Less able** students should be able to complete all the questions as they are grade C.

Plenary
- Write the sequence 2, 4, 8, 16 on the board. Ask students how it is built up. Can they find an *n*th term? It may help students if you write 2, 2 × 2, 2 × 2 × 2, etc.
- The *n*th term should be recognisable as 2^n.
- This principle similarly applies for the *n*th term of 10, 100, 1000, 10 000, etc.

Answers
Answers to Exercise 16B can be found at the end of this chapter.

© HarperCollins*Publishers* 2012

16.3 Special sequences

Curriculum references

Functional Skills standards
I2 Draw conclusions and provide mathematical justifications.

PLTS
Independent enquirers, Effective participators

APP
Using and applying mathematics L6.4

Collins references
- Student Book pages 401–404

Learning objectives
- Recognise and continue some special number sequences. [Grade C]

Learning outcomes
- Students **must** be able to use the formulas for special sequences of numbers.
  [Grade C]
- Students **should** be able to justify general statements about whether expressions are odd or even. [Grade C]
- Students **could** be able to find the nth term of a simple sequence involving square numbers. [Grade C]

Key words
- even
- prime
- odd
- square
- powers of 2
- triangular
- powers of 10

Prior knowledge
Students should be able to recognise number patterns, and they should be familiar with powers.

Common mistakes and remediation
Students often work out powers incorrectly when using calculators, e.g. they may calculate 109 instead of 1010. Encourage them to practise working with powers.

Useful tips
Suggest that students substitute appropriate values into expressions. They should also remember that a prime number may be odd or even, although the only even prime number is 2.

Functional maths and problem-solving help
Problem-solving (PS) questions 6, 7 and 9 of Exercise 16C involve combinations of odd, even and prime numbers.
Remind students to substitute appropriate values into the expressions, and that a prime number may be odd or even.
There are no functional mathematics (FM) questions in Exercise 16C.

© HarperCollins*Publishers* 2012

Lesson 16.3 Special sequences

Starter
- Recall the plenary from Lesson 16.1, relating to the 'silly' sequences and power sequences. Ask students if they have thought of or found any more unusual or silly sequences. If so, write these on the board and discuss them.
- If not, then other silly sequences include:
- 3, 3, 5, 4, 4, 3, … (5: number of letters in numbers as words)
- 15, 26, 40, 16, 37, … (58: sums of squares of digits in the previous number)
- Students may have found some other interesting sequences, such as Fibonacci or powers series.
- Remind students about 2, 4, 8, 16, … and 10, 100, 1000, … (powers of 2 and of 10)

Main lesson activity
- Refer to the Student Book. Go through the text about frequently occurring number sequences. Make sure students understand the nth terms for the odd and even number sequences.
- Students are not required to find the nth terms of quadratic sequences in examinations, but may need to substitute numbers into non-linear nth terms.
- The nth terms of the square numbers and the triangular numbers should be tested, as this is an essential part of FM skills. Ask students to test the square number rule for $n = 15$. Then show that the 15th square number is 225. ($1 \times 1 = 1$, $2 \times 2 = 4$, $3 \times 3 = 9$, $4 \times 4 = 16$, $5 \times 5 = 25$, …, $13 \times 13 = 169$, $14 \times 14 = 196$, $15 \times 15 = 225$)
- Ask students to extend the sequence of triangular numbers to 10 terms. Then test the rule for $n = 10$. Both answers should be 55. (1, 3, 6, 10, 15, 21, 28, 36, 45, 55)
- Test the rules for power sequences for $n = 10$; 210 = 1024 and 1010 = 10 000 000 000.
- Remind students about prime numbers. They have no pattern, but it is important to note that 2 is the only even prime number.
(2, 3, 5, 7, 11, 13, 17, 19, 23, 29, 31, 37, 41, 43, …)
- Work through Example 7, which is a typical examination question that tests knowledge of odd, even and prime numbers.
- Now students can do Exercise 16C, in which questions 6, 7 and 9 are PS and 4, 8 and 10 are assessing understanding (AU) questions. **Less able** students should be able to complete all of these grade C questions.

Plenary
- Put a copy of the current month's calendar on the screen or board.
- Highlight a row, column or diagonal and ask students to give the nth term. For example:
  - 2nd row $n + 5$
  - 3rd column $7n + 1$
  - Diagonal 1, 9, 17, 25 $8n - 7$
  - Reverse diagonal 4, 10, 16, 22, 28 $6n - 2$
- Ask students on what days the square numbers, triangular numbers, prime numbers and powers of 2 fall.

Answers
Answers to Exercise 16C can be found at the end of this chapter.

© HarperCollinsPublishers 2012

16.4 General rules from given patterns

Curriculum references

Functional Skills standards
I2 Draw conclusions and provide mathematical justifications.

PLTS
Independent enquirers, Effective participators

APP
Using and applying mathematics L6.4

Collins references
- Student Book pages 405–409
- Interactive Book: 10 Quick Questions

Learning objectives
- Find the nth term from practical problems. [Grade C]

Learning outcomes
- Students **must** be able to find the nth term from a sequence of patterns. [Grade C]
- Students **should** be able to use the formula for the nth term to solve problems. [Grade C]
- Students **could** be able to give a reasoned argument to support conclusions. [Grade C]

Key words
- difference
- rule
- pattern

Prior knowledge
Students should be able to substitute into simple algebraic expressions and have some experience in continuing a sequence. Having successfully completed Exercises 16B and 16C will help students enormously when they are working on Exercise 16D.

Common mistakes and remediation
Students often make the mistake of attempting to answer the question in their heads rather than writing down the sequence. Suggest that they sift through the words to get to the mathematics, write down the sequence and concentrate on the numbers rather than words such as 'fences' or 'pentagons'.

Useful tips
Always set up a table of results. This will help students to spot the pattern.

Functional maths and problem-solving help
Each question in Exercise 16D requires setting up a table and spotting the nth term. Once students find the nth term, they should test it. Checking results is an essential skill for functional maths (FM) at Higher level. **Less able** students may need help with problem-solving (PS) question 10 on Sierpinski's triangle. They may see the pattern, as the numerator is multiplied by 3 and the bottom is multiplied by 4, but they may not see the nth term.

© HarperCollins*Publishers* 2012

Lesson 16.4 General rules from given patterns

Starter
- Remind students how to find the next term and the *n*th term of a sequence. Provide two or three for practice,
  e.g. 7, 9, 11, 13, 15, ... 7, 17, 27, 37, ... 7, 107, 207, 307, ...

Main lesson activity
- Work through Example 8. The main point to emphasise is putting results in a table.
- Extend the table for a few columns (or rows). Test a generalisation to make sure the rule works.
- Initially, some **less able** students may need help with writing down a sequence.
- Now students can do Exercise 16D, in which questions 3, 6, 7, 8 and 12 are FM, question 10 is PS and 11 is an assessing understanding (AU) question. Allow **less able** students to work with a partner if you feel it is necessary.

Plenary
- Draw the following patterns on the board:

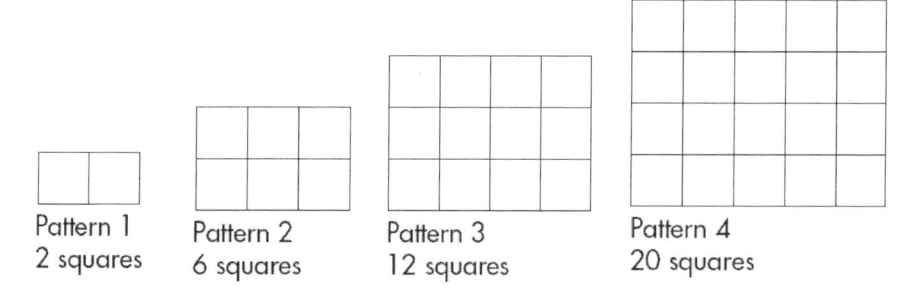

Pattern 1
2 squares

Pattern 2
6 squares

Pattern 3
12 squares

Pattern 4
20 squares

- Ask students how many squares there will be in pattern 5. (30)
- They may not spot this at first. You may need to prompt them.
- Ask how they can work it out. (The pattern goes up 4, 6, 8, 10, etc.)
- Ask if they could give a rule for the *n*th term.
- Look at the width (number of squares) of each block: 2, 3, 4, 5, ... *n* + 1.
- Look at the height (number of squares) of each block: 1, 2, 3, 4, ... *n*.
- The number of squares is: 1 × 2, 2 × 3, 3 × 4, 4 × 5, *n* × (*n* + 1).

Answers
Answers to Exercise 16D can be found at the end of this chapter.

Problem solving
Patterns in the pyramids

Curriculum references

Functional Skills standards
R3 Select a range of mathematics to find solutions.
A2 Use appropriate checking procedures and evaluate their effectiveness at each stage.

PLTS
Self-managers, Creative thinkers

APP
Algebra L6.3; Using and applying mathematics L6.3

Collins references
- Student Book pages 414–415

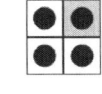

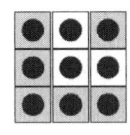

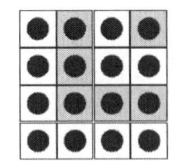

Learning objectives
- Simplify the situation or problem in order to represent it mathematically.
- Make connections within mathematics.
- Visualise and work with dynamic images.
- Calculate accurately and using approximations.
- Recognise patterns.
- Make and test generalisations.

Key words
- odd number
- Pythagorean triple
- square number

Functional maths help
This is an investigation, so students must record results, draw diagrams accurately and work systematically. This investigation could lead to a poster displaying a number of triples, or a written project about Pythagoras and why Pythagorean triples are important.

Lesson plans
Build
Briefly discuss the introductory paragraph, and then lead students to the first activity in the 'Getting started' text. For the benefit of **less able** students, work through the Pythagorean triple (3, 4, 5).

Activity 1
Refer students to number 1 of 'Getting started'. In pairs, they should create diagrams of odd numbers. **More able** students should be able to do this independently, but to help **less able** students, draw the diagrams as a class on the board.

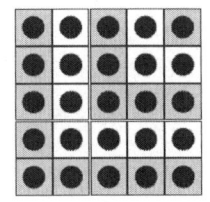

Activity 2
Guide students in using these diagrams to represent the patterns: $1^2, 2^2, 3^2, 4^2, 5^2$. They should develop the following sequence of squares.
Make sure all students have the right connection linking the last odd number and the square number:

$$((\text{last odd} + 1) \div 2)^2$$
$$1 + 3 + 5 + 7 + 9$$
$$(9 + 1) = 10$$
$$10 \div 2 = 5$$
$$5^2 = 25$$

© HarperCollins*Publishers* 2012

Problem solving Patterns in the pyramids

Activity 3
Ask students to prove that 5, 12, 13 is a Pythagorean triple. Students should draw their own diagrams accurately on squared paper.

First, they should draw an L-shaped square of 52; the square number formed is 25.
Then they should draw an L-shaped square of 122; the square number formed is 144.
Finally, they should draw the L-square of 132; the square number formed is 169.
Note: These two diagrams will be similar to the 52 L-shape, but larger in both cases.
Students' diagrams should indicate that 52 + 122 = 132, which is a Pythagorean triple, because 25 + 144 = 169. This is a sequence of three numbers where the squares of the first two numbers added together result in the square of the third number. **More able** students should be able to prove this triple independently and will only need confirmation that their diagrams are correct. Guide **less able** students through this description, step by step.

Activity 4
Ask students to work on their own to show the diagram for the triple (7, 24, 25), i.e. that 72 + 242 = 252 or 49 + 576 = 625. (Correct) Allow **less able** students to work in pairs, and if necessary, provide the clue that the odd numbers that form the outside 'L' are 25, 49, 81 and so on.
Now ask students to see if they can find other triples. Encourage students to test each triple they try, using a calculator to see if it works.

Activity 5
As a class, work through the sequences in number 2 of 'Your task' in the Student Book. Then let students, individually or in pairs, generate some sequences of their own.

Activity 6
Ask the class to use the internet to do some research on Pythagoras, his theorem and Euler's Formula.

Apply
Students working at this level should be able to represent numbers diagrammatically, and use these to complete patterns with minimal support.
Once they are ready to begin the main task, ask students to decide how they could approach the investigation to prove that (5, 12, 13) is a Pythagorean triple. Discuss ideas as a class and make sure students realise that they must construct diagrams for each part of the triple. Allow students to complete this process independently. Then to help **less able** students, work through the process as a class on the board, with individual students coming up to draw one stage of the process. Students should be able to find at least three more Pythagorean triples, using the skills they have built up. Discuss the sequences and how they can be related to real-life situations.
If there is time, ask students to research Pythagoras, his theorem and Euler's Formula (this formula states that, for any real number x, $eix = \cos x + i \sin x$). This research will encourage students to connect numerical sequences with patterns and shapes.

Master
At this level students should be able to simplify the problems and represent the solution mathematically. In order to demonstrate mastery of the learning objectives, students should be able to:
- Take a simple starting point and plan their work independently.
- Represent numerical patterns as accurate diagrams.
- Recognise patterns, make generalisations and test them.
- Be able to describe how they find Pythagorean triples.
- Independently research Pythagoras, his theorem and Euler's Formula and recognise the cross-mathematical value of the research.

Plenary
Check that students have correctly found their own Pythagorean triples.
To increase students' awareness of the links across mathematics, demonstrate with one of the triples found that they form the sides of a right-angled triangle.

© HarperCollins*Publishers* 2012

Answers Lessons 16.1 – 16.4

Quick check
1. **a** 17, 20, 23
   b 49, 64, 81

2. **a** 1 **b** 4 **c** 7

16.1 Number sequences
Exercise 16A

1. **a** 21, 34: add previous 2 terms
   b 49, 64: next square number
   c 47, 76: add previous 2 terms
2. 15, 21, 28, 36
3. 61, 91, 127
4. $\frac{1}{2}, \frac{3}{5}, \frac{2}{3}, \frac{5}{7}, \frac{3}{4}$
5. **a** 6, 10, 15, 21, 28
   b It is the sums of the natural numbers, or the numbers in Pascal's triangle or the triangular numbers.
6. **a** 2, 6, 24, 720 **b** 69!
7. X. There are 351 (1 + 2 + ... + 25 + 26) letters from A to Z. 3 × 351 = 1053. 1053 – 26 = 1027, 1027 – 25 = 1002, so, as Z and Y are eliminated, the 1000th letter must be X.
8. 364: Daily totals are 1, 3, 6, 10, 15, 21, 28, 36, 45, 55, 66, 78. (These are the triangular numbers). Cumulative totals are: 1, 4, 10, 20, 35, 56, 84, 120, 165, 220, 286, 364.
9. 29 and 41
10. No, because in the first sequence, the terms are always one less than in the 2nd sequence
11. $4n - 2 = 3n + 7$ rearranges as $4n - 3n = 7 + 2$, so $n = 9$.

16.2 Finding the *n*th term of a linear sequence
Exercise 16B

1. **a** 13, 15, $2n + 1$ **b** 25, 29, $4n + 1$
   c 33, 38, $5n + 3$ **d** 32, 38, $6n - 4$
   e 20, 23, $3n + 2$ **f** 37, 44, $7n - 5$
   g 21, 25, $4n - 3$ **h** 23, 27, $4n - 1$
   i 17, 20, $3n - 1$ **j** 42, 52, $10n - 8$
   k 24, 28, $4n + 4$ **l** 29, 34, $5n - 1$

2. **a** $3n + 1$, 151 **b** $2n + 5$, 105
   c $5n - 2$, 248 **d** $4n - 3$, 197
   e $8n - 6$, 394 **f** $n + 4$, 54
   g $5n + 1$, 251 **h** $8n - 5$, 395
   i $3n - 2$, 148 **j** $3n + 18$, 168
   k $7n + 5$, 355 **l** $8n - 7$, 393

3. **a** 33rd **b** 30th
   c 100th = 499

4. **a** **i** $4n + 1$ **ii** 401 **iii** 101, 25th
   b **i** $2n + 1$ **ii** 201 **iii** 99 or 101, 49th and 50th
   c **i** $3n + 1$ **ii** 301 **iii** 100, 33rd
   d **i** $2n + 6$ **ii** 206 **iii** 100, 47th
   e **i** $4n + 5$ **ii** 405 **iii** 101, 24th
   f **i** $5n + 1$ **ii** 501 **iii** 101, 20th
   g **i** $3n - 3$ **ii** 297 **iii** 99, 34th
   h **i** $6n - 4$ **ii** 596 **iii** 98, 17th
   i **i** $8n - 1$ **ii** 799 **iii** 103, 13th
   j **i** $2n + 23$ **ii** 223 **iii** 99 or 101, 38th and 39th

5. **a** $\frac{2n+1}{3n+1}$
   b Getting closer to $\frac{2}{3}$ (0.$\dot{6}$)
   c **i** 0.667 774 (6dp) **ii** 0.666 778 (6dp)
   d 0.666 678 (6dp), 0.666 667 (6dp)

6. **a** $\frac{4n-1}{5n+1}$
   b Getting closer to $\frac{4}{5}$ (0.8)
   c **i** 0.796 407 (6dp) **ii** 0.799 640 (6dp)
   d 0.799 964 (6dp), 0.799 9996 (7dp)

7. **a** £305 **b** £600 **c** 3 **d** 5

8. **a** $\frac{3}{4}, \frac{5}{7}, \frac{7}{10}$
   b **i** 0.666 666 777 8 **ii** $\frac{2}{3}$
   c for n, $\frac{2n-1}{3n-1} \approx \frac{2n}{3n} = \frac{2}{3}$

9. **a** $8n + 2$ **b** $8n + 1$
   c $8n$ **d** £8

10. **a** Sequence goes up in 2s; first term is 2 + 29
    b $n + 108$
    c Because it ends up as $2n \div n$
    d 79th

16.3 Special sequences
Exercise 16C

1. **a** 64, 128, 256, 512, 1024
   b **i** $2n - 1$ **ii** $2n + 1$ **iii** 3×2^n

2. **a** The number of zeros equals the power.
   b 6
   c **i** $10^n - 1$ **ii** 2×10^n

3. **a** Even,

+	Odd	Even
Odd	Even	Odd
Even	Odd	Even

   b Odd,

×	Odd	Even
Odd	Odd	Even
Even	Even	Even

4. **a** $1 + 3 + 5 + 7 = 16 = 4^2$,
   $1 + 3 + 5 + 7 + 9 = 25 = 5^2$
   b **i** 100 **ii** 56

5 a 28, 36, 45, 55, 66
 b i 210 **ii** 5050
 c You get the square numbers.
6 a Even **b** Odd **c** Odd **d** Odd **e** Odd
 f Odd **g** Even **h** Odd **i** Odd
7 a Odd or even **b** Odd or even
 c Odd or even **d** Odd
 e Odd or even **f** Even
8 a i Odd **ii** Even **iii** Even
 b Any valid answer, e.g. $x(y + z)$
9 11th triangular number is 66,
 18th triangle number is 171
10 a 36, 49, 64, 81, 100
 b i $n^2 + 1$ **ii** $2n^2$ **iii** $n^2 - 1$

16.4 General rules from given patterns
Exercise 16D
1 a
 b $4n - 3$
 c 97
 d 50th diagram
2 a
 b $2n + 1$
 c 121
 d 49th set
3 a 18 **b** $4n + 2$ **c** 12
4 a i 24 **ii** $5n - 1$ **iii** 224
 b 25
5 a i 20 cm **ii** $(3n + 2)$ cm **iii** 152 cm
 b 332
6 a i 20 **ii** 162
 b 79.8 km
7 a i 14 **ii** $3n + 2$ **iii** 41
 b 66
8 a i 5 **ii** n **iii** 18
 b 20 tins
9 a 2^n
 b i $100 \times 2^{n-1}$ ml **ii** 1600 ml
10 The nth term is $\left(\dfrac{3}{4}\right)^n$, so as n gets very large, the unshaded area gets smaller and smaller and eventually it will be zero; so the shaded area will eventually cover the triangle.
11 Yes, as the number of matches is 12, 21, 30, 39, ... which is $9n + 3$; so he will need $9 \times 20 + 3 = 183$ matches for the 20th step and he has $5 \times 42 = 210$ matches.
12 a 20 **b** 120

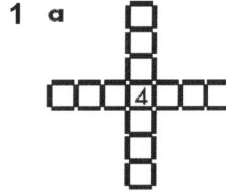

Examination questions
1 a $5n - 1$
 b 1.4, 1.5, 1.57...
2 a Always even
 b Could be either odd or even.
3 6, 9, 14
4 a i 5, 9, 13
 ii All numbers are odd
 b $(1 + 3)^2 - 9 = 4^2 - 9 = 16 - 9 = 7$
5 a Even
 b Odd
6 a i 25 **b** square numbers **c** 100
7 The square of any positive fraction less than 1, for example.
8 a For any value of n, $2n$ is even, so $2n + 1$ is odd.
 b The square of an odd number will be odd, the square of an even number will be even.
9 S, A, N

Chapter 17 Algebra: Linear graphs and equations

Overview

17.1	Linear graphs	17.3	Finding the equation of a line from its graph
17.2	Drawing graphs by the gradient-intercept method	17.4	3D coordinates

This chapter covers methods for drawing linear graphs and finding their equations. It also covers some of the ways in which they are used. The final section is an introduction to coordinates in three dimensions.

Context
Much of our knowledge and use of science can be displayed in graph form – everything from braking distances of cars to break-even points of a company's economic future.

Curriculum references
KS4 Programme of Study references
1.3b Understanding that mathematics is used as a tool in a wide range of contexts.
1.4a Knowing that mathematics is essentially abstract and can be used to model, interpret or represent situations.
2.1c Simplify the situation or problem in order to represent it mathematically, using appropriate variables, symbols, diagrams and models.
2.2k Make accurate mathematical diagrams, graphs and constructions on paper and on screen.
3.1e Linear, quadratic and other expressions and equations.

Linear specification
N6.4 Recognise and plot equations that correspond to straight-line graphs in the coordinate plane, including finding their gradients.
N6.5h Understand that the form $y = mx + c$ represents a straight line and that m is the gradient of the line and c is the value of the y-intercept.
N6.6h Understand the gradients of parallel lines.
N6.11 Construct linear functions from real-life problems and plot their corresponding graphs.

Functional Skills standards
R1 Understand routine and non-routine problems in familiar and unfamiliar contexts and situations.
R2 Identify the situation or problems and identify the mathematical methods needed to solve them.
A1 Apply a range of mathematics to find solutions.
A2 Use appropriate checking procedures and evaluate their effectiveness at each stage.
I2 Draw conclusions and provide mathematical justifications.

PLTS
Independent enquirers identify questions to answer and problems to resolve; explore issues, events or problems from different perspectives. **Creative thinkers** generate ideas and explore possibilities.

APP
Algebra L6.4 Plot the graphs of linear functions, where y is given explicitly in terms of x; recognise that equations of the form $y = mx + c$ correspond to straight-line graphs. **L7.2** Use algebraic and graphical methods to solve simultaneous linear equations in two variables. **Using and applying mathematics L7.3** Justify generalisations, arguments or solutions.

© HarperCollins*Publishers* 2012

Chapter 17 Algebra: Linear graphs and equations

Route mapping

Exercise	Grades				
	D	C	B	A	A*
A	1–11	12–13			
B		all			
C		1–5	6		
D		all			
E			all		
F		4	3	1	2

Overview test
Questions 1–5 are grade E.

1 Write down the coordinated of the points shown on the grid.

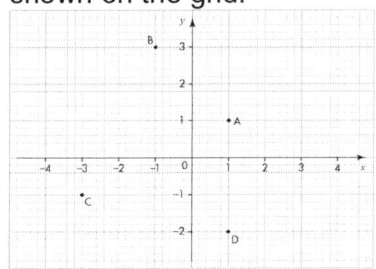

2 Find the value of y when $x = 3$, using the rule $y = 2x - 2$.

3 Find the value of y when $x = -3$, using the rule $y = 2x + 1$.

4 Plots the points given in the table on the grid below. Join the points with a straight line.

x	–2	0	2	4
y	–2	–1	0	1

5 Plots the points (–4, 6), (–1, 3), (0, 2), (2, 0) on the grid below. Join the points with a straight line.

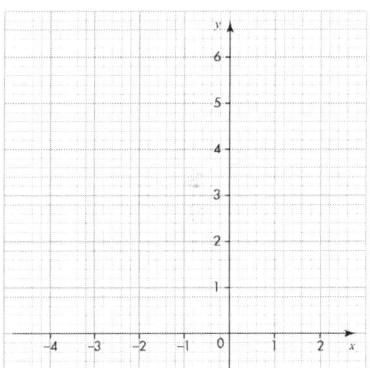

Answers to overview test
1 A(1, 1), B(–1, 3), C(–3, –1), D(1, –2)
2 $y = 4$
3 $y = -5$
4 ($y = \frac{1}{2}x - 1$)
5 ($y = -x + 2$)

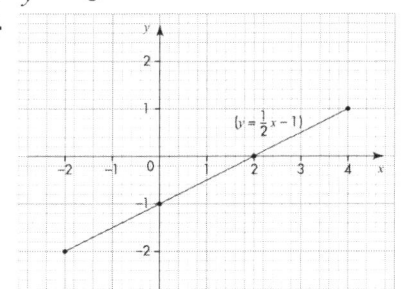

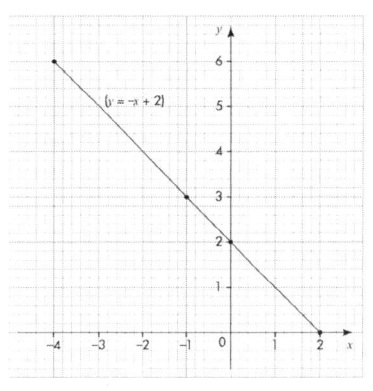

Why this chapter matters
Cross-curricular: This work ties in with the sciences, Geography and Physical Education.
Introduction: Linear graphs are fundamental to many areas of mathematics and they have many applications. Drawing linear graphs and identifying their equations is a basic skill.
Discussion points: What does it mean if a graph is parallel to the y-axis (or x-axis)? What does the gradient of a line tell you? How can you use a graph to illustrate information, e.g. how to work out a taxi fare or cook a roast?
Plenary: Using some examples from the introductory page, challenge students to draw the shapes without lifting their pencils from the paper. This type of puzzle lays the foundations for the mathematical study of graphs.

Answers to the quick check test can be found at the end of the chapter.

© HarperCollins*Publishers* 2012

17.1 Linear graphs

Curriculum references

Functional Skills standards
R2 Identify the situation or problems and identify the mathematical methods needed to solve them.

PLTS
Independent enquirers

APP
Algebra L6.4

Collins references
- Student Book pages 418–425
- Interactive Book: Multiple Choice Questions

Learning objectives
- Draw linear graphs without using flow diagrams. [Grade D–C]

Learning outcomes
- Students **must** be able to draw a linear graph from its equation. [Grade D]
- Students **should** be able to calculate the gradient of a line. [Grade C]
- Students **could** be able to interpret the gradient of a non-linear graph. [Grade C]

Key words
- axis (pl: axes)
- scale
- linear graphs

Prior knowledge
Students must be able to read and plot coordinates (even if they are given in table form). They must also be able to substitute into simple algebraic functions (Chapter 7, Lesson 7.1a).

Common mistakes and remediation
Students often work out two points and draw the line, rather than checking a third point to see that the line is correct. Inadequate algebra skills can also lead to mistakes. Students often get the x and y distances the wrong way round when calculating gradients. Tell them to remember that: x is a cross, and goes across.

Useful tips
Remind students that they should always use a sharp pencil when drawing linear graphs. Mention that if a point does not fit on a straight line, then it has been incorrectly calculated or plotted.

Functional maths and problem-solving help
Functional maths (FM) question 11 in Exercise 17A needs at least two values for H put into each formula. The obvious values to choose are 0 and 8, although students must identify the situation and the mathematics needed. In problem-solving (PS) question 13 students must link two graphs. Provide **less able** students with an example, e.g. $x = 1$ gives $y = 2$, and $y = 2$ gives $z = 3$, so $x = 1$ gives $z = 3$. Assessing understanding (AU) question 12 is structured. The answer will be the same even if the x and y lines are drawn incorrectly, so remind students that y lines are horizontal and x lines are vertical.
FM question 4 in Exercise 17B is structured, and so should be accessible. Students will need to apply a range of mathematics to answer the question and estimate the height and distance. PS question 8 requires interpretation of the information. Explain this in detail to **less able** students, and guide them towards a ratio or fraction that they can work out as a decimal, using a calculator.

© HarperCollins*Publishers* 2012

Lesson 17.1 Linear graphs

Starter
- Write a rule on the board, e.g. × 2 + 1. Ask for the results when this rule is applied to 0, 1, 2, 3 and 4. Give other rules and find the results, e.g. × 4 − 5, ÷ 2 − 3.
- Give other input numbers for these rules, e.g. −3, −2, −1.

Main lesson activity
  Drawing graphs by finding points
- Discuss the opening section of this lesson, to familiarise students with the concept of linear graphs.
  Stress these points.
    - To draw a straight line requires a minimum of two points, but three is better, as one point can be used as a check.
    - Use a sharp pencil. Drawing thick lines makes it more difficult to read values accurately. Students may be penalised for drawing thick lines in an examination.
    - Look directly down at the paper, to draw points and lines accurately. One millimetre is the tolerance allowed in an examination.
- Read this section with students and work through Example 1.
- Now students can complete Exercise 17A, in which question 11b is FM, 12 is AU and 13 is PS. **Less able** students can complete questions 1–11a; but may need support for 11b, 12 and 13. They may benefit from being given the sizes of the axes they must draw for each question. (Remind students that unlike at Foundation level, where axes are always provided in examinations, at Higher level they should not depend on this information being provided.)
- **Less able** students may also benefit from using a table for each question with extreme x-values and a 'central' x-value, so that they can work out the equivalent y-values and obtain the necessary three points needed to draw a straight line.
  Gradient, Drawing lines with a certain gradient
- Work through these sections with students. Before referring to Examples 2 and 3, stress these points. When drawing a triangle to measure the gradient, always use grid points so that the values are integers. Always divide the distance in the y-direction by the distance in the x-direction. Lines that slope from top left to bottom right have a negative gradient.
- Now students can complete Exercise 17B, in which question 4 is FM, 7 is AU and 8 is PS. **Less able** students may need support for the later questions. Consider providing photocopies of question 1 on which these students can draw the 'triangles'.

Plenary
- Draw a set of axes on the board. Ask a volunteer to draw the line with the equation $y = 2x + 1$. If it seems correct, ask for the (0, ?) and (?, 0) coordinates (and other points if suitable).
- Repeat with other volunteers and other lines, e.g. $y = 3x - 2$, $y = 6x$, $y = \frac{1}{2}x + 4$.
- Ask students to look at the first graph they drew in Exercise 17A and to tell you the gradient and the intercept of the line with the y-axis. Can they see a connection? If students spot it, test this on the second graph. If not, look at the second graph, and so on.

Answers
Answers to Exercises 17A and 17B can be found at the end of this chapter.

© HarperCollins*Publishers* 2012

17.2 Drawing graphs by the gradient-intercept method

Curriculum references

Functional Skills standards
A2 Use appropriate checking procedures and evaluate their effectiveness at each stage.

PLTS
Independent enquirers

APP
Not in assessment criteria.

Collins references
- Student Book pages 425–430
- Interactive Book: Paper Animation

Learning objectives
- Draw graphs using the gradient-intercept method. [Grade C]

Learning outcomes
- Students **must** be able to draw a linear graph by a variety of methods. [Grade B]
- Students **should** be able to select the most appropriate method to use when drawing a linear graph. [Grade B]
- Students **could** be able to describe properties of a linear graph from its equation. [Grade B]

Key words
- coefficient gradient-intercept
- constant term $y = mx + c$
- cover-up method

Prior knowledge
Students must be able to read and plot coordinates (even if they are given in table form). They must also be able to substitute into simple algebraic functions (Chapter 7, Section 7.1a).

Common mistakes and remediation
Students often calculate gradients incorrectly, e.g. they calculate x step ÷ y step. They may leave out the minus sign from a negative gradient. When using the cover-up method, students may misinterpret the intercept when the constant term is negative, or misunderstand and plot a point on the wrong axis.
To remediate these problems, provide additional practice and worked examples plus mnemonic devices such as 'x is a cross and goes across'.

Useful tips
Remind students always to start with the intercept. Drawing a line passing through the intercept may count as a mark in an examination, even if the gradient is wrong.

Functional maths and problem-solving help
There are no functional maths (FM) questions in this lesson. The problem-solving (PS) questions require students to read the information carefully. Problem-solving (PS) question 6 and assessing understanding (AU) question 5 in Exercises 17C and 17D require students to describe things that lines have in common. Students should look for a simple connection but, in both cases, it helps to draw the lines on the same grid; the common properties should then be obvious. The other questions explore relationships between lines. Encourage students to look for patterns and connections.

© HarperCollins*Publishers* 2012

Lesson 17.2 Drawing graphs by the gradient-intercept method

Starter
- Refer to the plenary in the previous lesson. Show some graphs or refer to graphs students have already seen.
- Link the equation to the gradient of the line and the intercept.
- Students should spot that the number in front of x (the coefficient) is the same as the gradient, and the point where the graph crosses the y-axis is given by the constant term.

Main lesson activity
- The starter covers the basic principle of the first part of this lesson, so go straight to Example 4 and work through it with the class.
- Now students can complete Exercise 17C, in which question 5 is an AU question and 6 is PS. **Less able** students can complete questions 1–5 first; they may need support with 6.
- All students should be able to find the intercept on the y-axis. **Less able** students may have trouble drawing the gradients. A printed sheet of the necessary gradients showing the x-step and the y-step may be helpful.

Cover-up method for drawing graphs
- Show students a set of axes. Ask for a series of points on the y-axis. Write these on the board and ask students what they have in common.
- They should spot that the x-value is 0. Recall that the y-axis has the equation $x = 0$.
- Repeat with points on the x-axis or ask students what the equation of this line is. Recall that the x-axis has the equation $y = 0$. Work through this section and Examples 5 and 6.
- Now students can complete Exercise 17D, in which question 5 is AU and 6 is PS. **Less able** students may need support. They may have difficulty when the equation involves a minus value. Remind them that, when dividing a positive by a negative, the result is negative, and that when dividing a negative by a negative, the result is positive. (This only occurs in question 1i.)
- Challenge **more able** students to find the equation, in the form $ax + by = c$, for the line passing through two points such as (5, 0) and (0, -3).

Plenary
- Write this equation on the board: $4x + 3y = 12$. Use the cover-up rule to plot the graph.
- Work through the following equation. Students may be able to solve this if they have experience of rearranging formulae.

$$3y = -4x + 12$$
$$y = -\frac{4}{3}x + 4$$

- Link this to the graph. It should have the same gradient and intercept.
- Point out that whatever form the equation is in, if it can be arranged in the form $y = mx + c$, then students can always use the gradient-intercept method to draw it.

Answers
Answers to Exercises 17C and 17D can be found at the end of this chapter.

17.3 Finding the equation of a line from its graph

Curriculum references

Functional Skills standards
I2 Draw conclusions and provide mathematical justifications.

PLTS
Independent enquirers

APP
Not in assessment criteria.

Collins references
- Student Book pages 430–432
- Interactive Book: Multiple Choice Questions

Learning objectives
- Find the equation of a line, using its gradient and intercept. [Grade B]

Learning outcomes
- Students **must** be able to find the equation of a line from its graph. [Grade B]
- Students **should** be able to find the equation of a line from knowing two points on it. [Grade B]
- Students **could** be able to find similarities between sets of linear graphs. [Grade B]

Key words
- coefficient
- intercept
- gradient

Prior knowledge
Students should know how to draw a line, using the gradient-intercept method. They should also know how to find the gradient of a straight line.

Common mistakes and remediation
Students often fail to recognise a negative gradient. Remind them (if necessary, every time they begin to answer questions about a graph) that lines that slope from top left to bottom right always have a negative gradient: forward slope = positive, backward slope = negative.

Useful tips
Suggest that students write down the equation as $y = mx + c$, then substitute the y-intercept value as the constant and work from there.

Functional maths and problem-solving help
There are no functional maths (FM) questions in Exercise 17E. All the problem-solving (PS) questions explore relationships. For example, the gradients in questions 2, 4 and 5 have values in common. In question 5, the coordinates can be used to identify a pattern. Encourage students to look for patterns and connections.

Lesson 17.3 Finding the equation of a line from its graph

Starter

- Write on the board: $y = 2x - 1$, $4x - 2y = 2$, $x = \frac{y+1}{2}$.
- Ask students what method they would use to plot the graphs of these equations.
- The first looks as if the gradient-intercept method would be most appropriate. The second looks as if the cover-up method would be best. The third doesn't look like anything with which students are familiar, but choosing y-values and calculating x-values to get coordinates would be one method, e.g. $y = 3, x = 2; y = 9, x = 5$.
- If students plot these lines, they will find that they are all the same.

Main lesson activity

The equation y = mx +c

- Show an accurate graph, e.g. $y = 3x + 1$ (unlabelled). Ask students if they can say what the equation of the line is.
- By now, students should be able to reverse the gradient-intercept method. If not, prompt them by asking them for the standard form of the equation of a line ($y = mx + c$).
- Eventually, students should link the gradient to m and the intercept to c. Work them out and write down the equation for this line.
- Work through the text and Example 7 to consolidate this.
- Now students can complete Exercise 17E, in which question 3 is an AU question and questions 2 and 5 are PS. **Less able** students may need support. Provide them with copies of the grids so that they can draw 'triangles' to find the gradients.

Plenary

- Show a sketch of a 'gas bill' graph, which consists of a basic charge and a price per unit. Do not put any values on the graph.

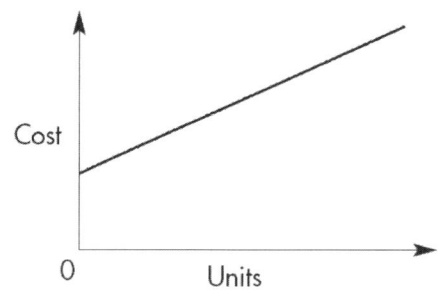

- Ask students to say how they can find the basic charge from the graph (the intercept). Then ask how they can find the price per unit (the gradient).

Answers

Answers to Exercise 17E can be found at the end of this chapter.

17.4 3D coordinates

Curriculum references

Functional Skills standards

R2 Identify the situation or problems and identify the mathematical methods needed to solve them.

I2 Draw conclusions and provide mathematical justifications.

PLTS

Independent enquirers, Creative thinkers, Self-managers

APP

Algebra L6; Using and applying mathematics L7

Collins references

- Student Book pages 433 – 436

Learning objectives

- Read and use 3D coordinates from a diagram. [C–A*]

Learning outcomes

- Students must be able to read 3D coordinates from a diagram. [Grade C–B]
- Students should be familiar with and ordered pair, an ordered triple and the right-hand rule. [Grade B]
- Students could be confident in using coordinates. [Grade B–A]

Key words

- 3D coordinates
- ordered pair
- ordered triple
- right-hand rule

Prior knowledge

Students should be familiar with coordinates in all four quadrants in two dimensions.

Common mistakes and remediation

Less able students may find this topic difficult. Spend extra time with these students. You could also pair them with more able students.

Students may misread diagrams. Practice will provide remediation. Also, working through one of the many dynamic demonstrations on the internet, of plotting 3D points, should help students to understand the concept.

Useful tips

Tell students to remember that in order to be able to specify the coordinates they want, they should remember the right-hand rule.

Students should also remember that coordinates in three dimensions are expressed as a triple of values, or an ordered triple.

Functional maths and problem-solving help

This topic lends itself to functional maths, in that it has to do with everyday life, from, for example, the classroom to a corner of the classroom or where something or someone is placed. Students should look around them and think of the coordinates in practical terms.

© HarperCollins*Publishers* 2012

Lesson 17.4 3D coordinates

Starter
- Take a corner of the classroom and call it the origin. Mark one horizontal edge as the x-axis, and the other horizontal edge as the y-axis.
- Take a unit to be approximately one metre (or one stride).
- Now ask students to identify their position relative to the origin. This may be something like (3, 5) for a student who is sitting approximately three metres from the origin in the *x* direction and five metres from the origin in the *y* direction.
- Now identify the vertical edge from the origin as the z-axis.
- Ask some students, including a very tall student, to stand up.
- Ask the class if they can identify the position of these students' heads as coordinates in all three directions. For example, if the tallest student is about two metres tall, his or her head may be at (3, 5, 2).
- Repeat for other students and objects in the room.
- Identify the far corner of the room as a 3D coordinate.

Main lesson activity
- The starter should have given students an idea of 3D coordinates.
- Explain that a number line is one-dimensional and any point on the line can be identified by a single value.
- Add that a typical graph is two-dimensional and any point on the grid can be identified by two values: an x- and a y-coordinate. However, we live in a three-dimensional world, and to identify points in space we need to give three values.
- Draw this diagram on the board or put it on screen.
- Explain that this is the standard convention for showing 3D coordinates.
- Explain the right-hand rule.
- There are many dynamic demonstrations on the internet of plotting 3D points. As has been mentioned under 'common mistakes and remediation', this is a difficult concept. Points in very different places can appear to be in the same place when plotted on a 3D diagram.
- Now work through the text and examples 8 and 9 in the 3D coordinates lesson.
- Students can complete the questions in Exercise 17F.

Plenary
- Refer back to the room marked out as a 3D space.
- Ask students to identify points in the room as 3D coordinates.

Answers
Answers to Exercise 17F can be found at the end of this chapter

© HarperCollins*Publishers* 2012

Problem solving
Traverse the network

Curriculum references

Functional Skills standards
R1 Understand routine and non-routine problems in familiar and unfamiliar contexts and situations.
I2 Draw conclusions and provide mathematical justifications.

PLTS
Independent enquirers, Creative thinkers

APP
Using and applying mathematics L7.3

Collins references
Student Book pages 440–441

Learning objectives
- Problem solving

Key words
- nodes
- arcs
- networks
- traversable

Functional maths help
Advise students not to spend too much time on one network, as this may disillusion or frustrate them. Students may need some help with the definitions in the Student Book. **Less able** students will benefit if you label the networks shown in 'Getting started' as arcs, nodes, etc. to reinforce these definitions.

Lesson plans
Build
Refer students to the 'Facts' text in the Student Book. Discuss the definitions of 'arcs', 'nodes' and 'the degree of a node'. Tell students they need to know these to achieve mastery of the topic.

Activity 1
Refer students to the 'Getting started' text in the Student Book.
More able students can work individually to look at the networks and decide if they are traversable. Allow **less able** students to work in small groups to do this. (first and third networks are possible)
This concept is likely to be new, so you may need to stop at various times during class to discuss their findings. Encourage students to tabulate the degree of the nodes and the traversable tendency of the network.

Activity 2
Refer students to the first question of 'Your task'.
Ask them to investigate the number of odd nodes and even nodes of each shape to work out which ones are traversable networks. (1st, 2nd, 5th, 7th and 9th; the rule: a variety of diagrammatical approaches could be used.).
More able students could try devising and summarising the rules for traversability.

Write the rules of traversability, below, on the board, and go through them with students.
Rules for traversable networks
1. A network with exactly two odd vertices is traversable: the starting point may be at either odd vertex; the end point is the other odd vertex.
2. A network with no odd vertices is traversable: any vertex may be the starting point; the same vertex will also be the end point.
3. A network with more than two odd vertices is not traversable.

© HarperCollins*Publishers* 2012

Problem solving Traverse the network

Activity 3
Now refer students to the second question of 'Your task'. Say that they must use their rule to trace a route through the house so that they pass through each door only once.
Students who are building skills may need help while doing this. It may be a good idea to simplify the building to scaffold learning. (See the 'Apply' text.)
Then ask students to work in pairs to look at the building and turn it into a network diagram.

Activity 4
Ask students to recall what they have learned about the Königsberg Bridge problems. Then ask them to research this topic further on the internet, or by using any other sources for information.

Apply
When applying skills, most students should be able to work through the definitions and Activity 1 independently. Students may need some help in finding a route through the building as this is an unfamiliar task. Scaffold this by asking students to consider rooms within their own home or within the school.
Students may use the internet or other sources to investigate the Königsberg Bridge problems.

Master
For students to show mastery in this activity they must be able to:
- Understand the definitions for traversing a network.
- Apply their knowledge of the definitions to the networks shown.

Plenary
Ask students the following questions.
- Can you think of any other uses for networks?
- Where might you see them?
- Which professions might be interested in analysing routes and looking at travelling, without revisiting places?

Have a class discussion about deliveries or sales people who look for the most cost-effective routes for their businesses. Why would they need to find such routes?

Answers Lessons 17.1 – 17.4

Quick check

1 a 13 **b** [graph of $y = 2x + 3$]

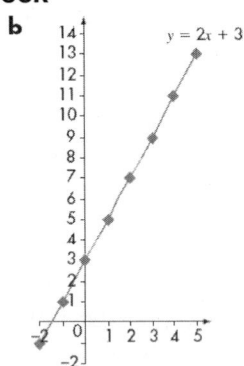

17.1 Linear graphs
Exercise 17A

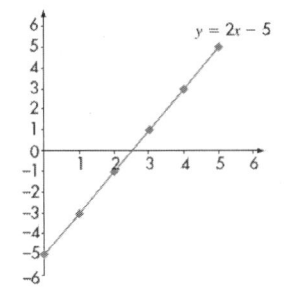

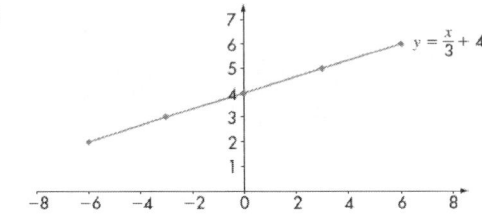

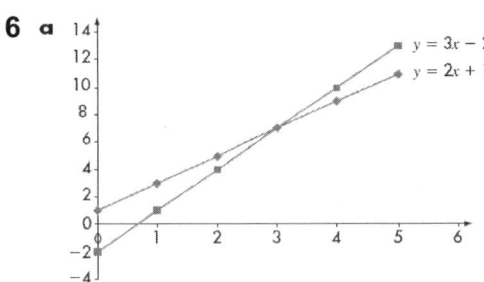

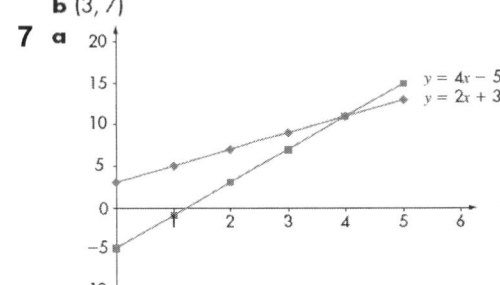

1 [graph of $y = 3x + 4$]

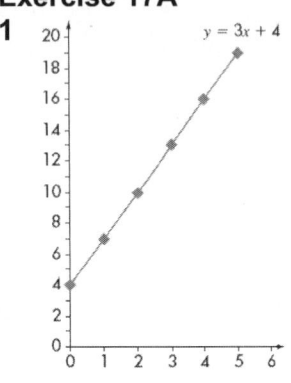

2 [graph of $y = 2x - 5$]

3 [graph of $y = \frac{x}{2} - 3$]

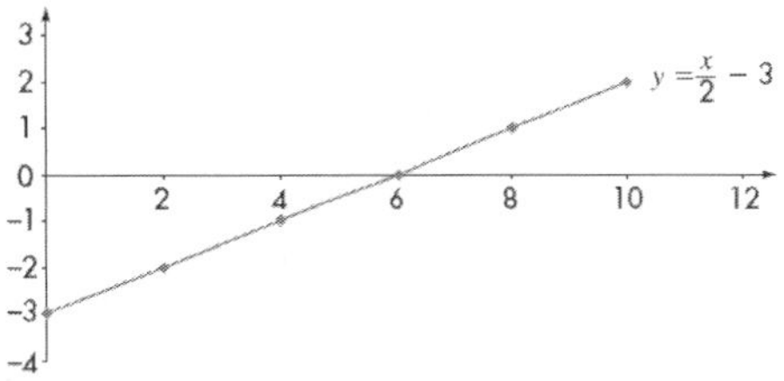

4 [graph of $y = 3x + 5$]

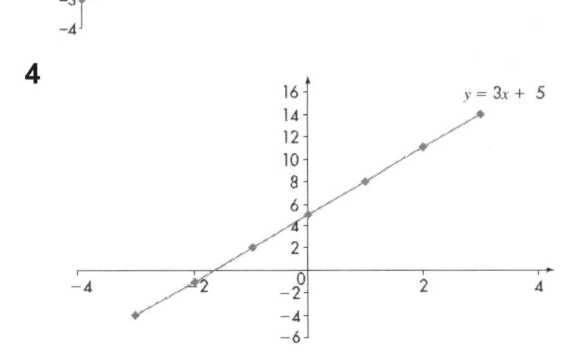

5 [graph of $y = \frac{x}{3} + 4$]

6 a [graphs of $y = 3x - 2$ and $y = 2x + 1$]
b (3, 7)

7 a [graphs of $y = 4x - 5$ and $y = 2x + 3$]
b (4, 11)

8 a [graphs of $y = \frac{x}{2} - 2$ and $y = \frac{x}{3} - 1$]
b (6, 1)

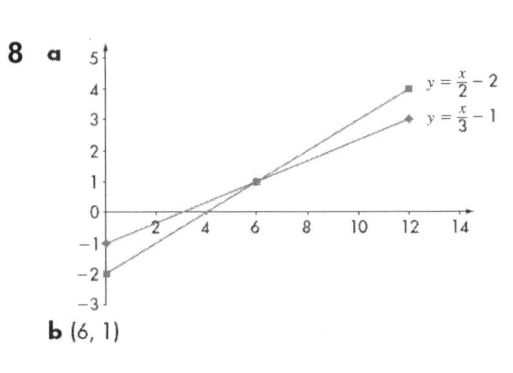

© HarperCollins*Publishers* 2012

280

Answers Lessons 17.1 – 17.4

9 a [graph of $y = 3x+1$ and $y = 3x-2$]
b No, because the lines are parallel.

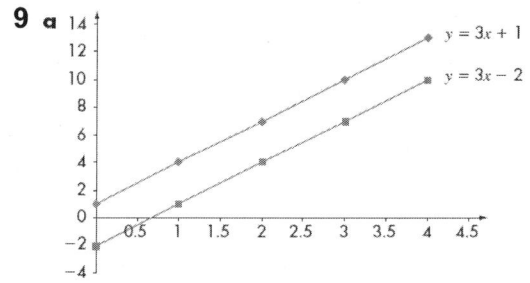

10

x	0	1	2	3	4	5
y	5	4	3	2	1	0

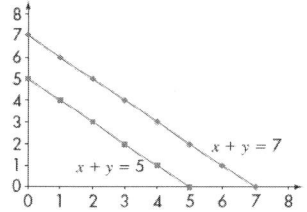

[graph of $x+y=5$ and $x+y=7$]

11 a [graph showing Ian and John lines]
b Ian, as Ian only charges £85, whilst John charges £90 for a 2-hour job.

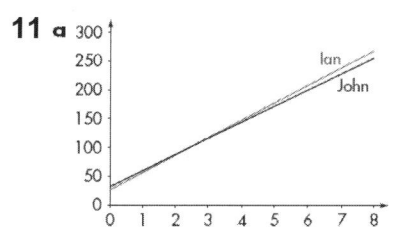

12 a [graph] **b** 4.5 units squared

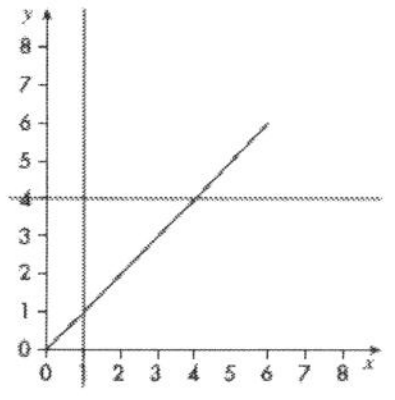

Exercise 17B
1 a 2 **b** $\frac{1}{3}$ **c** −3 **d** 1 **e** −2
f $-\frac{1}{3}$ **g** 5 **h** −5 **i** $\frac{1}{5}$ **j** $-\frac{3}{4}$

2 [grid showing lines a–f]

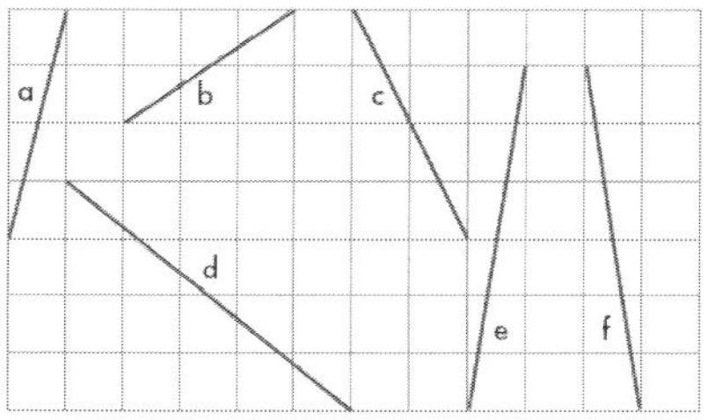

3 a 1
b −1. They are perpendicular and symmetrical about the axes.

4 a Approximately 320 feet in half a mile (2640 feet), so gradient is about 0.12.
b i Because the line on the graph has the steepest gradient.
ii Approximately 550 feet in half a mile so gradient is about -0.21
c BM. Approximately 1200 feet of climbing in 6.3 miles ≈ 190 feet of ascent on average

5 a 0.5 **b** 0.4 **c** 0.2 **d** 0.1 **e** 0
6 a $1\frac{2}{3}$ **b** 2 **c** $3\frac{1}{3}$ **d** 10 **e** ∞
7 Raisa has misread the scales. The second line has four times the gradient (2.4) of the first (0.6).
8 6 : 4, 8 : 5, 5 : 3, 11 : 6, 2 : 1, 5 : 2

17.2 Drawing graphs by the gradient-intercept method
Exercise 17C
1 a, b, c, d [graph]

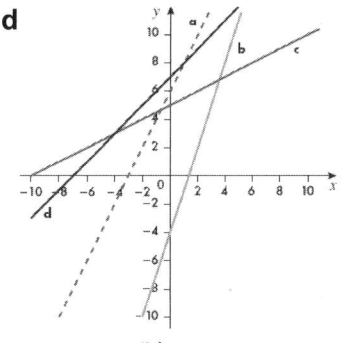

e, f, g, h [graph]

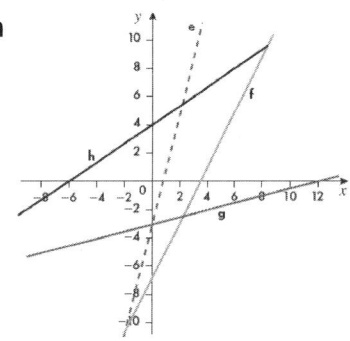

i, j, k, l [graph]

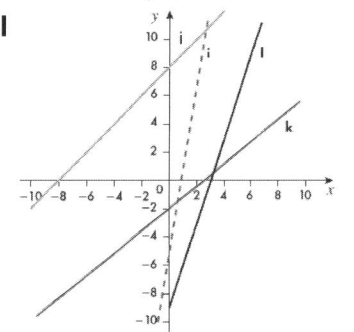

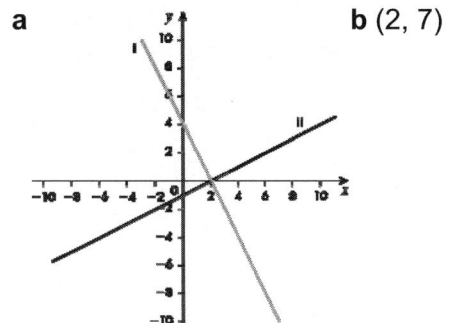

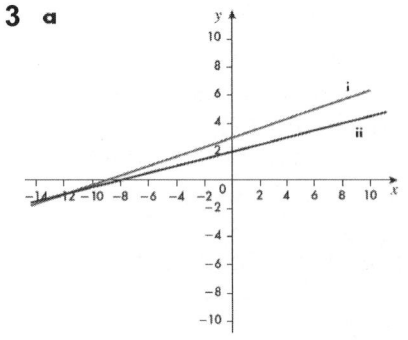

Answers Lessons 17.1 – 17.4

2 a [graph] **b** (2, 7)

3 a [graph]
b (−12, −1)

4 a [graph]
b (3, 6)

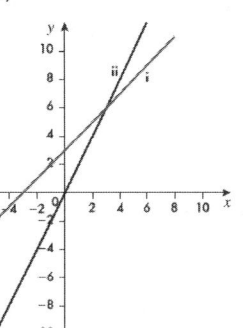

5 a They have the same gradient (3).
b They intercept the y-axis at the same point (0, −2).
c (−1, −4)

6 a −2 **b** $\frac{1}{2}$ **c** 90°
d Negative reciprocal **e** $-\frac{1}{3}$

Exercise 17D

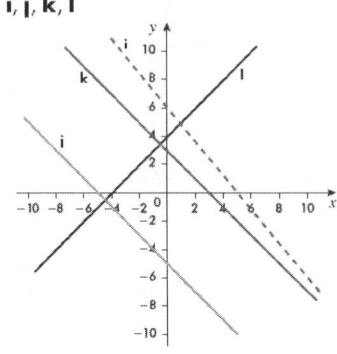

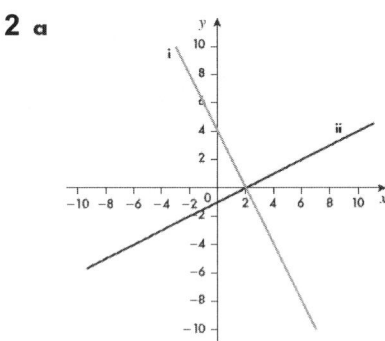

1 a, b, c, d [graph]

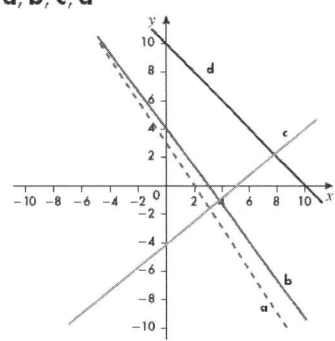

e, f, g, h [graph]

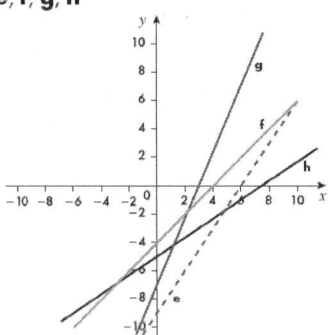

i, j, k, l [graph]

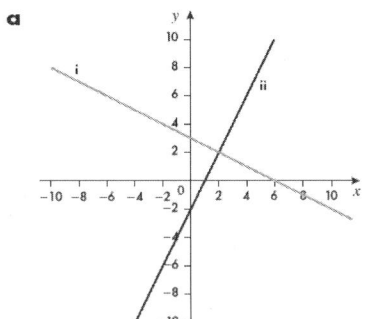

2 a [graph]
b (2, 0)

3 a [graph]
b (2, 2)

Answers Lessons 17.1 – 17.4

4 a

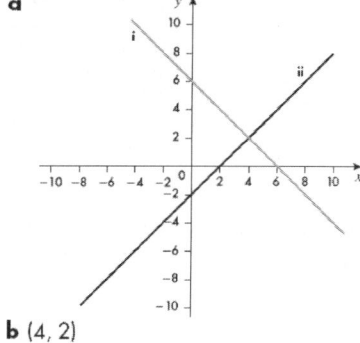

b (4, 2)

5 a Intersect at (6, 0)
  b Intersect at (0, –3)
  c Parallel

6 a i $x = 3$ **ii** $x - y = 4$
  iii $y = -3$ **iv** $x + y = -4$
  v $x = -3$ **vi** $y = x + 4$
  b i -3 **ii** $\frac{1}{3}$ **iii** $-\frac{1}{3}$

17.3 Finding the equation of a line from its graph
Exercise 17E

1 a $y = \frac{4}{3}x - 2$ or $3y = 4x - 6$ **b** $y = x + 1$
  c $y = 2x - 3$ **d** $2y = x + 6$
  e $y = x$ **f** $y = 2x$

2 a i $y = 2x + 1, y = -2x + 1$
     ii Reflection in y-axis (and $y = 1$)
     iii Different sign
  b i $5y = 2x - 5, 5y = -2x - 5$
     ii Reflection in y-axis (and $y = -1$)
     iii Different sign
  c i $y = x + 1, y = -x + 1$
     ii Reflection in y-axis (and $y = 1$)
     iii Different sign

3 a x-coordinates go from $2 \to 1 \to 0$
  and y-coordinates go from
  $5 \to 3 \to 1$.
  b x-step between the points is 1 and y-step is 2.
  c $y = 3x + 2$

4 a $y = -2x + 1$
  b $2y = -x$
  c $y = -x + 1$
  d $5y = -2x - 5$
  e $y = -\frac{3}{2}x - 3$ or $2y = -3x - 6$

5 a i $2y = -x + 1, y = -2x + 1$
     ii Reflection in $x = y$
     iii Reciprocal of each other
  b i $2y = 5x + 5, 5y = 2x - 5$
     ii Reflection in $x = y$
     iii Reciprocal of each other
  c i $y = 2, x = 2$
     ii Reflection in $x = y$
     iii Reciprocal of each other (reciprocal of zero is infinity)

17.4 3-D coordinates
Exercise 17F

1 a A(0, 0, 0), B(0, 3, 0), C(0, 3, 3),
  D(0, 0, 3), E(3, 0, 0), F(3, 0, 3),
  G(3, 3, 3), H(3, 3, 0)
  b i (0, 1.5, 1.5) **ii** (1.5, 3, 1.5)
  c (1.5, 1.5, 1.5)
  d i $\sqrt{18} = 4.24$ **ii** $\sqrt{27} = 5.2$

2 a i (0, 3, 0) **ii** (3, 3, 0) **iii** (3, 0, 0)
  b i (0, 3, 2) **ii** (–3, 3, 2) **iii** (–3, 3, 0)
  c $\tan^{-1}(2 \div \sqrt{18}) = 28.1°$

3 a (2, 2, 2) **b** (2, 2, 3) **c** (0, 2, –1)
  d (–4, 3, 3) **e** (7, 0, 3) **f** (4, –3, 2)

4 a (0, 4, 2) **b** (2, –3, 0)
  c 28 cubic units

Examination questions

1 a

  b 1.75

2

3 a $y = -3x + 9$ **b** –9
  c $\frac{1}{3}$

4 (6, 5)

5 Straight-line graph from (–4, –13) to (4, 12)

6 12.5 square units

7 a Straight line from (–4, –10) to (4, 14)
  b $y = -2x - 1$

8 a The line shown is $y = -x$
  b $y = x + 1$

Chapter 18 Algebra: More graphs and equations

Overview

| 18.1 Quadratic graphs | 18.2 The significant points of a quadratic graph |

This chapter covers more graph and equation work for the Higher tier. Note that transformation of graphs is covered in Chapter 15 of AQA Linear Higher Book 2. Students will learn how to recognise and draw quadratic graphs and read values and calculate the significant points from them.

Context
In this chapter, students will learn that graphs and equations have many uses in science, economics, architecture and other areas.

Curriculum references
KS4 Programme of Study references
1.2b Using existing mathematical knowledge to create solutions to unfamiliar problems.
2.2a Make connections within mathematics.
2.2k Make accurate mathematical diagrams, graphs and constructions on paper and on screen.
3.1e Linear, quadratic and other expressions and equations.
3.1f Graphs of exponential and trigonometric functions.

Linear specification
N6.13 Generate points and plot graphs of simple quadratic functions, and use these to find approximate solutions.

Functional Skills standards
R2 Identify the situation or problems and identify the mathematical methods needed to solve them.
R3 Select a range of mathematics to find solutions.
A1 Apply a range of mathematics to find solutions.
A2 Use appropriate checking procedures and evaluate their effectiveness at each stage.
I2 Draw conclusions and provide mathematical justifications.

PLTS
Independent enquirers identify questions to answer and problems to resolve; explore issues, events or problems from different perspectives. **Reflective learners** assess themselves and others, identifying opportunities and achievements. **Self-managers** work towards goals, showing initiative, commitment and perseverance. **Creative thinkers** connect their own and others' ideas and experiences in inventive ways. **Effective participators** discuss issues of concern, seeking resolution where needed. **Team workers** collaborate with others to work towards common goals.

APP
Using and applying mathematics L8.1 Develop and follow alternative methods and approaches.
Algebra L8.6 Sketch, interpret and identify graphs of linear, quadratic, cubic and reciprocal functions, and graphs that model real situations.

© HarperCollins*Publishers* 2012

Chapter 18 Algebra: More graphs and equations

Route mapping

Exercise	Grades C	Grades B	Grades A	Grades A*
A	1–9	10		
B	1–9	10		11–18

Overview test

Questions 1–4 are grade E, questions 5–7 are grade D, question 8 is grade B and question 9 is grade A*.

1. For the equation $y = x + 3$, find the value of y when $x = 3$.
2. For the equation $y = x^2 - 2$, find the value of y when $x = 3$.
3. For the equation $y = 2x$, find the value of y when $x = 3$.
4. For the equation $y = \frac{1}{2}x$, find the value of y when $x = 3$.
5. For the equation $y = 3x^2 - 2x + 1$, find the value of y when $x = 3$.
6. For the equation $y = x^3 + x^2 + x + 1$, find the value of y when $x = 3$.
7. For the equation $y = 2x^3 - 3x^2 + 4x - 10$, find the value of y y when $x = 3$.
8. For the equation $y = \sin x$, find the value of y when $x = 60°$.
9. Use the 'completing the square' method to solve $x^2 - 4x - 3 = 0$, giving your answers correct to two decimal places.

Answers to overview test

1. 6
2. 7
3. 6
4. 1.5
5. 22
6. 40
7. 29
8. 0.866
9. 4.65, −0.65

Why this chapter matters

Cross-curricular: This work has links with the sciences, Geography, History, Design and Technology.
Introduction: Quadratic graphs are essential to many areas of mathematics and they have many applications. Drawing these graphs and identifying their equations is a basic skill that leads to greater things. The task on the introductory page in the Student Book leads into the material in the chapter.
Discussion points: What does it mean if a graph is parallel to the y-axis (or x-axis)? What does the gradient of a line tell you? How do you measure the slope of a curved graph?
Plenary: Discuss the task with students. Refer to the notes under 'Why it works'. Ask them to suggest where they might see this type of curve in their environment.

Answers to the quick check test can be found at the end of this chapter.

© HarperCollins*Publishers* 2012

18.1 Quadratic graphs

Curriculum references

Functional Skills standards
A2 Use appropriate checking procedures and evaluate their effectiveness at each stage.

PLTS
Independent enquirers, Creative thinkers

APP
Not in assessment criteria.

Collins references
- Student Book pages 444–449
- Interactive Book: 10 Quick Questions

Learning objectives
- Draw and read values from quadratic graphs. [Grade C–B]

Learning outcomes
- Students **must** be able to calculate points on a quadratic graph. [Grade C]
- Students **should** be able plot a quadratic graph accurately. [Grade C]
- Students **could** be able to use symmetry to identify and correct errors in drawing a quadratic graph. [Grade C]

Key words
- parabola
- quadratic

Prior knowledge
Students should be able to read and plot coordinates (even if given in table form), and be able to substitute into algebraic functions.

Common mistakes and remediation
Students often produce poorly plotted and poorly drawn curves. Remind them that in the GCSE all points should be within small squares on the graph paper, and the curve should also pass within a small square. The curve should be one continuous line, with no gaps (or multiple attempts at drawing the line).

Useful tips
Remind students that they should always be careful only to use smooth, accurate, one-line curves, and that nothing else is acceptable.

Functional maths and problem-solving help
Problem-solving (PS) questions 8 and 9 in Exercise 18A are constructions of parabolas. Even the **more able** students may find these difficult at first. It is worth working through constructions of parabolas and having a pre-prepared sheet for both questions. Once students have grasped how to plot a few points, they should be able to complete the questions themselves. There are no functional maths (FM) questions in Exercise 18A.

Starter
- Ask students to draw a set of axes numbered −5 to +5 on the x-axis and 0 to 30 on the y-axis. Ask them to plot these points:

x	−5	−4	−3	−2	−1	0	1	2	3	4	5
y	25	16	9	4	1	0	1	4	9	16	25

Then ask them to join the points with a smooth curve. Walk around and check students' graphs, pointing out unacceptable curves and ways to improve them. Explain that this is the $y = x^2$ curve, which they need to learn and recognise.

© HarperCollins*Publishers* 2012

Lesson 18.1 Quadratic graphs

Main lesson activity
- Explain that a quadratic equation is an equation containing a term in x^2. Look at the examples provided.
- The most fundamental graph is $y = x^2$, and all quadratic graphs are based on this and have the same basic shape. This is the graph students were required to draw in the starter.
- The main points to emphasise are:
  - In examinations, a range of values of x, and axes, are always given.
  - Usually a table is provided, but this is just the x values and a line for the y values. Intermediate lines to aid with calculation are not given in examinations.
  - Most values will be filled in and only two or three will need to be calculated.
- Read and then discuss the information about drawing accurate graphs and relate this to some of the errors that students may have made in the starter.
- Remind students that they should always label the graph.
- Work through Examples 1 and 2. Example 1 has intermediate lines for calculating the different parts of the equation, but Example 2 only has x and y values.
- It is expected that students will use a calculator to work out y values. Calculating with a table is only helpful for **less able** students, who may need more structure.
- Make sure students know how to go from an x value to find the equivalent y value, and also how to go from a y value to find the equivalent x values. Note that usually there will be two x values to find.
- Now students can complete Exercise 18A, in which questions 8 and 9 are PS, and 10 is an assessing understanding (AU) question. **Less able** students can complete the questions in pairs, if necessary. They should find the structure of the table useful and may find it easier to have pre-prepared grids to work on.

Plenary
Ask the following questions:
- What shapes are $y = x^2$ and $y = -x^2$ graphs? Where do they intercept with the axes?
- What shapes are $y = 2x^2$, $y = 3x^2$, $y = 5x^2$ graphs? (Sketch them on the same axes.)
- Where do the following graphs intercept with the axes? $y = x^2 + 2$, $y = 2x^2 + 2$, $y = 3x^2 + 2$

Answers
Answers to Exercise 18A can be found at the end of this chapter.

© HarperCollins*Publishers* 2012

18.2 The significant points of a quadratic graph

Curriculum references

Functional Skills standards
I2 Draw conclusions and provide mathematical justifications.

PLTS
Independent enquirers, Creative thinkers

APP
Not in assessment criteria.

Collins references
- Student Book pages 450–454

Learning objectives
- Recognise and calculate the significant points of a quadratic graph. [Grade B–A*]

Learning outcomes
- Students **must** be able to use a quadratic graph to solve a quadratic equation. [Grade B]
- Students **should** be able to find a minimum or maximum point on a quadratic graph. [Grade B]
- Students **could** be able to find a relationship between a quadratic equation and the coordinates of a minimum point. [Grade A*]

Key words
- intercept
- maximum
- minimum
- roots
- vertex

Prior knowledge
Students should know how to solve a quadratic by the method of completing the square, and also how to draw quadratic graphs.

Common mistakes and remediation
Students often draw graphs poorly. Tell them they can avoid this by making sure they use the right equipment – sharp pencils, a compass, ruler, and so on. Miscalculating y values is a common mistake. To remediate this, encourage students to look vertically, or up the graph to see the value, as opposed to finding the x value, when one looks horizontally, or along the graph. To distinguish between the two axes, encourage students to remember the simple method of thinking of x as across (so you look across to get the value).

Functional maths and problem-solving help
There are no functional maths (FM) questions in Exercise 18B. Questions 3 and 7 are problem solving (PS), and grade B, questions 11 and 16 are grade A*, as are 17 and 18, which are assessing understanding (AU) questions. The grade A* questions, which may not be appropriate for all students, are designed to assess students' understanding of the work in this lesson. Work through the questions to help **less able** students, who will find the concepts difficult. **More able** students should be able to complete the questions, possibly with some support.

Starter
- Sketch a linear graph on the board.

- Ask students to give a possible equation for this line, e.g. $y = 2x$
- Ask what points we might be interested in as mathematicians.
- This would be the two points where the line crosses the axes.

© HarperCollins*Publishers* 2012

Lesson 18.2 The significant points of a quadratic graph

- For the equation provided, work out these points and discuss how to find them. [For $y = 2x + 1$ these are (0, 1) and (−0.5, 0)]
- The intersect of the y-axis is the constant term.
- The intersect on the x-axis is found by solving the equation $2x + 1 = 0$.

Main lesson activity
- Do not allow students to open their textbooks.
- Draw a quadratic on the board and ask what points would be of interest to mathematicians.

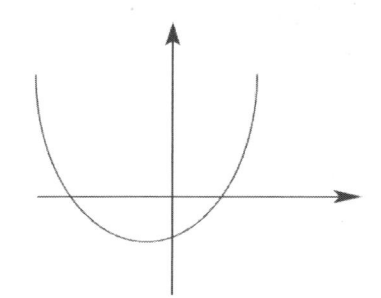

- Students will see the points where the graph crosses the axes, but may not spot the 'vertex'. Encourage them by saying: There are four points, what is the fourth one? Eventually they should identify the vertex.
- Now go through the text dealing with the roots of an equation, followed by Examples 3 and 4.
- Students can complete Exercise 18B, in which questions 3, 7, 11 and 16 are PS and 17 and 18 are AU.
- If students could draw the graphs in the previous exercise, then they will certainly manage questions 1, 2 and 4–6. **Less able** students can complete these questions first, then questions 8–15, and then move on to PS and AU. In PS questions 3 and 7 students must spot a connection. **Less able** students may need a clue, e.g. ask them to look at the equation in question 1 and the roots from the graph. Although there are many valid mathematical concepts covered in all the questions, **more able** students will find the latter questions challenging.

Plenary
- Ask students to use their calculators to work out 5^8. (390 625). They may need help in using the power button.
- Once they have figured this out, ask them to work out other powers, e.g. $2^{0.5}$ (1.414…) or $3^{0.8}$ (2.408..).
  Finish by raising a number to the power 0, e.g. 7^0 (= 1).
- Make sure students can use the power button and know that anything raised to the power zero is 1.

Answers
Answers to Exercise 18B can be found at the end of this chapter.

© HarperCollins*Publishers* 2012

Problem solving
Quadratics in bridges

Curriculum references

Functional Skills standards
R2 Identify and obtain necessary information to tackle the problem.
A1 Apply mathematics in an organised way to find solutions.

PLTS
Independent enquirers, Reflective learners

APP
Algebra L7

Learning objectives
- Model real life quadratics

Key words
- curve
- quadratic
- model

Collins references
- Student Book pages 458–459

Functional maths help
Students will not have yet generated quadratic equations and graphs from real-life situations, so even **more able** students will need guidance through this challenging task.
You may wish to direct students to an appropriate software package (if one is available), to help them to complete the task, particularly if the class is **less able**.

Build
Discuss the shape that bridges take. Include as many types as possible, including single beam, arches, multiple beam, cantilever and suspension. Propose that the shape of any bridge can be related to mathematical equations. In the case of suspension bridges, the curve of the cable or chain is relatively easy to describe mathematically. Recap on quadratic expressions and equations.

Activity 1
Familiarise students with quadratics in real-life situations by asking them to identify bridges and other landmarks that might be based on quadratics. Students will gain most benefit from this activity if they can identify landmarks in the local area. Have a few examples prepared, to give students some direction in their research.
Once students have found their examples, ask them to identify what they would need to know in order to be able to describe the shape of a bridge in terms of quadratic expressions or equations. Guide them to the following ideas:
- the width of the parabola
- where the parabola is centred
- the overall height of the parabola.

Activity 2
Direct students to the fact box about the Clifton Suspension Bridge. Ask students to construct a drawing of the Clifton Suspension Bridge, labelling the dimensions as appropriate. A **more able** class could construct an accurate scale drawing. Conversely, for a **less able** class, you may prefer to supply a blank diagram of the bridge, asking them to label the dimensions. You may need to hold a class discussion about the units that should be used.

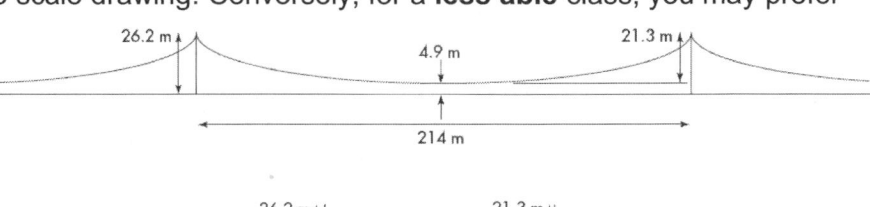

Check the span and pier heights on students' diagrams, which should look like these.

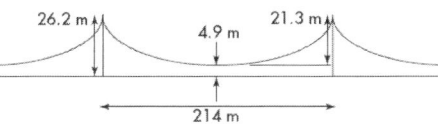

© HarperCollins*Publishers* 2012

Problem solving Quadratics in bridges

Activity 3
Ask students to work in pairs, using their diagrams to find the quadratic equation for the Clifton Suspension Bridge. Suggest that they measure the width of the bridge from the centre, counting in both directions. This will simplify the equation, allowing the quadratic to be symmetrical about the origin. Encourage students to state the assumptions that they are making, in order to solve the problem. Assuming that the bottom of the curve passes through the origin, the working should be like this.

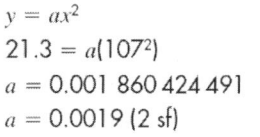

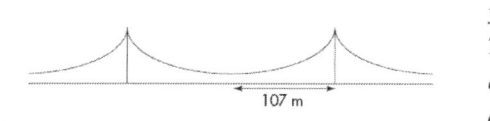

$y = ax^2$
$21.3 = a(107^2)$
$a = 0.001\,860\,424\,491$
$a = 0.0019$ (2 sf)

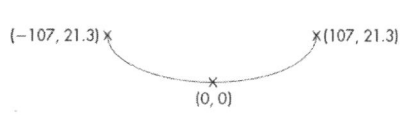

$(-107, 21.3)$, $(107, 21.3)$
$(0, 0)$

The equation is $y = 0.0019x^2$

More able students may choose to take the origin as the point along the deck below the minimum point of the curve. In this case, the equation will be $y = 0.0019x^2 + 4.9$.

Activity 4
Ask students to work in pairs to draw the graph for their quadratic equation. Students may need help in scaling and formulating the graph; for example, guide them to put the values of the width of the bridge on the x-axis and the height on the y-axis. After drawing the graph they can write their reports, explaining their problem-solving approach, the mathematical skills they have used, assumptions they made and how changing their assumptions would change their answer.

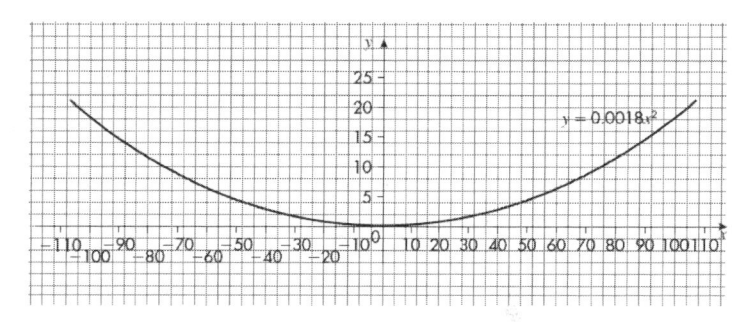

Activity 5
Ask pupils to think about these questions:
- Does the height of the deck above the water influence the equation of the bridge? (No.)
- What happens to the basic equation $y = x^2$ when you apply it to the bridge? (The equation is stretched along the x-axis.)

Apply
Students working at this level should be able work collaboratively, in small groups, to break down the problem. Group work will allow students to support themselves and each other in the application of modelling skills. They should be able to make their own assumptions and find the connection between
$y = x^2$ and the equation of the bridge, if the equation is centred at the origin. If the class is **less able**, use the activities as prompts to help the groups to find their own routes through the task.

Master
In order to demonstrate their mastery of the learning objective, students should be able to:
- Construct a scale drawing.
- Find a quadratic equation.
- Model quadratic equations.
- Use mathematical language to describe the investigation and state any assumptions that they have made.

If the class is **more able**, extend the task by asking students to investigate the quadratics of other bridges and similar structures. Alternatively, they could make up their own equations and see if they are suitable to be used for construction.

Plenary
Ask students to present their reports. Unless the students have already done so, hold a brief class discussion about where else they see quadratics in real life, for example, in arches in buildings, over railways and in historical architecture such as Roman baths.

© HarperCollins*Publishers* 2012

Answers Lessons 18.1 – 18.2

Quick check

1

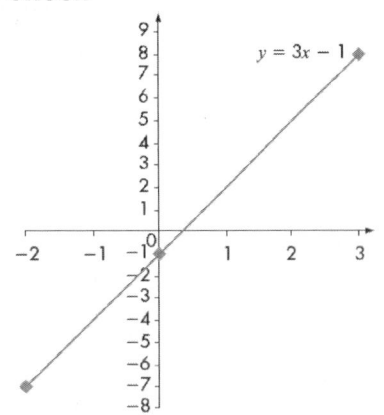

2 $y = 3x + 1$

18.1 Quadratic graphs
Exercise 18A
1 **a** Values of y: 27, 12, 3, 0, 3, 12, 27
   b 6.8 **c** 1.8 or −1.8
2 **a** Values of y: 27, 18, 11, 6, 3, 2, 3, 6, 11, 18, 27
   b 8.3 **c** 3.5 or −3.5
3 **a** Values of y: 27, 16, 7, 0, −5, −8, −9, −8, −5, 0, 7
   b −8.8 **c** 3.4 or −1.4
4 **a** Values of y: 2, −1, −2, −1, 2, 7, 14
   b 0.25 **c** 0.7 or −2.7
   d

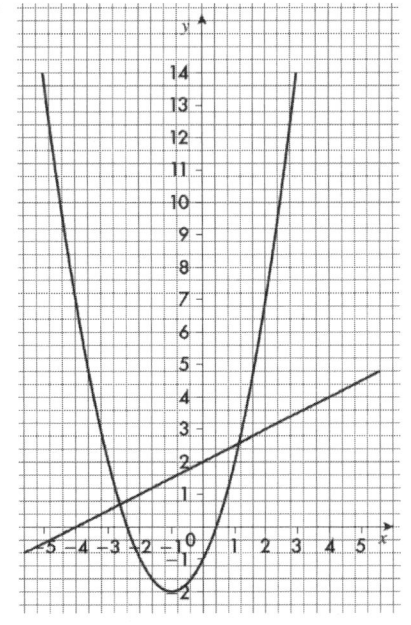

   e (1.1, 2.6) and (−2.6, 0.7)
5 **a** Values of y: 18, 12, 8, 6, 6, 8, 12
   b 9.75 **c** 2 or −1
   d Values of y: 14, 9, 6, 5, 6, 9, 14
   e (1, 6)
6 **a** Values of y: 4, 1, 0, 1, 4, 9, 16
   b 7.3 **c** 0.4 or −2.4

   d

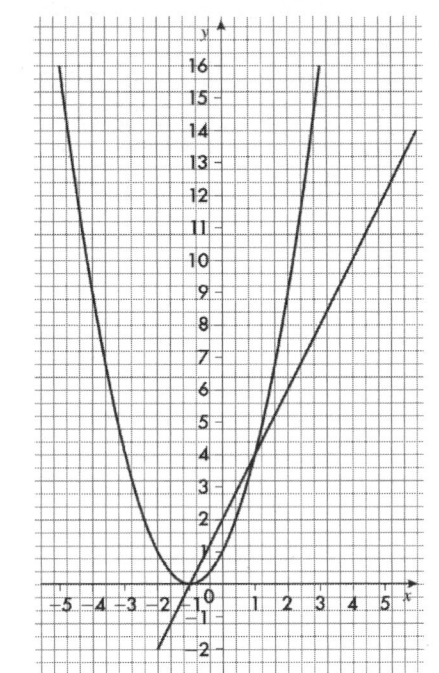

   e (1, 4) and (−1, 0)
7 **a** Values of y: 15, 9, 4, 0, −3, −5, −6, −6, −5, −3, 0, 4, 9
   b −0.5 and 3
8 Points plotted and joined should give parabolas.
9 Points plotted and joined should give a parabola.
10 Line A has a constant in front, so is 'thinner' than the rest.
   Line B has a negative in front, so is 'upside down'.
   Line C does not pass through the origin.

© HarperCollins*Publishers* 2012

292

18.2 The significant points of a quadratic graph
Exercise 18B
1. **a** Values of y: 12, 5, 0, −3, −4, −3, 0, 5, 12
   b 2 and −2
2. **a** Values of y: 7, 0, −5, −8, −9, −8, −5, 0, 7
   b 3 and −3
3. **a** The roots are positive and negative square roots of the constant term.
   b Check predictions.
   c Values of y: 15, 8, 3, 0, −1, 0, 3, 8, 15
   d Values of y: 11, 4, −1, −4, −5, −4, −1, 4, 11
   e 1 and −1, 2.2 and −2.2
4. **a** Values of y: 5, 0, −3, −4, −3, 0, 5, 12
   b −4 and 0
5. **a** Values of y: 16, 7, 0, −5, −8, −9, −8, −5, 0, 7, 16
   b 0 and 6
6. **a** Values of y: 10, 4, 0, −2, −2, 0, 4, 10, 18
   b −3 and 0
7. **a** The roots are 0 and the negative of the coefficient of x.
   b Check predictions.
   c Values of y: 10, 4, 0, −2, −2, 0, 4, 10
   d Values of y: 6, 0, −4, −6, −6, −4, 0, 6, 14
   e 0 and 3, −5 and 0
8. **a** Values of y: 9, 4, 1, 0, 1, 4, 9
   b −2
   c Only 1 root
9. **a** Values of y: 10, 3, −2, −5, −6, −5, −2, 3, 10
   b 0.6 and 5.4
10. **a** Values of y: 19, 6, −3, −8, −9, −6, 1, 12
    b 0.9 and −3.4
11. **a** (0, −4), (0, −9), (0, −1), (0, −5), (0, 0), (0, 0), (0, 0), (0, 0), (0, 0)
    b (0, −4), (0, −9), (0, −1), (0, −5), (−2, −4), (3, −9), (−1.5, −2.25), (1.5, −2.25), (−2.5, −6.25)
    c The y-intercept; the point where the x-value is the mean of the roots.
12. **a** $y = (x − 2)^2$ **b** 0
13. **a** $y = (x − 3)^2 − 6$ **b** 6
14. **a** $y = (x − 4)^2 − 14$ **b** −14
15. **a** $y = −(x − 1)^2 − 5$ **b** −5
16. **a** The minimum point is (a, b).
    b (−5, −28)
17. $y = (x − 3)^2 − 7$, $y = x^2 − 6x + 9 − 7$, $y = x^2 − 6x + 2$
18. **a** (−2, −7)
    b i $(a, 2b − a^2)$ **ii** $(2a, b − 4a^2)$

Examination questions
1. **a** −3, 7 **b** Correct graph
   c Read from the x-axis
2. **a** −2.2 to −2.3 **b** −1 and 2
3. **a** 7 and −2
   b Check students' graphs.
   c 0.6 and 3.4
4. **a** 0.55 **b** −4.25 **c** −6.25

Chapter 19 Algebra: Inequalities and regions

Overview

| 19.1 Solving inequalities | 19.2 Graphical inequalities |

This chapter deals with linear inequalities in one and two dimensions. These are used in the end-of-chapter activity to solve practical real-life problems.

Context
Inequalities are used to maximise profits and minimise wastage in a whole range of industrial situations. They are also used in economics, for example, to help forecast growth, profit and loss.

Curriculum references
KS4 Programme of Study references
1.1b Communicating mathematics effectively.
2.2a Make connections within mathematics.
2.2k Make accurate mathematical diagrams, graphs and constructions on paper and on screen.
3.1e Linear, quadratic and other expressions and equations.

Linear specification
N5.7 Solve linear inequalities in one variable and represent the solution set on a number line.
N5.7h Solve linear inequalities in two variables and represent the solution set on a suitable diagram.

Functional Skills standards
R2 Identify the situation or problems and identify the mathematical methods needed to solve them.
A1 Apply a range of mathematics to find solutions.
I2 Draw conclusions and provide mathematical justifications.

PLTS
Independent enquirers identify questions to answer and problems to resolve; explore issues, events or problems from different perspectives. **Creative thinkers** generate ideas and explore possibilities. **Effective participators** present a persuasive case for action. **Reflective learners** assess themselves and others, identifying opportunities and achievements.

APP
Algebra L7.3 Solve inequalities in one variable and represent the solution set on a number line.
L8.5 Solve inequalities in two variables and find the solution set. **Using and applying mathematics**
L8.5 Examine generalisations or solutions reached in an activity, commenting constructively on the reasoning and logic or the process employed, or the results obtained.

Route mapping

Exercise	Grades C	Grades B
A	1–5	6–9
B	1–6	7
C	1–4	5–16

© HarperCollins*Publishers* 2012

Chapter 19 Algebra: Inequalities and regions

Overview test
Questions 1 and 2 are grade D.
1 Solve the following equations.
  a $4x + 5 = 17$
  b $\frac{x}{3} - 2 = 4$
  c $2x + 3 = x + 6$
  d $3(x + 4) = 2(4x + 1)$

2 On the axes supplied, draw the graphs of the following equations.
  a $y = -6$
  b $x = 7$
  c $y = -x$
  d $y = 2x + 2$
  e $3x + 2y = 12$

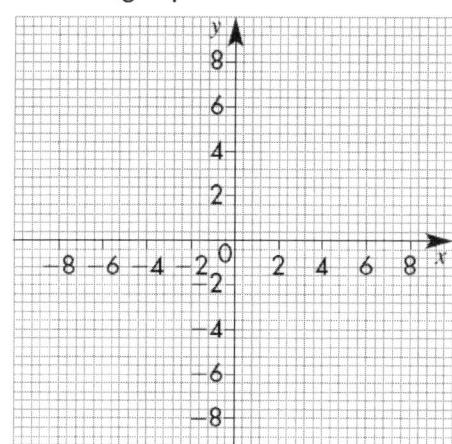

Answers to overview test
1 a $x = 3$ c $x = 3$
  b $x = 18$ d $x = 2$

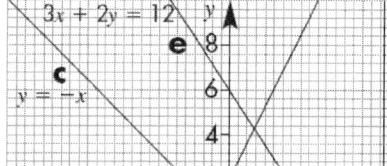

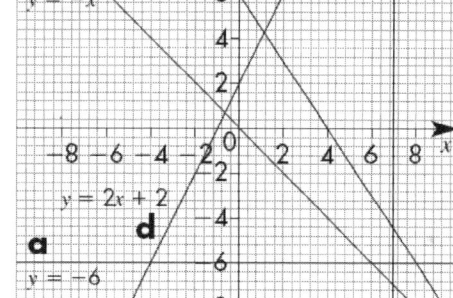

Why this chapter matters
Cross-curricular: This work is linked to the sciences, IT and other areas where the size of one value varies with the value of another.
Introduction: There are many situations that are controlled by limiting parameters. The precise value of a variable often depends on a number of factors. This idea has many applications in real life, especially in commercial contexts such as production of a minimum cost mix when blending animal feeds. The introductory page in the Student Book gives some background.
Discussion points: What mathematical symbol shows 'not equal to'? How are symbols used to show the nature of an inequality? How do we show whether the limiting value is included? How can we show this on a graph?
Plenary: Ask students if they have ever been to a theme park where there was a ride with a height restriction. Was there a set maximum height? Were people whose height was exactly this maximum allowed on?

Answers to the quick check test can be found at the end of the chapter.

© HarperCollinsPublishers 2012

19.1 Solving inequalities

Curriculum references

Functional Skills standards
A1 Apply a range of mathematics to find solutions.

PLTS
Independent enquirers

APP
Algebra L7.3

Collins references
- Student Book pages 462–467
- Interactive Book: Common Misconceptions

Learning objectives
- Solve a simple linear inequality. [Grade C–B]

Learning outcomes
- Students **must** be able to solve simple linear inequalities. [Grade C]
- Students **should** be able to represent the solution to an inequality on a number line. [Grade C]
- Students **could** be able to solve more complex linear inequalities. [Grade B]

Key words
- inclusive inequality
- number line
- inequality
- strict inequality

Prior knowledge
Students should be able to solve linear equations, including those with the variable occurring on both sides of the equation.

Common mistakes and remediation
Students often solve these as equations and do not put the inequality back in for the answer. Encourage students to keep the inequality in the solution at each step.
A grade C inequality question becomes a grade E equation when the inequality is replaced with an equals sign. Remind students that failure to recover the inequality in the final answer would result in zero marks, as the level of difficulty of the question has been changed.

Useful tips
Solve inequalities in the same way you solve equations, i.e. by rearrangement, but do not replace the inequality with an equals sign.

Functional maths and problem-solving help
Less able students find the concept of inequalities difficult to grasp, so these students will benefit from working through the functional maths (FM) and problem-solving (PS) questions of Exercises 19A and 19B (i.e. questions 4 and 6 in each). A set of cards (or similar) could be made to illustrate question 6 of Exercise 19A.

© HarperCollins*Publishers* 2012

Lesson 19.1 Solving inequalities

Starter
- Write on the board: $x < 5$. Ask students to say what it means. (x is less than 5)
- Then ask: What is the biggest value of x you can find that obeys the rule?
- Many will say 4, then 4.9. Until it is established that x can be as large as 4.999…, keep on saying, "No, there is a bigger number." Students may be familiar with the recurring decimal notation; otherwise they will be content with 4.9 followed by an infinite number of 9s.
- Now ask students for the smallest value that obeys $x > 3$. This is more difficult, and the answer is 3.00000…0000001, where there is an infinite number of zeros.
- Write the following on the board $x \leq 5$. Ask students to say what it means. (x is less than or equal to 5)
- Ask: What is the biggest value of x you can find that obeys the rule? (The answer is, of course, 5.)
- Now ask students: What is the smallest value that obeys $x \geq 3$? (The answer this time is 3.)

Main lesson activity
- Ask students if they can give the values of x that are true for $x + 3 > 7$.
- Say, "Clearly 5 will work, as will 4.5 – but what about 4? 4.1? 4.05, etc?" Establish that a value just bigger than 4 will work, so $x > 4$.
- Now show students how to solve this inequality.

$$x + 3 > 7$$
$$x > 7 - 3$$
$$x > 4$$

- Students should instantly recognise this as being the same basic method as solving an equation. Say that the methods are the same, but that the equals sign is replaced with an inequality sign.
- Work through the introductory text and Examples 1, 2, 3 and 4.
- Now students can do Exercise 19A, in which question 4 is FM, 5 is an assessing understanding (AU) question and 6 is a PS question. **Less able** students can complete questions 1–5. They may need support to finish the exercise.
- **Less able** students may find this topic difficult, but those who were able to solve equations should be able to access this. Provide students who do find this topic difficult, with more practice of simple equations until they are confident they can do them.

The number line
- Go through this section and Example 5. Most students, even those who are **less able**, will find this accessible.
- Now students can complete Exercise 19B, in which question 4 is FM, 5 is AU and 6 is PS. **Less able** students should be able to do questions 1 and 2. Pre-printed number lines would help them with question 3. Let **more able** students help **less able** students with the remaining questions in this exercise.

Plenary
- Write $2x + y \leq 7$ on the board and ask for values of x and y that make it true. For example, $x = 3, y = 1$ or $x = 0, y = 0$.
- Repeat with $3x + 4y > 12$.

Answers
Answers to Exercises 19A and 19B can be found at the end of this chapter.

© HarperCollins*Publishers* 2012

19.2 Graphical inequalities

Curriculum references

Functional Skills standards
A1 Apply a range of mathematics to find solutions.

PLTS
Independent enquirers

APP
Algebra L8.5

Collins references
- Student Book pages 467–471
- Interactive Book: Worked Exam Questions

Learning objectives
- Show a graphical inequality. [Grade B]
- Show how to find regions that satisfy more than one graphical inequality. [Grade C–B]

Learning outcomes
- Students **must** be able to show the region on a grid that satisfies an inequality with two variables. [Grade B]
- Students **should** be able to show the solution set for several inequalities on a grid. [Grade B]
- Students **could** be able to answers problems involving inequalities and points on a grid. [Grade B]

Key words
- boundary
- origin
- included
- region

Prior knowledge
Students will need to know how to substitute into simple expressions.

Common mistakes and remediation
Students often do not draw the correct line, or they are confused about whether the line should be dotted or solid. They also choose the incorrect side of the line as the one required. Testing a point to check should help to eliminate these mistakes. Also, students often plot points on the wrong axes, using the cover-up rule. They may also shade the wrong side of the line or not define clearly which side is required. To help to prevent these mistakes, work through any and point out the danger areas.

Useful tips
Tell students: Whatever the inequality, draw as though the inequality is an '=' sign. Then, if the sign is > or <, draw a dotted line, and if the sign is ≥ or ≤, draw a solid line. Think carefully about whether you want the region above or below the line.

Functional maths and problem-solving help
There are no functional maths (FM) questions in Exercise 19C, although the topic, as it is developed, does lend itself to this area. The problem-solving (PS) and assessing understanding (AU) questions test students' understanding of two-dimensional inequalities. There are many answers to AU question 13. Encourage students to draw inequalities that surround the points, and then define them. In PS question 14, it may not be obvious which conditions could be true, so encourage students to test them against the original inequality.

© HarperCollins*Publishers* 2012

Lesson 19.2 Graphical inequalities

Starter
- Recall the method of drawing linear graphs by the cover-up method.
- Write $2x + 3y = 6$ on the board. Remind students that the x-axis is the line $y = 0$, so, covering up the y term and solving the remaining linear equation, $2x = 6$ results in the intersect on the x-axis (3, 0). Similarly the y-axis is the line $x = 0$, leading to $y = (0, 2)$.
- Repeat with other similar lines if necessary. Note that there is no need for negative values, although there is no reason why it cannot be covered for **more able** students. **Less able** students will prefer positive values.

Main lesson activity
- Ask students to sketch the graph of $2x + 3y = 6$. Ask which side of the line would satisfy $2x = 3y = 6$.
- The natural instinct, as the sign is 'less than or equal to', is to go for the region under the line. This is correct.
- But ask students how they can be sure. The answer is to test a point. Suggest to students that there is an 'obvious' choice of a point to test.
- If they do not come up with the answer, tell them that the origin is the best point to test if it isn't on the line, so $2 \times 0 + 3 \times 0 \leq 6$. This means that the origin is on the side required.
- Make sure students realise that the answer to $2x + 3y \leq 6$ is a region and that it is infinitely big.
- Repeat with the inequality $x > 2$. Ask why this is different. Students should realise that there is only one variable and the inequality is a 'strict' inequality.
- Sketch the region and point out that it is still a region, and that the way we deal with the strict inequality is to draw the boundary line as a dashed line.
- Make sure students are aware of the steps involved in drawing a graphical inequality, as follows.
  Step 1: Draw the boundary line (dashed or solid as appropriate).
  Step 2: Test a point not on the line to establish which side of the line is required.
- A cause of confusion is which side to shade. Examination questions often say: Mark the region clearly with an R. There is no established convention. In the activity at the end of the chapter, it is better to shade the unwanted side. (Make sure students are aware of the need to mark the region clearly.)
- Work through Example 6 to make sure students have understood the principles.

More than one inequality
- Introduce the idea that some regions may be bound by more that one line. Draw a sketch of a set of coordinate axes, with two lines drawn on them. Indicate the area between the lines and ask students to identify some points in this region. Move on to use other pairs of lines.
- Use a similar method to show regions enclosed by three or more lines. Work through this section and Example 7.
- Now students can complete Exercise 19C, in which questions 13, 15 and 16 are AU and 14 is PS. **Less able** students can complete questions 1–6 first, then try the rest of the exercise. Note that questions 1–4 are grade C and 5–14 are grade B.
- All the graphs are straightforward $x = a$, $y = b$ types, or they use positive values in the cover-up rule, so **less able** students should be able to cope with drawing them. These students will benefit from having pre-printed graphs to complete.

Plenary
- Provide some simple inequalities such as $x \geq 2$, $y < 3$, $x > y$, $y \leq 2x + 1$ on the board, and ask volunteers to explain the steps they should take.

Answers
Answers to Exercises 19C can be found at the end of this chapter.

© HarperCollins*Publishers* 2012

Functional maths
Linear programming

Curriculum references

Functional Skills standards
R2 Identify the situation or problems and identify the mathematical methods needed to solve them.
I2 Draw conclusions and provide mathematical justifications.

PLTS
Independent enquirers, Reflective learners

APP
Algebra L8.5; Using and applying mathematics L8.5

Collins references
- Student Book pages 476–477

Learning objectives
- Use linear programming to solve a real life problem.

Key words
- condition
- equal to
- less than
- region

Functional maths help
In 'Your task' of the Student Book (Activity 2 in this guide), students may fail to see the connection between the use of the Cartesian plane and the problem. They may try to look at all possible combinations of stock and deal with each combination in its own right. This may lead to accurate solutions, but it is not an efficient method. Therefore, question students about alternative solutions.

Lesson plans

Build
Activity 1
The text under 'Getting started' requires students to place themselves in the real-life situation of going to the fair and having only a certain amount of money to spend.
- **More able** students may be able to work on the questions independently.
- However, to help **less able** students, it is likely that you will need to discuss, especially, question a iii as a class, emphasising that W and D stand for 'the number of …'

Activity 2
The 'Your task' activity requires students to use the information provided to investigate the number of sofas and beds the shop should stock.
- To help **less able** students, you may wish to break down the problem. You can look at each limiting factors one at a time, produce the inequality and then plot it.
- Students may need guidance through each criterion. Try highlighting the key points about space, stock and value.
- Take each point one at a time and discuss the mathematical representation of this factor.

Activity 3
In the 'Extension' in the Student Book, students are required to give some limiting factors of their own and display them diagrammatically.

Activity 4
- If there is time, or for **more able** students, give them these questions.

Mushtaq has to buy some apples and some pears. He has £3.00 to spend. Apples cost 30p each and pears cost 40p each. He must buy at least two apples and at least three pears, and at least seven fruits altogether. He buys x apples and y pears.

a Explain each of these inequalities:
  i $3x + 4y \leqslant 30$ (Cost $30x + 40y \leqslant 300 \Rightarrow 3x + 4y = 300$)
  ii $x \geqslant 2$ (At least 2 apples, so $x \geqslant 2$)
  iii $y \geqslant 3$ (At least 3 pears, so $y \geqslant 3$)
  iv $x + y \geqslant 7$ (At least 7 fruits, so $x + y \geqslant 7$)
b Which of these combinations satisfy all of the above inequalities?
  i   Three apples and three pears (No)
  ii Four apples and five pears (No)
  iii No apples and seven pears (No)
  iv Three apples and five pears (Yes)

Apply
The first activity is relatively straightforward. The answers are as follows:
a i Because $4 \times 1.5 = 6$ and the boy only has £6, he cannot buy more than four rides.
  ii Similarly, the boy cannot afford more than three hotdogs.
  iii If you purchase W rides and D hotdogs you will spend a total of:
     $1.5W + 2D \leqslant 6 \qquad 3W + 4D \leqslant 12$
b $D \leqslant 2$
c i Possible **ii** Not possible **iii** Not possible **iv** Possible

When applying skills in the 'Your task' activity, students should be able to break down the problem into manageable chunks. They should be able to draw out the information about cost, stock and space. However, they may need guidance with transferring this information into mathematical representations.
Ask students how they can represent the inequalities graphically. **Less able** students may need to work in groups to produce the graph. **More able** students may be able to work independently at this task.
Ensure that students see the eloquence of the solution when graphs are used. (Students' graphs should be as shown here.)

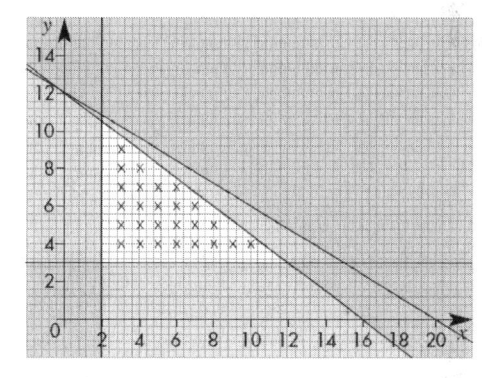

Master
In order to demonstrate mastery of the learning objectives, students should be able to:
- Recognise the information about stock, space and value.
- Represent this information using inequalities and graphs.
- Produce the required diagram and interpret acceptable points.
- Use the suggested extension to develop this task further.

Plenary
Students should think about the following questions.
- What other restrictions may the shop face?
- How efficient was your method?
- Was any of the information supplied surplus, or unnecessary, to the requirements?
- If the shop were to stock a third line, such as tables, would you still be able to use linear programming to solve the stock issues?

© HarperCollins*Publishers* 2012

Answers Lessons 19.1 – 19.2

Quick check
1 a 8 b 10

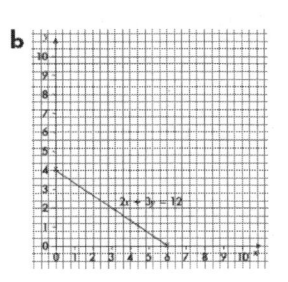

2 a [graph of y = 3x + 1]   b [graph of 2x + 3y = 12]

19.1 Solving inequalities
Exercise 19A
1 a $x < 3$ b $t > 8$ c $p \geq 10$ d $x < 5$
   e $y \leq 3$ f $t > 5$ g $x < 6$ h $y \leq 15$
   i $t \geq 18$ j $x < 7$ k $x \leq 3$ l $t \geq 5$

2 a 8 b 6 c 16 d 3 e 7

3 a 11 b 16 c 16

4 $2x + 3 < 20$, $x < 8.50$, so the most each could cost is £8.49

5 a Because $3 + 4 = 7$, which is less than the third side of length 8
   b $x + x + 2 > 10$, $2x + 2 > 10$, $2x > 8$, $x > 4$, so smallest value of x is 5

6 a $x = 6$ and $x < 3$ scores -1 (nothing in common), $x < 3$ and $x > 0$ scores $+1$ (1 in common for example), $x > 0$ and $x = 2$ scores $+1$ (2 in common), $x = 2$ and $x \geq 4$ scores -1 (nothing in common), so we get $-1 + 1 + 1 - 1 = 0$
   b $x > 0$ and $x = 6$ scores $+1$ (6 in common), $x = 6$ and $x \geq 4$ scores $+1$ (6 in common), $x \geq 4$ and $x = 2$ scores -1 (nothing in common), $x = 2$ and $x < 3$ scores $+1$ (2 in common). $+ 1 + 1 - 1 + 1 = 2$
   c Any acceptable combination, e.g. $x = 2$, $x < 3$, $x > 0$, $x \geq 4$, $x = 6$

7 a $x \geq -6$ b $t \leq \frac{8}{3}$ c $y \leq 4$
   d $x \geq -2$ e $w \leq 5.5$ f $x \leq \frac{14}{5}$

8 a $x \leq 2$ b $x > 38$ c $x < 6\frac{1}{2}$
   d $x \geq 7$ e $t > 15$ f $y \leq \frac{7}{5}$

9 a $3 < x < 6$ b $2 < x < 5$ c $-1 < x \leq 3$
   d $1 \leq x < 4$ e $2 \leq x < 4$ f $0 \leq x \leq 5$

Exercise 19B
1 a $x > 1$ d $x \geq -1$
   b $x \leq 3$ e $x \leq -1$
   c $x < 2$ f $x \geq 1$

2 a–h [number line diagrams]

3 a $x \geq 4$ e $x < 1.5$
   b $x < -2$ f $x \leq -2$
   c $x \geq 3.5$ g $x > 50$
   d $x < -1$ h $x \geq -6$

4 a Because 3 apples plus the chocolate bar cost more that £1.20: $x > 22$.
   b Because 2 apples plus the chocolate bar left Max with at least 16p change: $x \leq 25$.
   c [number line diagram]
   d Apples could cost 23p, 24p or 25p.

5 Any two inequalities that overlap only on the integers −1, 0, 1 and 2, for example, $x \geq -1$ and $x < 3$.

6 1 and 4

7 a $x > 2$ c $x \leq -1$
   b $x \geq 6$ d $x \geq -4$

© HarperCollins*Publishers* 2012

302

Answers Lessons 19.1 – 19.2

19.2 Graphical inequalities
Exercise 19C

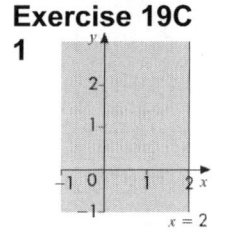

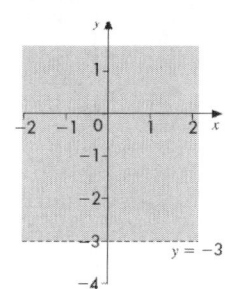

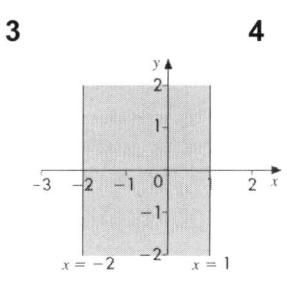

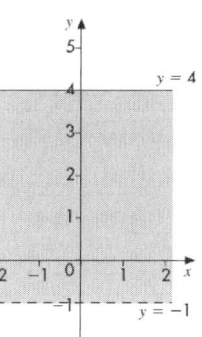

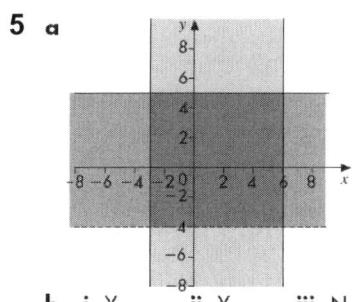

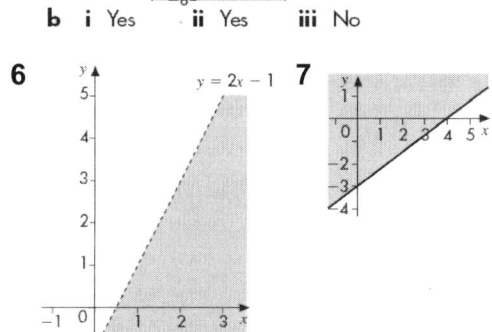

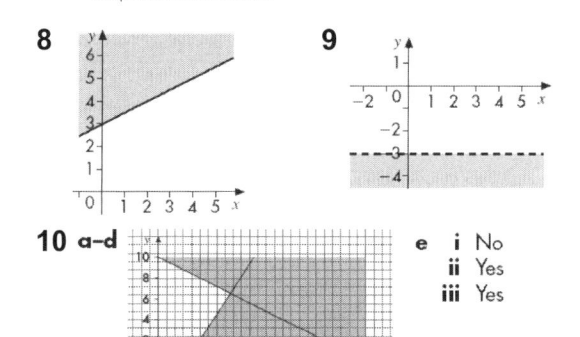

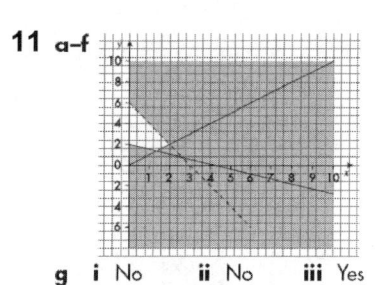

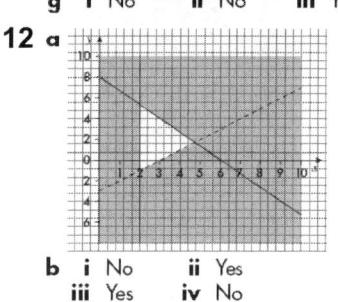

1, 2, 3, 4 [graphs showing shaded regions for $x = 2$, $y = -3$, $x = -2$ and $x = 1$, $y = 4$ and $y = -1$]

5 a [graph]
   b i Yes   ii Yes   iii No

6 [graph of $y = 2x - 1$]

7 [graph]

8, 9 [graphs]
   e i No   ii Yes   iii Yes

10 a–d [graph]

11 a–f [graph]
   g i No   ii No   iii Yes

12 a [graph]
   b i No   ii Yes
     iii Yes iv No

13 For example, $x \geq 1$, $y \leq 3$ and $y \geq x + 1$. There are many other valid answers.

14 May be true: a, c, d, g
   Must be false: b, e
   Must be true: f, h

15 Test a point such as the origin (0, 0), so $0 < 0 + 2$, which is true. So the side that includes the origin is the required side.

16 a (3, 0) b (4, 5)

Examination questions

1 $x \leq 1.4$

2 a $x \leq 5$
  b A line with a solid circle from –2 to 3.

3 a $x \leq -1$ b $y > -3$
  c Top side shaded; test a point such as the origin $0 \leq 12$.

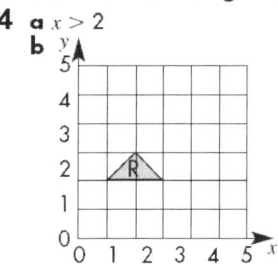

4 a $x > 2$
  b [graph with region R]

5 Region 1 is graph D, Region 2 is graph C, Region 3 is graph E, Region 4 is graph A.

Collins Schemes of Work

We have put together a suggested Scheme of Work for 2 years and 3 years. The overviews for planning are printed here and the detailed breakdowns are included on the accompanying CD-ROM. The Schemes of Work are based on a minimum of 3 GCSE Maths lessons per week and follow the content in the *Collins New GCSE Maths* course for AQA Linear Higher Book 1 and 2. Suggestions about revision, extra recap topics and mock exams are given in italics.

The 3-year Scheme of Work takes you up to the end of term 1 of the third year. As there is more flexibility in this scheme, you can supplement the textbook with more activities from Collins' Interactive Books, revision guides and functional skills resources.

2-Year Scheme of Work: Overview

		Book	Chapter	Total no. of teaching hours	Comments/suggestions
Year 1	Term 1	1	Ch 1: Number: Number skills	4	
		1	Ch 2: Number: More number	5	
		1	Ch 3: Number: Fractions, percentges and ratios	8	
			HALF TERM		
		1	Ch 4: Number: Proportions	5	
		1	Ch 5: Geometry: Shapes	6	
		1	Ch 6: Algebra: Expressions and equations 1	6	
		1	Ch 7: Algebra: Expressions and equations 2 (part 1)	3	
			End of term assessment	1	
			Assessment review	1	
	Term 2	1	Ch 7: Algebra: Expressions and equations 2 (part 2)	3	
		1	Ch 8: Geometry: Pythagoras and trigonometry	5	
		1	Ch 9: Geometry: Angles	3	
		1	Ch 10: Geometry: Constructions	5	
			HALF TERM		
		1	Ch 11: Geometry: Transformation geometry	7	
		1	Ch 12: Statistics: Data handling	12	
		1	Ch 13: Algebra: Real-life graphs	3	

© HarperCollins*Publishers* 2012

Year 2	Term 3		End of term assessment	1
			Assessment review	1
		1	Ch 14: Statistics: Statistical representation	4
		1	Ch 15: Probability: Calculating probabilities	7
		1	Ch 16: Algebra: Algebraic methods	6
		1	Ch 17: Algebra: Linear graphs and equations	3
		1	Ch 18: Algebra: More graphs and equations	3
		1	Ch 19: Algebra: Inequalities and regions	3
			HALF TERM	
		2	Ch 1: Number: Using a calculator	2
		2	Ch 2: Geometry: Shapes	4
		2	Ch 3: Geometry and trigonometry (part 1)	4
		2	Ch 3: Geometry and trigonometry (part 2)	4
			Revision	2
			End of term assessment	1
			Assessment review	1
	Term 1		Number recap	2
			Algebra recap	3
			Data recap	2
			Geometry recap	3
		2	Ch 4: Geometry: Angles and properties of circles	5
		2	Ch 5: Number: Powers, standard form and surds	6
			HALF TERM	
		2	Ch 6: Algebra: Quadratic equations	7
		2	Ch 7: Measures: Similarity	3
		2	Ch 8: Geometry: Trigonometry	7
			Revision	2
			Mock exams	2
			Mock exam review	2

305

© HarperCollins*Publishers* 2012

Term 2	2	Ch 9: Algebra: More graphs and equations	6
	2	Ch 10: Statistics: Statistical representation	7
	2	Ch 11: Algebra: Algebraic methods	6
		HALF TERM	
	2	Ch 12: Number: Variation	4
	2	Ch 13: Number: Number and limits of accuracy	3
	2	Ch 14: Geometry: Vectors	5
	2	Ch 15: Algebra: Transformations of graphs	3
		Number recap	3
		Algebra recap	3
		End of term assessment	1
Term 3		Assessment review	2
		Geometry recap	3
		Statistics and probability recap	3
		Revision and exam preparation	12
		HALF TERM	
		Final Exam – TBC	

306

© HarperCollins*Publishers* 2012

3-Year Scheme of Work: Overview

	Book	Chapter	Total no. of teaching hours	Comments/suggestions
Year 1				
Term 1	1	Ch 1: Number: Number skills	4	
	1	Ch 2: Number: More number (part 1)	4	
	1	Ch 3: Number: Fractions, percentages and ratios (part 1)	5	
		HALF TERM		
	1	Ch 3: Number: Fractions, percentages and ratios (part 2)	4	
	1	Ch 2: Number: More number (part 2)	3	
	1	Ch 4: Number: Proportions	6	
		Revision	2	
		End of term assessment	1	
		Assessment review	1	
Term 2	1	Ch 5: Geometry: Shapes	8	
	1	Ch 6: Algebra: Expressions and equations 1	7	
	1	Ch 7: Algebra: Expressions and equations 2	3	
		Number recap	2	
		HALF TERM		
	1	Ch 8: Geometry: Pythagoras and trigonometry	6	
	1	Ch 9: Geometry: Angles (part 1)	1	
	1	Ch 10: Geometry: Constructions	5	
		Revision	1	
		End of term assessment	1	
		Assessment review	1	
Term 3	1	Ch 9: Geometry: Angles (part 2)	2	
	1	Ch 11: Geometry: Transformation geometry	8	
	1	Ch 12: Statistics: Data handling	11	
		HALF TERM		
	1	Ch 13: Algebra: Real-life graphs	3	
		Number recap	3	

Year 2	**Term 1**		*Geometry recap*	3
			Statistics recap	3
			Algebra recap	3
			End of term assessment	1
			Assessment review	1
		2	Ch 1: Number: Using a calculator	4
		2	Ch 2: Geometry: Shapes	4
		1	Ch 14: Statistics: Statistical representation	4
		1	Ch 15: Probability: Calculating probabilities	7
			HALF TERM	
		1	Ch 16: Algebra: Algebraic methods	5
		1	Ch 17: Algebra: Linear graphs and equations	2
		1	Ch 18: Algebra: More graphs and equations	3
		1	Ch 19: Inequalities and regions	3
		2	Ch 3: Geometry and trigonometry 1	2
		2	Ch 3: Geometry and trigonometry 2	8
			End of term assessment	1
			Assessment review	1
	Term 2	2	Ch 4: Geometry: Angles and properties of circles	5
		2	Ch 5: Number: Powers, standard form and surds	7
		2	Ch 6: Algebra: Quadratic equations (part 1 – factorising)	5
			HALF TERM	
		2	Ch 7: Measures: Similarity	4
		2	Ch 6: Algebra: Quadratic equations (part 2 – formula, etc.)	3
		2	Ch 8: Geometry: Trigonometry (part 1)	6
		2	Ch 8: Geometry: Trigonometry (part 2)	2
			Revision	1
			End of term assessment	1
			Assessment review	1

© HarperCollins*Publishers* 2012

308

Year 3	**Term 3**	2	Ch 9: Algebra: More graphs and equations	9
		2	Ch 10: Statistics: Statistical representation	7
		2	Ch 11: Algebra: Algebraic methods	8
			HALF TERM	
		2	Ch 12: Number: Variation	4
			Number recap	2
			Geometry recap	2
			Statistics recap	2
			Algebra recap	2
			End of term assessment	1
			Assessment review	1
	Term 1		Number recap	2
			Geometry recap	2
			Statistics and probability recap	3
		2	Ch 13: Number: Number and limits of accuracy	3
			Algebra recap	2
			MOCK EXAM	2
			Mock exam review	2
			Algebra recap – graphs	2
			HALF TERM	
		2	Ch 14: Geometry: Vectors	4
		2	Ch 15: Algebra; Transformation of graphs	3
			Revision and exam preparation	6
			GCSE MATHEMATICS EXAM (TBC)	

309

© HarperCollins*Publishers* 2012

William Collins' dream of knowledge for all began with the publication of his first book in 1819. A self-educated mill worker, he not only enriched millions of lives, but also founded a flourishing publishing house. Today, staying true to this spirit, Collins books are packed with inspiration, innovation and practical expertise. They place you at the centre of a world of possibility and give you exactly what you need to explore it.

Published by Collins
An imprint of HarperCollins*Publishers*
77–85 Fulham Palace Road
Hammersmith
London
W6 8JB

Browse the complete Collins catalogue at
www.collinseducation.com

© HarperCollins*Publishers* Limited 2012

10 9 8 7 6 5 4 3 2 1

ISBN 978-0-00-748936-7

Brian Speed, Keith Gordon, Kevin Evans, Trevor Senior and Chris Pearce assert their moral rights to be identified as the authors of this work. All rights reserved. No part of this publication may be reproduced, stored in a retrieval system, or transmitted in any form or by any means, electronic, mechanical, photocopying, recording or otherwise, without the prior written permission of the Publisher or a licence permitting restricted copying in the United Kingdom issued by the Copyright Licensing Agency Ltd., 90 Tottenham Court Road, London W1T 4LP.

British Library Cataloguing in Publication Data
A Catalogue record for this publication is available from the British Library

Commissioned by Katie Sergeant
Suggested Schemes of Work and Grade progression maps by Claire Powis, Maths AST; Children, Families and Education, Kent
Edited and proofread by Brian Asbury, Joan Miller, Margaret Shepherd, Sarah Vittachi and
Karen Westall
Artwork by Ann Paganuzzi
Answer check by Amanda Dickson
Cover design by Angela English
Word conversion by Hart McLeod, Grace Glendinning and Jocelyn Whitmore
PDF conversion by Mark Walker
Production by Rebecca Evans
Printed and bound by Martins the Printers, Berwick-upon-Tweed

Acknowledgements
With thanks to Andy Edmonds, Claire Beckett, Samantha Burns,
Anton Bush (Gloucester High School for Girls), Matthew Pennington
(Wirral Grammar School for Girls), James Toyer (The Buckingham School),
Gordon Starkey (Brockhill Park Performing Arts College), Laura Radford and
Alan Rees (Wolfreton School) and Mark Foster (Sedgefield Community College).
Chris Curtis and Chris Pearce.

MIX
Paper from responsible sources
FSC
www.fsc.org
FSC™ C007454

© HarperCollins*Publishers* 2012

© HarperCollins*Publishers* 2012

© HarperCollins*Publishers* 2012